Rehabilitation
of Rivers

LANDSCAPE ECOLOGY SERIES

Ecology and Management of Invasive Riverside Plants
Edited by Louise C. de Waal, Lois E. Child, P. Max Wade and John H. Brock

European Wet Grasslands
Edited by Chris B. Joyce and P. Max Wade

Rehabilitation of Rivers
Edited by Louise C. de Waal, Andrew R. G. Large and P. Max Wade

Rehabilitation of Rivers

Principles and Implementation

Edited by

LOUISE C. DE WAAL
Division of Environmental and Analytical Sciences, School of Applied Sciences, University of Wolverhampton, UK

ANDREW R. G. LARGE
Department of Geography, University of Newcastle, UK

and

P. MAX WADE
Department of Environmental Sciences, University of Hertfordshire, UK

JOHN WILEY & SONS
Chichester · New York · Weinheim · Brisbane · Singapore · Toronto

Copyright © 1998 by John Wiley & Sons Ltd,
Baffins Lane, Chichester,
West Sussex PO19 1UD, England

National 01243 779777
International (+44) 1243 779777
e-mail (for orders and customer service enquiries): cs-books@wiley.co.uk
Visit our Home Page on http://www.wiley.co.uk
or http://www.wiley.com

All rights reserved. No part of this publication may be reproduced, stored in a retrieval system, or transmitted, in any form or by any means, electronic, mechanical, photocopying, recording, scanning or otherwise, except under the terms of the Copyright, Designs and Patents Act 1988 or under the terms of a licence issued by the Copyright Licensing Agency, 90 Tottenham Court Road, London, UK W1P 9HE, without the permission in writing of John Wiley and Sons Ltd., Baffins Lane, Chichester, West Sussex PO19 1UD, UK.

Other Wiley Editorial Offices

John Wiley & Sons, Inc., 605 Third Avenue,
New York, NY 10158–0012, USA

WILEY-VCH Verlag GmbH, Pappelallee 3,
D-69469 Weinheim, Germany

Jacaranda Wiley Ltd, 33 Park Road, Milton,
Queensland 4064, Australia

John Wiley & Sons (Canada) Ltd, 22 Worcester Road,
Rexdale, Ontario M9W 1L1, Canada

John Wiley & Sons (Asia) Pte Ltd, 2 Clementi Loop #02-01,
Jin Xing Distripark, Singapore 129809

British Library Cataloguing in Publication Data

A catalogue record for this book is available from the British Library

ISBN 0-471-95753-4

Typeset in 10/12pt Times from the authors' disks by Vision Typesetting, Manchester, UK
Printed and bound in Great Britain by Biddles Ltd, Guildford and Kings Lynn

This book is printed on acid-free paper responsibly manufactured from sustainable forestry, in which at least two trees are planted for each one used for paper production.

Contents

List of Contributors   vii

Preface   xi

1   **Rehabilitation of Degraded River Habitat: An Introduction**   1
   P. Max Wade, Andrew R. G. Large and Louise C. de Waal

Part One   PRINCIPLES

2   **Policy Networks and the Success of Lowland Stream Rehabilitation Projects in the Netherlands**   13
   Peter Jasperse

3   **An Approach to Classification of Natural Streams and Floodplains in South-west Germany**   31
   Rolf Bostelmann, Ulrich Braukmann, Elmar Briem, Thomas Fleischhacker, Georg Humborg, Ina Nadolny, Karl Scheurlen and Uwe Weibel

4   **A Systematic Approach to Ecologically Sound River Bank Management**   57
   Jennie Simons and René Boeters

5   **The Influence of Riparian Ecotones on the Dynamics of Riverine Fish Communities**   87
   Maciej Zalewski and Piotr Frankiewicz

6   **Rehabilitation of Rivers by Using Wet Meadows as Nutrient Filters**   97
   Ann Fuglsang

7   **Practical Approaches for Nature Development: Let Nature Do Its Own Thing Again**   113
   Jan P. M. van Rijen

Part Two   IMPLEMENTATION: STRATEGIC APPROACHES

8   **The River Restoration Project and Its Demonstration Sites**   133
   Nigel T. H. Holmes

9   **Strategic Approaches to River Rehabilitation: The River Leen and the River Derwent, UK**   149
   David Hickie

10 **Problems Associated with the Degradation of Rivers in the North East Region, England, and Initiatives to Achieve Rehabilitation** 159
John R. Pygott and Andrew R. G. Large

11 **Rehabilitation of the North American Great Lakes Watershed: Past and Future** 171
Frank Butterworth and Robbin Hough

12 **Four Decades of Sustained Use, of Degradation and of Rehabilitation in Various Streams of Toronto, Canada** 189
Gordon A. Wichert and Henry A. Regier

13 **Rehabilitation of the River Murray, Australia: Identifying Causes of Degradation and Options for Bringing the Environment into the Management Equation** 215
Anne Jensen

Part Three   IMPLEMENTATION: CASE STUDIES

14 **Efforts for In-stream Fish Habitat Restoration within the River Iijoki, Finland – Goals, Methods and Test Results** 239
Timo Yrjänä

15 **Rehabilitation of the Acque Alte Drainage Canal on the River Po Alluvial Plain, Italy** 251
Bruna Gumiero, Gianpaolo Salmoiraghi, Marco Rizzoli and Roberta Santini

16 **Degradation and Rehabilitation of Waterways in Australia and New Zealand** 269
Christopher J. Gippel and Kevin J. Collier

17 **Rehabilitation of Japan's Waterways** 301
Christopher J. Gippel and Shubun Fukutome

Index 319

Contributors

René Boeters
Ministry of Transport, Public Works and Water Management, Directorate-General for Public Works and Water Management, Road and Hydraulic Engineering Division, PO Box 5044, 2600 GA Delft, The Netherlands

Rolf Bostelmann
c/o ALAND-Arbeitsgemeinschaft Landschaftsökologie, Vorholzstrasse 36, D-76137 Karlsruhe, Germany

Ulrich Braukmann
Landesanstalt für Umweltschutz Baden-Württemberg, Postfach, D-76157 Karlsruhe, Germany

Elmar Briem
Am Kolmerberg 7, D-76889, Dörrenbach, Germany

Frank Butterworth
Institute for River Research International, 920 Ironwood Drive, Suite 344, Rochester, MI 48309-1333, USA

Kevin J. Collier
National Institute of Water and Atmosphere, Hamilton, New Zealand

Thomas Fleischhacker
Institute für Geographie und Geoökologie, Universität Karlsruhe, Karlsruhe, Germany

Piotr Frankiewicz
Department of Applied Ecology, University of Łódz, Banacha 12/16, PL-90-237 Łódz, Poland

Ann Fuglsang
Funen County Council, Department of Nature and Water Environment, Ørbækvej 100, DK-5220 Odense SØ, Denmark

Shubun Fukutome
Nishinihon Institute of Technology, Wakamatsu-cho, Kochi City 780, Kochi-ken, Japan

Christopher J. Gippel
Centre for Environmental Applied Hydrology, Department of Civil and Environmental Engineering, The University of Melbourne, Parkville, Victoria 3052, Australia

Bruna Gumiero
Department of Biology, University of Bologna, via Selmi 3, 40126 Bologna BO, Italy

David Hickie
The Environment Agency, Midlands Region, Sapphire East, 550 Streetsbrook Road, Solihull B91 1QT, UK

Nigel T. H. Holmes
River Restoration Centre, Silsoe Campus, Silsoe, Bedfordshire, UK

Robbin Hough
Great Lakes Institute for Systems Research, Oakland University, Rochester, Michigan 48309-4401, USA

Georg Humborg
Franz-Heinrich-Strasse 10, D-95100 Selb, Germany

Peter Jasperse
Rÿnweide 102, 3451 VK Vleuten, The Netherlands

Anne Jensen
Department of Environment and Natural Resources, Resource Management Branch, PO Box 1047, Adelaide, SA 5001, Australia

Andrew R. G. Large
Department of Geography, Daysh Building, University of Newcastle upon Tyne, Newcastle NE1 7RU, UK

Ina Nadolny
Institute für Wasserbau und Kulturtechnik, Universität Karlsruhe, Theodor-Rehbock-Laboratorium, Kaïserstraße 12, D-76128 Karlsruhe, Germany

John R. Pygott
The Environment Agency, North East Region, Olympia House, Gelderd Road, Leeds LS12 6DD, UK

Henry A. Regier
Institute of Environmental Studies, University of Toronto, Toronto, Ontario, Canada M5S 1A4

Jan P. M. van Rijen
Ministry of Agriculture, Nature Management and Fisheries, Regional Policy Department South, PO Box 6111, 5600 HC Eindhoven, The Netherlands

CONTRIBUTORS

Marco Rizzoli
Provincia di Bologna, via Zamboni 13, 40126 Bologna, Italy

Gianpaolo Salmoiraghi
Department of Biology, University of Bologna, via Selmi 3, 40126 Bologna BO, Italy

Roberta Santini
Unitá Sanitaria Locale 26, Corso Italia 58, 40017 San Giovanni in Persiceto (BO), Italy

Karl Scheurlen
Institut für Umweltstudien, Weisser & Ness GmbH, Waldhofer Strasse 104, D-69123, Heidelberg, Germany

Jennie Simons
Centre for Civil Engineering Research, Codes and Specifications, PO Box 420, 2800 AK Gouda, The Netherlands

Louise C. de Waal
School of Applied Sciences, University of Wolverhampton, Wulfruna Street, Wolverhampton, WV1 1SB, UK

P. Max Wade
Department of Environmental Sciences, University of Hertfordshire, Hatfield, AL10 9AB, UK

Uwe Weibel
Institut für Umweltstudien, Weisser & Ness GmbH, Waldhofer Strasse 104, D-69123, Heidelberg, Germany

Gordon A. Wichert
Faculty of Environmental Science, University of Guelph, Guelph, Ontario, Canada N1G 2W1

Timo Yrjänä
North Ostrobothnia Regional Environment Centre, PO Box 124, 90101 Oulu, Finland

Maciej Zalewski
Department of Applied Ecology, University of Łódz, Banacha 12/16, PL-90-237 Łódz, Poland

Preface

Landscape ecology concerns the science that focuses on the interrelationships between the various components of the landscape, i.e. flora, fauna, soil, water, air and human impact. This rapidly expanding science needs to take into account both the growing knowledge concerned with the processes that underlie the landscape, for example ecological, geomorphological and hydrological processes, and the experience gained by those managing the landscape, such as engineers, conservation officers and planners. The last decade has seen the emergence of a series of river rehabilitation projects aimed at the enhancement of degraded river habitats and improvement of the wider river landscape. These rivers have been damaged in the past by over-engineering, pollution, over-abstraction and unsympathetic management. Each river rehabilitation project extends our knowledge of the rehabilitation process, whether it be from a planning, ecological, hydrological or engineering viewpoint. Some of this work has resulted from opportunities for practical river rehabilitation work associated with flood defence, water resources and pollution control activities as well as via the consultations on planning and development control by local authorities often in conjunction with voluntary groups or local businesses, and others are research-based initiatives derived from ongoing programmes investigating the river environment. A survey of a range of river rehabilitation projects undertaken across Europe by the International Centre of Landscape Ecology, Loughborough University, identified the need to bring together collective experience and to share the emerging knowledge of how to ensure that the effort being put into river rehabilitation was used efficiently and effectively. This survey, funded by the former National Rivers Authority (R&D Project 477), also identified a range of constraints on river rehabilitation. Those most commonly cited involved the inability to allow floodplain inundation because of the requirement to provide flood control and the lack of control over floodplain land as a result of land ownership or because of urban areas occupying the floodplain area. Constraints upon river rehabilitation can be of a scientific, natural, social, financial or political nature or a combination of all five. The challenge of rehabilitating rivers is therefore one which truly shows the importance of landscape ecology as an emerging science. At one end of the spectrum the interrelations of animals, plants, water and sediment must not only be understood but knowledge must be acquired as to how to manipulate these processes to achieve the long-term aim of putting rivers back to a better working order. At the other end, a more geomorphological and hydrological perspective is required setting the river in its channel, floodplain, corridor and even catchment in order to ensure that these manipulations at the local scale are sustainable on a regional scale and into the future.

This book has resulted from an international workshop entitled 'Rehabilitation of Rivers' organised by the International Centre of Landscape Ecology and held at Loughborough University. The aims of the workshop were to determine the necessity for river rehabilitation, to identify methods for effective rehabilitation, including ecological and morphological considerations, and to identify the likely limitations in rehabilitating rivers. The workshop brought together professionals from 12 different countries: Australia, Denmark, Finland, France, Germany, Hungary, Italy, the Netherlands, Norway, Poland, Sweden and the United Kingdom. Their backgrounds varied from academics to people from environmental consultancies, local authorities, the former National Rivers Authority (now the Environment Agency) and other water authorities, nature conservancy councils, including Scottish Natural Heritage, Dutch Stichting Het Utrechts Landschap and the Norwegian Directorate for Nature Management, and people from several European ministries. Although a number of professionals from Australia, Canada, Japan and the USA were not able to attend the workshop itself, their contributions to this text are invaluable in providing a worldwide overview of the rehabilitation of rivers.

This book aims to bring together an array of specialists who deal with the various facets of the rehabilitation process: ecologists, hydologists, geomorphologists, nature conservationists, planners, river managers and landscape architects. The first section (Chapters 2–7) deals with the principles underlying the enhancement of the river environment, recognising not only the need to consider the river in its broadest sense (channel, floodplain and catchment), but also the policy networks which underpin success. The second section deals with the implementation of rehabilitation projects which is considered from the strategic viewpoint (Chapters 8–13) and through four case studies from Finland, Italy, Australia and New Zealand, and Japan (Chapters 14–17).

Max Wade, Louise de Waal and Andy Large
February 1998

1 Rehabilitation of Degraded River Habitat: An Introduction

P. MAX WADE,[1] ANDREW R. G. LARGE[2] and LOUISE C. DE WAAL[3]
[1]*Department of Environmental Sciences, University of Hertfordshire, UK*
[2]*Department of Geography, Daysh Building, University of Newcastle upon Tyne, UK*
[3]*School of Applied Sciences, University of Wolverhampton, UK*

INTRODUCTION

Over the past 6000 years, humans have altered river corridors by over-engineering, pollution, over-abstraction and unsympathetic management. In Europe, for example, water supply and land drainage schemes were implemented as early as 3200 BC with the construction of embankments and weirs. Later, rivers were channelised for the purpose of navigation and flood defence. From the early 1900s major dam building activities started for both hydroelectric power and drinking water supply. Petts and Wood (1988) suggested that, as a result of these activities, around 89% of the rivers in the UK are regulated to some extent. The continued increase in the use of water for domestic, industrial and agricultural consumption, and for hydroelectric power will lead to further development of river regulation, water abstraction and inter-basin transfer schemes (Petts *et al.*, 1995). At the same time, environmental concern from public and scientific communities regarding stressed and degraded river systems is growing rapidly. The increased cost of river management, the reduced opportunities and at the same time increased demand for sites for recreation, and the nature conservation imperative have all primed changes in river management policies.

LEGISLATION AND RIVER REHABILITATION

In the UK, the Environment Agency has a statutory duty under the Water Resources Act 1991 (Section 2), 'generally to promote the conservation and enhancement of the natural beauty and amenity of inland waters and of land associated with such waters and the conservation of flora and fauna which are dependent on an aquatic environment'. The Wildlife and Countryside Act 1981 empowers the government's statutory adviser on nature conservation, English Nature, the Countryside Council for Wales and Scottish Natural Heritage, to select and notify areas of land or water containing plants, animals, geological or landform features as Sites of Special Scientific Interest (SSSIs). According to Boon (1995), the quantity of this type of legislation that incorporates at least an element of ecology is still increasing from year to year. Examples of international environmental

---

Rehabilitation of Rivers: Principles and Implementation. Edited by L. C. de Waal, A. R. G. Large and P. M. Wade.
© 1998 John Wiley & Sons Ltd.

legislation include the EC Directive on the Conservation of Natural Habitats and of Wild Fauna and Flora (European Communities, 1992) and the statutory requirement for Environmental Impact Assessment (European Communities, 1985). These changes in environmental legislation have, for example, led to changes in river management objectives towards undertaking statutory operational functions whilst maintaining and, if possible, enhancing the natural river environment (Harper et al., 1995).

TOWARDS SUCCESSFUL RIVER REHABILITATION

Returning a river to a former or more semi-natural state is a challenging prospect. In its purest sense, restoration means the full structural and functional return of a river to a pre-disturbance state, an opportunity which rarely occurs for the whole of a river. Indeed, this is even difficult to achieve for discrete sections or reaches of a river (Jasperse, 1998), because the conditions prior to the disturbance may not be fully known. Even if the natural conditions are known, they may not fit the present-day hydrological conditions as a result of, for example, land-use change and regulation in the upstream reaches. In most cases rehabilitation (a more achievable objective) indicates a process which can be defined as the partial functional and/or structural return to a former or pre-degradation condition or putting back to good working order. This challenge of rehabilitating rivers is one which truly shows the importance of emerging sciences such as landscape ecology. It involves at one end of the spectrum the understanding of interrelations between animals, plants, water and sediment and acquiring knowledge as to how to manipulate the processes involved. At the other end, it involves a more geomorphological and hydrological perspective, setting the river channel in its floodplain and catchment in order to ensure that manipulations at the local scale are sustainable into the future.

Successful river rehabilitation relies on bringing together the skills and determination of specialists from a range of disciplines at all stages of the process, from conception and planning through to implementation and appraisal (de Waal et al., 1995). The co-operation of a range of practitioners enables them not only to achieve high standards within their own sphere of specialism but also to be appreciative and sensitive to the needs of other parts of the landscape and other specialists who are trying to manage them. The range of experts includes geomorphologists and hydrologists (Bostelmann et al., 1998; Gippel and Collier, 1998), ecologists (Simons and Boeters, 1998), landscape architects (Hickie, 1998), and planners (Jasperse, 1998). Successful examples of projects integrating these skills, and highlighted in this volume, include the River Leen, Nottingham, UK (Hickie, 1998), the River Murray, Adelaide, Australia (Jensen, 1998) and the Aque Alte Canal, Italy (Gumiero et al., 1998).

STATING THE AIM

At the outset, probably the single most important stage in the rehabilitation process is to establish and agree the aim or aims which are both achievable within the time-scale of the project and sustainable thereafter. Different parties will have different expectations, and if a large measure of all-round satisfaction is to be achieved, it is important not to encourage false hopes. In order to return a river to a condition of better working order it is necessary to decide on the scope of the rehabilitation: what is the type of river, i.e. upland/lowland, rural/urban, and what does it include, i.e. channel, riparian zone, floodplain, catchment? What length is to be rehabilitated? What is meant by 'working order' and what level of rehabilitation is to be achieved, i.e. visual, ecological, water quality, biodiversity or any combination of these? The stated aims need to describe the intent and the direction in which a project will go. The more specific details can be defined through a number of secondary objectives.

A common aim of rehabilitation projects is to make the river more natural, and 'naturalness' has become an important element of the enhancement ethos. Primarily it is controlled by river type, structure and geographical position in the landscape. Indications and insight into naturalness can be achieved through personal experience of the river and the region, information provided through river or water authorities, and data from local biological records centre(s) along with old maps and other relevant archives. A common term that is used in the rehabilitation process is that of the reference point, i.e. the former state or condition of the river or any of its attributes, or the vision of the return of the river to a better working order. These reference points will vary widely depending on the landscape type and its cultural history, and will be unique for each rehabilitation scheme embarked upon. One drawback of this is the difficulty of standardising methodologies between different rehabilitation schemes. Techniques that will work in one scheme will not necessarily be of use in another project. Despite this, some degree of strategic planning is necessary for all rehabilitation schemes as they are invariably being undertaken in landscapes subject to a wide range of other uses, both consumptive, e.g. water abstraction and irrigation, and non-consumptive, e.g. recreation and landscape aesthetics.

THE NEED TO INCORPORATE REHABILITATION INTO THE PLANNING PROCESS

There are many hundreds of kilometres of rivers in Europe alone which are badly in need of rehabilitation and the likelihood of their receiving appropriate attention depends on the success today's schemes achieve and the support they generate. However, the future health of rivers cannot depend solely on the successful implementation of one-off initiatives or trial pioneering projects. Rehabilitation needs to be built into the planning process for both town and country (Wade *et al.*, 1993). Much has been achieved within the rural environment: a recent review of 66 rehabilitation projects across Europe identified 68% as being within the rural context; 11% of the projects were purely urban in character with 21% being both urban and rural (de Waal *et al.*, 1995).

Town and city councils have for some time recognised the need to treat rivers as more than just conduits carrying waste and drainage away from the urban environment and have subsequently started to incorporate rivers into the broader urban planning process. As early as medieval times the people of Cambridge, UK, identified the importance of improving the foul nature of the River Cam running through the city centre (Benstead, 1968). The development of this river and its floodplain to create the present-day landscape of this university city must be one of the earliest successes in river rehabilitation. This example has included significant lengths of the river channel, whereas other more recent projects are more concerned with prestigious waterfront developments such as the River Medlock, Manchester, UK (Central Manchester Development Corporation, 1990). In both cases, however, the water body itself has been identified as the focus of attention within the wider landscape and has served to encourage broad co-operation between a range of agencies and the contribution of substantial funding to ensure the financing of the project. In the case of the River Leen, Nottingham, UK, the National Rivers Authority (now the Environment Agency) co-ordinated the project and the emphasis was put upon the river as a linear feature in the urban landscape (Hickie, 1998). Its rehabilitation enabled the local community to gain access to the waterfront by means of riverside parks and for housing estates to benefit from river views. Overall, it achieved an improved value of the local amenity in Nottingham. This project involved co-operation between landscape architects, planners (in both the Environment Agency and the local authority), ecologists and river engineers. In contrast, the Manchester Development Corporation aimed to create an urban landscape centred on the River Medlock which would encourage industry and commerce to locate in an inner city situation previously blighted not only by river degradation, but by overall environmental decline. This project necessitated involvement from landscape architects, geomorphologists, ecologists, planners, civil and hydraulic engineers, and quantity and structural surveyors, and a substantial financial input from the local authority and local businesses. Both cities have recognised these rehabilitation projects in local plans, the former as a completed project with maintenance needs into the future, the latter as the intention for river improvement into the future.

A planning feature which is included in many town or city plans in the UK is the green or wildlife corridor or urban greenway, and many of these have a river as the core element. In London, the London Ecology Unit has recommended that Borough Councils should confer a planning status on such corridors, encouraging habitat conservation and riparian habitat continuity, and to limit development which might damage the corridor. Whilst the ecological role of such corridors underpins their designation and inclusion in planning documents, town and city authorities also emphasise the importance of public access at points along these linear features.

Rivers in the urban environment are also frequently designated as sites of nature conservation interest, again conferring protection into the future. Such sites are usually of local interest and, again, part of their value is that they can be enjoyed and appreciated by the local community. This contrasts with those sites of nature conservation value in the rural environment which, because of their more natural condition, often have a higher relative value placed upon them. This can result in the public being discouraged from using these sites as amenity areas. Rivers in the urban environment can also serve a navigation role and are often linked into a wider navigation network, e.g. the Amstel in the Netherlands. This integrates the corridor and inner city development strategies, bringing people into the urban environment and making this a pleasurable and interesting experience.

As described earlier, the Environment Agency in England and Wales has a responsibility to take the opportunity to enhance rivers and has undertaken a number of rehabilitation projects and is supporting the River Restoration Project (Holmes, 1998). Other UK government agencies have also developed roles in relation to rivers and their rehabilitation. The stated policy of the Forestry Commission (1991) in the UK is to conserve not only wildlife but also soil, water quality and fish. Such a policy can incorporate a more sympathetic use of the land and also encourages a more integrated approach to rivers and forestry, for example the planting of more appropriate trees and the creation of buffer zones. A scheme in Scotland effectively returned a stream to a more natural condition by allowing the riparian vegetation to regenerate naturally after the Norway spruce (*Picea abies*), which had been planted right up to the edge of the stream, was felled (Adamson, 1993). Alterations in the light regime encourage the recovery of ground flora and stream invertebrates with the expectation of a decrease in the rate of erosion and the consequent benefits for downstream habitats. A recent initiative promoted by English Nature and the Ministry of Agriculture, Fisheries and Food in England has encouraged farmers and other landowners to produce a water-level management plan for the watercourses passing through their land. This is a form of rehabilitiation in its own right, but has also provided the stimulus to develop more substantial rehabilitation of watercourses in mostly rural and lowland environments.

Buffer zones are of high potential value in any rehabilitation scheme as they provide a flexible tool for acting as a filter between activities in the catchment and the water body. Water bodies reflect those land uses to varying degrees via the degradation of their water quality and wildlife habitat value. There is a growing amount of evidence to suggest that riparian buffer zones can play a major role in proactive landscape management and that they are a potentially valuable tool for ecologically sound management of river catchments (Large and Petts, 1994). While the primary focus of buffer zones would be on the river corridor itself, use of these features enables more holistic catchment approaches to be adopted. Whilst a riparian zone, comprising no more that the river or stream bank, can have considerable ecological importance, buffers can be situated as far back as the floodplain–terrace boundary. This allows full use to be made of the potential of buffer zones to act as links between patches of habitat and provide valuable wildlife corridors through the river–floodplain landscape. The buffer zone does not necessarily need to be positioned at the stream-side to be effective. However, design, management and maintenance of these features will depend on the geographical setting. Commitment to habitat maintenance is importance, as the rehabilitation value may progressively decline through lack of planned aftercare. Planning is necessary to ensure that such maintenance is minimal and set in a long-term context.

Changes in agricultural policies experienced in the European Union have created opportunities for a range of interested parties, from governments to individual landowners, to devise schemes to conserve habitats within the rural environment. Good examples of these initiatives in the UK are the Countryside Stewardship Scheme and the designation of Environmentally Sensitive Areas. Such initiatives typically incorporate an element of financial support to return to a more traditional form of land use, including watercourse management.

Another good example of a more integrated approach towards sustainable management of river environments are the catchment management plans compiled by agencies such as the Environment Agency in England and Wales, and the Ministry of Agriculture, Nature Management and Fishery in the Netherlands (Jaspere, 1998). As Newson (1992) states, this is the result of two aspects of system thinking – firstly the interdependence of constituent parts, and secondly time-dependent behaviour – becoming relevant in many river issues. Pollution and conservation have assumed prominance amongst river management issues in the last two decades. The challenge therefore will be to recognise that the entire river corridor has potential for rehabilitation and, as Gardiner (1992) recommends, the corridor should be treated as a single unit with emphasis on both its conservation restoration and public access.

THE NEED FOR EFFECTIVE COMMUNICATION

As stated earlier, in any rehabilitation effort there is a need for an integrated approach. With a wide range of expertise being involved in rehabilitation, the principal requirement becomes one of effective communication, not only between the specialists, but also between specialists and the general public. This is perhaps one of the greatest challenges facing river workers as we approach the beginning of the 21st century. On the institutional side, the agencies directly involved in rehabilitation initiatives should include a combination of local authorities, nature conservation bodies, water authorities, farmers union and fisheries. On the consultant's side, an interdisciplinary team should be involved, including civil engineers, landscape engineers, biologists/ecologists, sedimentologists, hydrologists and geomorphologists. Brookes (1990) has also identified the need to form a partnership between the public organisation carrying out the works and the affected landowners. In the initial stages a plan should outline the broad option for rehabilitation and form a basis for landowner negotiation and public consultation. The River Leitha, Burgenland, Austria, is a good example of a successful rehabilitation project that arose from intensive liaison between water authorities and the local community. This has subsequently led to the reinstatement of a meander and the creation of a wetland on a section of community-owned land. However, further negotiation with the farming community will be necessary in order to complete the scheme, and to hydrologically reconnect other sections of the floodplain with remnant forest by removal of sections of the floodbank (de Waal *et al.*, 1995).

That there is also a need to educate people and to make people aware of the need for, and the existence of, river rehabilitation projects. This education should be targeted not only at planners, decision-makers, politicians, farmers/landowners, practitioners, water managers/engineers, pressure groups and scientists, but again also at the public (all age groups) and this requires a marketing approach. The best way of educating people is showing not only the local community but also all relevant user groups the benefits of river rehabilitation through practical examples, by making use of visual material, as well as by involving these people directly in rehabilitation projects. The River Restoration Project, UK, is one example that established international demonstration projects and developed methods for establishing partnerships for structured collaboration for river rehabilitation between institutions and interested landholders with differing powers, resources and responsibilities who share a common aim of improving rivers (Holmes, 1998). However, conflicting arguments remain as to whether projects should be 'people-led' or 'project-led' or for that matter 'bottom-up' or 'top-down' respectively. While we need to develop an integrated approach, principally by developing a strategic framework for rehabilitation, increased public awareness can be viewed as being tantamount to support.

THE NEED FOR A REALISTIC PROJECT SCALE AND TIME-SCALE

Integrated approaches are more likely to succeed if work is undertaken at a realistic scale. This will make quantification of true costs and benefits easier to achieve. It will also make it easier to concentrate on natural, self-sustaining regimes. The majority of the rehabilitation schemes carried out in Europe thus far involve river sector lengths of 5 km or less (de Waal *et al.*, 1995). However, these so-called small-scale rehabilitation projects often arise from opportunities which fall outside the remit of strategic initiatives, for example opportunities resulting from road development or as a planning gain for industrial development. While these opportunistic schemes are important providers of demonstration sites, they may not be sustainable in the long term. Upstream and/or downstream of the rehabilitated stretch the ecology, hydrology and/or geomorphology of the river is still degraded, a factor which can have negative feedback on the rehabilitated stretch. Locally the stretch may be improved in the short term, though the long-term impact is very unpredictable. For example, a 200 m long stretch of a degraded reach of the River Bush, Northern Ireland, was rehabilitated using different techniques to introduce salmonid spawning and nursery habitat. Shortly after the rehabilitation works were undertaken, the introduced stones were displaced and the newly created habitat destroyed by high winter flows (de Waal et al., 1995).

In terms of ecological issues, specific questions may be difficult to answer in the immediate time-scale for any rehabilitation scheme. While ecological trends are fairly well understood, real numbers may be difficult to provide. Harper et al. (1995) highlight the circular nature of this problem, suggesting that information collection and interpretation by ecologists is focused on both structural and functional aspects of river ecosystems and their constituent habitats, species and communities. The collection of functional information is expensive and therefore the pattern has become one of collection from a few sites and extrapolation to other sites of similar structure. Management, however, has tended to use ecological information biased towards the more structural aspects of habitats and their component species and communities. In some cases, this has led to difficulties when management schemes have failed to incorporate function. Despite this, planners often demand extremely specific answers which can lead to problems when attempting to incorporate both average and extreme conditions into project specifications.

Another factor difficult to specify is the quantification of the time-scales of recovery from both degradation and rehabilitation efforts themselves. Other questions that are posed are 'do we have enough science?' or 'do we have the right science?' In calling for the case for river conservation to be made more effectively, Boon (1992) pointed out that conservation management is often ineffective because the requirements of important species or communities are not known. There is, undoubtedly, a real need for more emphasis on this currently unfashionable area of the science.

Offsetting these drawbacks, however, are the significant advances made in stream ecology since the early 1980s, principally through the river continuum, serial discontinuity and flood pulse concepts. These have highlighted the need for holistic approaches in our efforts to promote catchment management. The aim must be long-term sustainable management and our vision for rehabilitation also needs to be in balance with the context of today, a feature recognised by the *Leitbild* (or vision) concept as put forward by Kern (1992). This recognises that the *Leitbild* is an ideal solution for the stream and its floodplain, but that the final project design will be one that is optimal under present conditions, opportunities and constraints.

SUMMARY

The first and possibly most important step in any rehabilitation scheme is to define a *Leitbild* and associated project objectives. Once the principal objectives have been chosen, the site selected should ensure (i) the maximum probability that the objective(s) can be met, (ii) that rehabilitation can be carried out at a reasonable cost, and (iii) that the long-term maintenance costs of the system are not excessive or hidden. The surrounding land use must be considered, as well as existing planning strategies for the river and its landscape. It must also be borne in mind that projects designed to mimic natural processes in the fluvial landscape will be more likely to yield less expensive and more satisfactory solutions in the long term. Mitsch and Gosselink (1993) point out that one problem with this approach however is that naturally designed schemes may not develop as predictably as more artificially designed systems may. They have also highlighted the general principles of eco-technology which can be applied to the creation of wetlands; principles which also have direct relevance when discussing the rehabilitation of rivers. These include designing systems that have the following characteristics:

1. involve minimum maintenance costs in the long term;
2. use the potential energy of the river as a natural subsidy;
3. are suitable for the existing climate and hydrological regime;
4. can cope with, and indeed are designed for, extreme events as well as average conditions;
5. incorporate the various interested parties, i.e. involve an integrated approach;
6. have a long-term approach (rehabilitation schemes will not become functional overnight);
7. are designed for function and not for form;
8. are not over-engineered, but are as natural as possible.

In addition, rehabilitation schemes will require consideration of the ethos of the project, i.e. is rehabilitation, enhancement or restoration the required course of action? To aid this decision-making and the rehabilitation planning process, there must be (i) a search for relevant documentation and appropriate data regarding the site selected prior to implementation of the scheme, (ii) inclusion of a detailed budget which can be compared with actual costs, and (iii) the establishment before implementation has taken place of a programme for post-project appraisal. In all rehabilitation schemes there is a need for appraisal and evaluation methods to ascertain the success of integrated efforts at rehabilitation. This evaluation needs to be tailored to the objectives of individual schemes, but requires at the same time a standard approach in order to permit comparison between schemes. Post-project appraisal and evaluation can be expensive.

Four primary objectives can be identified in order to achieve a more effective implementation of strategic frameworks. Communication between interest groups needs to be initiated and optimised. At the same time, public education and targeting specific user groups is essential. The public should be made aware of the rehabilitation implications, e.g. the new risks involved with floodplain inundation and the untidiness of nature, which will promote realistic expectations. By increasing the educational component of the activities, political goodwill and secure funding for future works can be created. To that end it is imperative that information should be shared wherever possible and a multidisciplinary approach to river rehabilitation should be adopted. In advocating this, Morgan (1978) has stated: 'Good policy analysis recognises that the physical truth may be poorly or incompletely known. Its objective is to evaluate, order and structure incomplete knowledge so as to allow decisions to be made with as complete an understanding of the current state of knowledge.'

Thus far, one of the principal shortcomings of many rehabilitation schemes has been the 'disappearance' of the results and information gained from pre- and post-project appraisal into unpublished reports, i.e. the so-called 'grey' literature, held by government and other organisations. Consequently, there is a great need to increase communication via mainstream international journals, whilst ensuring that the scientific base of our endeavours remains rigorous. Balancing these requirements will provide a major challenge to workers in the field of river rehabilitation as we approach the 21st century.

REFERENCES

Adamson, J. (1993) Revegetation of streamsides following felling, monitoring at Kirk Burn. In Glimmerveen, I. and Ritchie, A. (eds) *What is the Value of River Woodlands?* Institute of Chartered Foresters and British Ecological Society, pp. 9–12.

Benstead, C.R. (1968) *Portrait of Cambridge*. Robert Hale, London.

Boon, P.J. (1992) Essential elements in the case for river conservation. In Boon, P.J., Calow, P. and Petts, G.E. (eds) *River Conservation and Management*. Wiley, Chichester, pp. 11–34.

Boon, P.J. (1995) The relevance of ecology to the statutory protection of British Rivers. In Harper, D.M. and Ferguson, A.J.D. (eds) *The Ecological Basis for River Management*. Wiley, Chichester, pp. 239–250.

Bostelmann, R., Braukmann, U., Briem, E., Fleischhacker, T., Humborg, G., Nadolny, I., Scheurlen, K. and Weibel, U. (1998) An approach to classification of natural streams and floodplains in south-west Germany. In de Waal, L.C., Large, A.R.G. and Wade, P.M. (eds) *Rehabilitation of Rivers: Principles and Implementation*. Wiley, Chichester, pp. 31–55.

Brookes, A. (1990) Restoration and enhancement of engineered river channels: some European experiences. *Regulated Rivers: Research and Management*, **5**, 45–56.

Central Manchester Development Corporation (1990) The River Medlock improvement scheme. CMDC, Manchester.

European Communities (1985) Council Directive of 27 June 1985 on the assessment of the effects of certain public and private projects on the environment (85/337/EEC). *Official Journal of the European Communities*, **L175**, 40–48.

European Communities (1992) Council Directive of 21 May 1992 on the conservation of natural habitats and of wild fauna and flora (92/43/EEC). *Official Journal of the European Communities*, **L206**, 7–50.

Forestry Commission (1991) *Forests and Water Guidelines*. Her Majesty's Stationery Office, London.

Gardiner, J.L. (1992) Catchment planning: the way forward for river protection in the UK. In Boon, P.J., Calow, P. and Petts, G.E. (eds) *River Conservation and Management*. Wiley, Chichester, pp. 397–406.

Gippel, C.J. and Collier, K. (1998) Rehabilitation of waterways in Australia and New Zealand. In de Waal, L.C., Large, A.R.G. and Wade, P.M. (eds) *Rehabilitation of Rivers: Principles and Implementation*. Wiley, Chichester, pp. 269–300.

Gumiero, B., Rizzoli, M., Salmoiraghi, G. and Santini, R. (1998) Rehabilitation of a lowland drainage canal on the Po River Plain. In de Waal, L.C., Large, A.R.G. and Wade, P.M. (eds) *Rehabilitation of Rivers: Principles and Implementation*. Wiley, Chichester, pp. 251–267.

Harper, D., Smith, C., Barham, P. and Howell, R. (1995). The ecological basis for the management of the natural river environment. In Harper, D.M. and Ferguson, A.J.D. (eds) *The Ecological Basis for River Management*. Wiley, Chichester, pp. 219–238.

Hickie, D. (1998) Strategic approaches to river rehabilitation: River Leen and River Derwent. In de Waal, L.C., Large, A.R.G. and Wade, P.M. (eds) *Rehabilitation of Rivers: Principles and Implementation*. Wiley, Chichester, pp. 149–158.

Holmes, N.T.H. (1998) The River Restoration Project, UK. In de Waal, L.C., Large, A.R.G. and Wade, P.M. (eds), *Rehabilitation of Rivers: Principles and Implementation*. Wiley, Chichester, pp. 133–148.

Jasperse, P. (1998). Policy networks and the success of lowland stream rehabilitation projects in the Netherlands. In de Waal, L.C., Large, A.R.G. and Wade, P.M. (eds) *Rehabilitation of Rivers: Principles and Implementation*. Wiley, Chichester, pp. 13–29.

Jensen, A. (1998) Rehabilitation of the River Murray: identifying causes of degradation and options for bringing the environment into the management equation. In de Waal, L.C., Large, A.R.G. and Wade, P.M. (eds), *Rehabilitation of Rivers: Principles and Implementation*. Wiley, Chichester, pp. 215–236.

Kern, K. (1992) Rehabilitation of streams in south-west Germany. In Boon, P.J., Calow, P. and Petts, G.E. (eds) *River Conservation and Management*. Wiley, Chichester, pp. 0–0.

Large, A.R.G. and Petts, G.E. (1994) Rehabilitation of river margins. In Calow, P. and Petts, G.E. (eds) *The Rivers Handbook*, volume 2. Blackwell Scientific Publications, Oxford, pp. 401–418.

Mitsch, W.J. and Gosselink, J.G. (1993) *Wetlands*, 2nd edition. Van Nostrand Reinhold, New York.

Morgan, M.G. (1978) Bad science and good policy analysis. *Science*, **201**, 971.

Newson, M.D. (1992) River conservation and catchment management: a UK perspective. In Boon, P.J., Calow, P. and Petts, G.E. (eds) *River Conservation and Management*. Wiley, Chichester, pp. 386–396.

Petts, G.E. and Wood, R. (eds) (1988) Regulated rivers in the United Kingdom. *Regulated Rivers, Special Issue*, 2.

Petts, G.E., Maddock, I., Bickerton, M. and Ferguson, A.J.D. (1995) Linking hydrology and ecology: the scientific basis for river management. In Harper, D.M. and Ferguson, A.J.D. (eds) *The Ecological Basis for River Management*. Wiley, Chichester, pp. 1–16.

Simons, J. and Boeters, R. (1998) Ecologically sound approaches to river bank management. In de Waal, L.C., Large, A.R.G. and Wade, P.M. (eds) *Rehabilitation of Rivers: Principles and Implementation*. Wiley, Chichester, pp. 57–85.

Waal, L.C. de, Large, A.R.G., Gippel, C.J. and Wade, P.M. (1995) River and floodplain rehabilitation in Western Europe: opportunities and constraints. *Archive für Hydrobiologie Suppl. 101, Large Rivers 9*, **3/4**, 1–15.

Wade, P.M., Waal, L.C. de and Haukeland, A. (1993) Integrating river rehabilitation into the planning process – a comparison of Norway and England. In Wolf, P. (ed.) Ökologische Gewässersanierung im Spannungsfeld zwischen Natur und Kultur. *Wasser, Abwasser, Abfall*, Schriftenreihe des Fachgebietes Siedlungswasserwirtschaft. Universität Gesamthochschule Kassel, Germany, **11**, 85–92.

Part One
PRINCIPLES

2 Policy Networks and the Success of Lowland Stream Rehabilitation Projects in the Netherlands

PETER JASPERSE
Previously University of Twente, Civil Engineering and Management, The Netherlands

INTRODUCTION

Since the 1930s most lowland streams in the Netherlands have been regulated for the purpose of land reclamation and drainage schemes. In the 1970s, regulation was undertaken to lower the groundwater levels in large areas. Meanders were cut off, bends straightened, channels widened and deepened, bank vegetation removed and weirs were placed at regular intervals to regulate flow, reduce flooding and sediment transport, and to retain minimum water levels in summer for irrigation purposes (Tolkamp, 1980). At present, 96% of all stream kilometres in the Netherlands are degraded to some degree, with only 4% remaining as semi-natural lowland streams.

The degradation of stream systems has a number of related problems, including water quality (adversely affected by regional economic development), water quantity (low flows in summer, peak flows in winter) and ecological problems (a decrease in native species diversity). The water systems can no longer function in a sound ecological way and water user requirements with respect to quantity and quality cannot be met. This necessitates the development of rehabilitation policies for the regional water systems. Some efforts have already been made in this direction (e.g. Van der Hoek and Higler, 1993).

Rehabilitation of lowland streams should therefore create a situation where the necessary conditions of an unimpaired hydrology, a restored groundwater flow and protection of seepage areas are present (Verdonschot *et al.*, 1993), facilitating the natural development of lowland streams.

Until recently, the Dutch water manager was concerned with a limited stretch of a stream (Higler, 1993), focusing the rehabilitation activities on local conditions only. In these situations the inputs of water and nutrients are not under the control of the rehabilitation project. Verdonschot *et al.* (1993) consider that efforts should be made to rehabilitate lowland streams with an emphasis on the larger systems. The main objective in such rehabilitation schemes should be large-scale nature development. However, catchment rehabilitation is seldom undertaken, which makes river rehabilitation somewhat cosmetic.

THEORETICAL FRAMEWORK: INTEGRATED WATER MANAGEMENT AND POLICY NETWORKS

DEGRADATION OF WATER SYSTEMS

In recent decades three major changes have taken place in Dutch water management. Firstly, problems related to the functioning of the water system due to the abovementioned events have led to degradation of the water system (Table 2.1). Secondly, the number of interest and user groups has increased and subsequently the use of water has changed. Thirdly, after decades of degradation of stream systems, their natural function has been recognised as an important one, shifting the emphasis to nature conservation and, more importantly, to natural development in the catchment area of lowland streams.

The problems indicated in Table 2.1 can be characterised by the dimensions of complexity and uncertainty. The problems are complex: each problem has a growing number of causes and the causes and/or outcomes of a problem are all interrelated. The cause or outcome of one problem can weaken or strengthen another (an example is discussed in a later section) and obviously the solution to one problem depends on the solution to another. Uncertainty arises mostly from a lack of scientific knowledge about cause–effect relations. This means that the effects of certain measures are not known before they are put into practice, e.g. the lack of knowledge about diffuse transport processes and the behaviour of nitrates and phosphates in the soil. On the basis of these two dimensions, four types of problems can be recognised (Figure 2.1).

In lowland stream rehabilitation, a shift can be perceived in problem solving from a simple and certain problem to a complex and uncertain problem. It becomes clear that all problems affect the natural function of a stream (Table 2.1) and only if rehabilitation is able to solve these problems can the natural function be restored.

INTEGRATED WATER MANAGEMENT

A more integrated approach is needed in water management as the individual water manager, now facing a complexity of problems, can no longer deal with these problems separately. This need for an integrated water management approach was recognised by the Dutch government and in 1985 led to a new policy concept: *integrated water management* – a concept aimed at problem solving. The idea of integrated water management is based on an increased knowledge of the relationship between the different water functions, the management activities and the interest groups. Many government agencies have responsibilities for different aspects of river management and consequently the problem solving capacity of the Dutch government is distributed over many departments.

In the Netherlands a distinction can be made between two types of rehabilitation policies for regional water systems: the natural lowland stream rehabilitation policy and the area directed policy. The natural lowland stream rehabilitation policy dates back to the early 1970s and the area directed policy was first mentioned in 1984 in the Government Note 'More than the Sum of the Parts'. In recent years, area directed policy at a provincial level has become almost synonymous with integrated water management.

The main difference between natural lowland stream rehabilitation policy and area

Table 2.1 Events, problems, causes and affected functions in the Dutch regional water systems

Events	Problems	Causes	Affected functions
Supply of nutrients	Eutrophication, acidification, diffuse pollution	Agriculture, industry, urbanisation, recreation	Nature conservation, drinking water, fisheries, recreation
Supply of organic micro-pollution/heavy metals	Point-source pollution	Industry, agriculture, urbanisation	Nature conservation, drinking water, fisheries, recreation, agriculture
	Diffuse-source pollution	Industry, agriculture, urbanisation, recreation	Nature conservation, drinking water, fisheries, recreation
	Pollution of water bed	Industry, agriculture, urbanisation	Nature conservation, fisheries, drinking water, recreation, agriculture
Drought	Drought (lowering of groundwater level)	Water abstraction for agriculture and drinking water, urbanisation, industry	Nature conservation, urban areas, agriculture

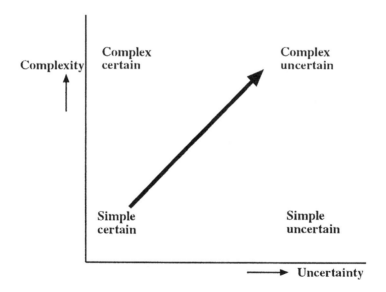

Figure 2.1 Problems characterised by their complexity and uncertainty

directed policy is one of scale. Natural lowland stream rehabilitation is merely aimed at the rehabilitation of factors such as dimensions, current velocity, acidity, load of organic material and nutrient load. Area directed policy is aimed at the rehabilitation of the hydrology and water quality at the catchment scale. Both policies are aimed at solving problems but the problems differ in complexity and uncertainty. The problems at the catchment level are relatively complex and uncertain, while the problems concerning factors such as dimensions, current velocity and so on are relatively simple and certain. Consequently, implementation of an area directed policy is far more uncertain compared to natural lowland stream rehabilitation. Clearly, implementation of natural lowland stream rehabilitation cannot solve problems which appear at the catchment scale. Similarly, if an area directed policy is executed to solve simple problems an overkill in problem solving capacity can occur. Natural lowland stream rehabilitation policy is more appropriate for relatively simple and certain problems and area directed policy for relatively complex and uncertain problems. The general hypothesis is that a specific type of rehabilitation policy should be developed to solve a specific type of problem and that government agencies should realise, before implementing a certain type of rehabilitation policy, what type of problem it is that they want to solve.

POLICY NETWORKS

Area directed policy and natural lowland stream rehabilitation policy also differ in the size of the policy network, defined as the total number of agencies responsible for solving a particular problem (Glasbergen and Driessen, 1993). Area directed policies usually have larger policy networks than natural lowland stream rehabilitation policies. This difference in size is due to the difference in scale and the complexity of the problems. As the problem solving capacity is distributed over an increasing number of government and non-government agencies, effective and efficient problem solving becomes less likely.

The problem solving capacity could be increased if the water boards, who are primarily responsible for the Dutch regional surface waters, would work together with those other public and private agencies which are also part of the problem solving capacity, that is by creating a policy network. At present, an agency is only responsible for one or more aspects of the water system complex but not for solving the problem-complex as a whole. Besides, all agencies have different interests and objectives regarding management of the water system and the objectives are likely to conflict at least in part (Glasbergen, 1989). A policy network should consist of both public and private agencies; the latter because economic activities can be a cause of the problems and if they could be changed or stopped, a part of the problem-complex would be solved. The consequences of a low problem solving capacity are at least a delay in the project and at worst its cancellation. The remaining questions concern how a policy network should be created and what the driving force is behind the creation of a policy network. Mobilisation means participation of relevant agencies in solving the problem-complex and the activation of the policy network. But mobilisation itself does not determine the eventual size of the policy network. The eventual size of the network is determined by the nature and extent of the problem. The policy network is complete when the problem solving capacity cannot be further increased. Therefore, if the initiator (probably a water board or a province) of a policy network fails to mobilise the complete policy network, the problem solving capacity will consequently be less and the rehabilitation policy could fail.

There might already be a consensus within a policy network, but this is rather rare. A non-consensus on subjects such as problem definitions, objectives, solutions, measures and financial aspects is more likely, and consultation between the members of the policy network in order to achieve a consensus on these subjects will be very important with regard to the problem solving capacity. However, the more agencies there are in a policy network, the harder it is to reach a consensus. This is a common situation, especially with area directed policies (Glasbergen and Driessen, 1993) when no consensus exists. The balance of power between the agencies can play an important role (Driessen, 1990; Van Heffen, 1993). This balance is made up by those agencies with the power 'to do' and those with the power 'to stop' (Maarse, 1991). In the Netherlands, for example, the individual rights of landowners are better protected by laws than the collective rights of the environment. Therefore the power of landowners is larger than that of nature conservation agencies (Driessen, 1990). Table 2.2 summarises problem solving capacity in relation to mobilisation and consensus.

Glasbergen et al. (1992) identify several other factors that contribute to the problem solving capacity. These factors include the following:

- the initiative taken by a government agency which, on the basis of a legitimate task, has the main responsibility for the problem;
- a system of project management;
- an arrangement for feedback on decision points to all agencies;
- a legal status for the plan, committing participating agencies to the plan;
- a time schedule;
- an organisation established or commissioned to conduct and supervise the implementation of the measures.

With good project management, an open planning process and, as a result, a clear plan, the chances of success are high (Glasbergen et al., 1992).

AN OVERVIEW OF DUTCH LOWLAND STREAM REHABILITATION PROJECTS

Lowland stream rehabilitation has been included in several plans of the national government including the National Nature Plan (1990) and the third National Water Management Plan (1989). The main focus in the National Nature Plan is the so-called Ecological Main Structure (EMS), implemented in 1991. The main goal of the EMS is nature conservation on a large scale in the Netherlands. The EMS consists of ecological corridors and nature development areas, and it has given nature conservation in the Netherlands a new incentive. Lowland streams are an important part of the EMS and

Table 2.2 Mobilisation and consensus in relation to the resulting problem solving capacity

Policy network	Problem solving capacity
Incomplete mobilised network	Moderate/low
Complete mobilised network	
Consensus among agencies	High
No consensus, but agencies will compromise	Moderate
No consensus and the opponents do not have enough power to influence or to stop the implementation	High
No consensus but the opponents do have enough power to influence or to stop the implementation	Low

72% of 171 lowland stream rehabilitation projects are being carried out as part of the EMS (Hermens and Wassink, 1992).

The focus of the third National Water Management Plan is the development of ecologically sound water systems which should guarantee the sustainable use of water systems. This implies the rehabilitation of regional lowland water systems and rehabilitation to improve the natural function, a current issue in Dutch water management. The awareness of the need for lowland stream rehabilitation which has grown in recent years was also stimulated by the introduction of the policy concept of integrated water management in 1985 and by extra state-aided funding, the so-called REGIWA subsidy (REGional Integrated WAter management). In these REGIWA projects, 50% of all the costs are paid by the national government and the remaining 50% by local authorities and water boards. These projects are meant to stimulate in particular, the implementation of the EMS, integrated water management and the creation of ecologically sound river banks (Simons and Boeters, 1998). Analysis of 55 rehabilitation projects undertaken in the Netherlands (Jasperse, 1994) shows the importance of REGIWA projects, of which there are 25 in total (Table 2.3). Without this financial incentive, it is unlikely that these rehabilitation projects would ever have been implemented. The significance of this REGIWA subsidy is also shown in Table 2.4, the number of rehabilitation projects increasing significantly in 1992 and 1993. (The subsidy commenced in 1991 and was stopped in 1995.)

Examination of the phase reached by these projects (Table 2.5) shows that many were still under development, with only 33% completed in 1995. Some projects are in more

Table 2.3 Types of stream rehabilitation projects, 1977–1995

Rehabilitation project type	No. of projects ($n = 55$)
REGIWA projects	25
Land reclamation projects	7
National Nature Plan projects	13
Others	10

Table 2.4 Number of rehabilitation projects, 1977–1995

Year	Number of rehabilitation projects (completed or under way) ($n = 55$)
1977–1985	4
1986	4
1987	3
1988	3
1989	5
1990	2
1991	4
1992	14
1993	10
1994	2
1995	1
Unknown	3

Table 2.5 Current phases of stream rehabilitation projects undertaken between 1977 and 1995

Project phase	No. of rehabilitation projects ($n = 55$)
Planned	9
Development	17
Implementation	11
Completed	18
Unknown	12

Table 2.6 Distribution of stream rehabilitation projects in eastern and southern part of the Netherlands, 1977–1995

Province	No. of rehabilitation projects ($n = 55$)
Groningen	1
Drenthe	10
Overijssel	13
Gelderland	18
Utrecht	1
Noord-Brabant	10
Limburg	5

than one project phase. The geographical distribution of projects varies (Table 2.6) with lowland stream rehabilitation projects being restricted to the eastern and southern parts of the Netherlands, the regions in which most of the lowland streams in the Netherlands are found.

Tables 2.7 and 2.8 give an insight into the degree of participation of agencies in the projects and the number of agencies within the policy networks. The water boards are legally responsible for the management of regional surface waters (including many lowland streams) and have therefore taken the initiative in most rehabilitation projects (Table 2.7). The Rijkswaterstaat (the national water authority) took the initiative for the rehabilitation project 'Overijsselse Vecht'. The Rijkswaterstaat is primarily concerned with the national waterways and this is the only regional rehabilitation project where it has taken the initiative.

The degree of participation for the different agencies varies, e.g. that of the municipalities

Table 2.7 Degree of participation of agencies in lowland stream rehabilitation projects ($n = 47$)

Agencies	Degree of participation
Water Board	100%
Water Purification Board	47%
NBLF (Department of Nature, Forest, Landscape and Wildlife Management)	47%
Landinrichtingsdienst (Planning Department)	85%
State Forestry Services	38%
Provincial and national government agencies	77%
Municipalities	40%

Table 2.8 Number of other agencies besides the initiator in policy networks in lowland stream rehabilitation projects

Number of participants	1	2	3	4	5	6	7	8	9	10	11	> 11
Number of projects with policy network ($n = 47$)	–	1	2	2	4	8	6	5	3	1	5	8

is rather low, whereas that of the Landinrichtingsdienst is very high due to their expertise and financial support. The involvement of municipalities is important because of their responsibility for development plans. This planning document has to be changed before a rehabilitation project can be implemented.

Most rehabilitation projects have policy networks with six to eight participants (Table 2.8). Rehabilitation projects with smaller policy networks ($n < 5$) are rare due to the introduction of integrated water management. In eight rehabilitation projects the number of participants exceeds 11, some of which are associated with area directed policies.

TWO EXAMPLES OF DUTCH LOWLAND STREAM REHABILITATION PROJECTS

THE TONGELREEP: PROJECT ACHELSE KLUIS

The rehabilitation of the lowland stream Tongelreep is part of the nature development project called Achelse Kluis. The total area consists of 107 ha of agricultural land and woodland areas purchased by the Ministry of Agriculture, Nature Management and Fisheries (LNV) in 1989 (Figure 2.2). The Tongelreep flows through the area over a length of 1.4 km and was channelised in 1890. The original planform can still be recognised on the map in the boundaries of the municipalities. The Tongelreep is part of the catchment of the River Dommel, with a total catchment of 9625 ha of which 6870 ha is situated in Belgium.

As a result of land-use changes such as land-ownership changes, and changes in soil structure, drainage and urban developments, the original discharge pattern of the Tongelreep has been significantly altered. The consequences are substantial fluctuations in discharge (average $0.85 \text{ m}^3 \text{ s}^{-1}$ and peak discharges as high as $3.5 \text{ m}^3 \text{ s}^{-1}$). From 1985 onwards the water quality of the Tongelreep deteriorated as a result of discharges of waste water from the Belgian municipality of Achel.

The corridor of this stream is part of the EMS. Rehabilitation by means of spontaneous natural processes, such as erosion, sedimentation, inundation, seepage and succession of vegetation, is the main objective of the project. The implementation will create a meandering middle reach of a lowland stream (Figure 2.3) with areas of bank erosion and sedimentation, with overbank flow occurring three or four times a year.

The initiative for the rehabilitation of the Tongelreep was taken by the Ministry of Agriculture, Nature Management and Fisheries, especially the Department of Nature, Forest, Landscape and Wildlife Management (NBLF) (formerly the NMF – Nature, Environment and Fauna), in co-operation with the State Forestry Service (Staatsbosbeheer). The policy network is shown in Figure 2.4.

LOWLAND STREAM REHABILITATION IN THE NETHERLANDS

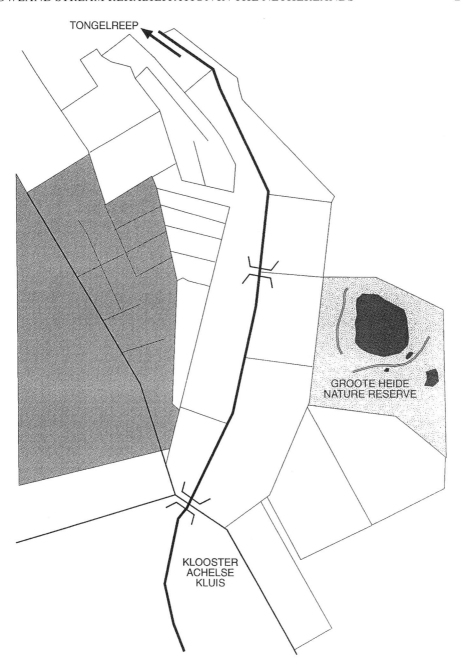

Figure 2.2 Channelised Tongelreep in the area of Achelse Kluis

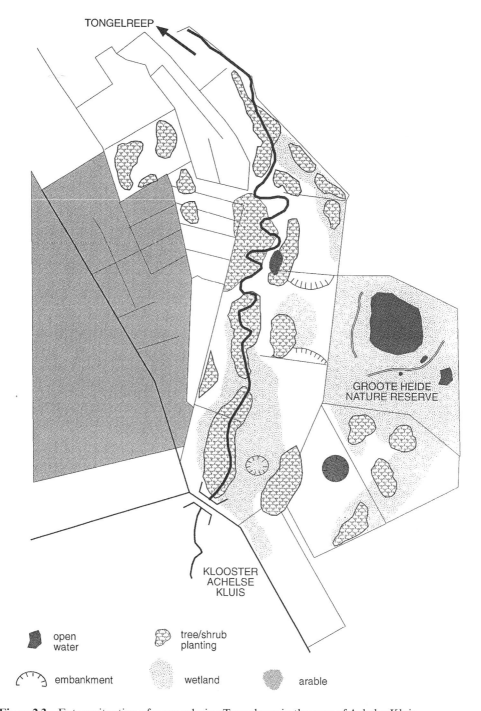

Figure 2.3 Future situation of a meandering Tongelreep in the area of Achelse Kluis

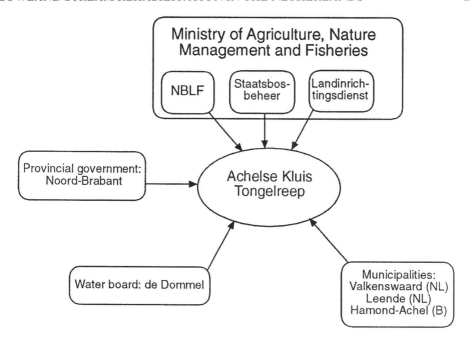

Figure 2.4 Policy network for the Achelse Kluis project (NL, the Netherlands; B, Belgium)

The water board, De Dommel, the two Dutch municipalities, Leende and Valkenswaard, and the Belgian municipality, Hamond-Achel, participated but did not contribute to the project financially. Initially the Landinrichtingsdienst was not invited but became involved at a later stage because of their expertise and good relationship with the water board which became very important (Ambting, 1992). In the early stages of the project a conflict existed between the State Forestry Service and the other participants regarding the objectives and management of the project. Both Dutch municipalities were able to achieve limited influence of the policy process in relation to the functioning of the rehabilitated stretch of the river. Research was undertaken by two engineering consultants, Heidemij and Oranjewoud, resulting in a plan for a meandering stretch of the Tongelreep (Figure 2.3). The planform of the stream is influenced by the present discharge pattern in which inundation is prevented because of the polluted nature of the water. Even in its rehabilitated condition, inundation will not be allowed and a diversion canal is planned to cope with peak discharges without the bank vegetation being adversely affected by the eutrophic water. Whilst this rehabilitation project will not improve the water quality, it is expected that in the future the quality will improve and inundations will then be allowed for approximately 20 days a year. The post-project sedimentation pattern downstream should remain the same as the pre-project pattern.

The implementation of the rehabilitation project has been delayed by two local farmers even though all permits and licences have been issued. The development of a meandering stream will affect the groundwater levels in the surrounding area and therefore improve natural habitat and species diversity as well as processes in a wider terrestrial area. Unfortunately, the farmers view this as a constraint on their modern agricultural business.

The purchase of land along the Tongelreep stream has been an important factor contributing to the success so far, improving the opportunities for nature development, reducing the constraint of incorporating the plan into the development plans of the local municipalities and thereby eliminating a possible factor of delay. However, ownership of the land will not solve the water quality problem of the stream. It can only solve the water quantity problem (i.e. discharge fluctuations) for a restricted stretch of the Tongelreep. Although a meandering stream will improve the nature conservation value of this particular area, there are some doubts regarding ecological recovery of the whole stream, including up- and downstream stretches.

The functioning of this specific policy network has been successful. However, some agencies were not invited who were essential in the solving of certain problems. This shortcoming was partly corrected later in the process. For example, the obstruction by two local farmers might have been avoided if representatives of the local farmer agencies had been invited to join the policy network at an earlier stage. In terms of the theoretical framework put forward in this chapter, this represents a failure in the mobilisation of a complete policy network, resulting in only a moderate problem solving capacity. It explains the delay of the rehabilitation project.

REHABILITATION OF THE DINKEL

The rehabilitation of the Dinkel system is an example of lowland stream rehabilitation within an area directed policy. The Dinkel system consists of the River Dinkel and its tributaries (Figure 2.5), with the whole Dinkel valley as part of the north-east Twente area directed policy. The River Dinkel is also part of the EMS. The Dinkel flows through Germany and the Netherlands, with a total length of about 84 km. The middle reach is situated in the Netherlands, in the region of Twente in the eastern part of the province of Overijssel, with a catchment area of 22 400 ha. The Dinkel flows in a wide valley, the west flank of which is an ice-pushed ridge, a remnant of the last ice age, with an attitude of up to 70 m. The eastern part of the valley is lower but the gradient is steep. Three main environmental problems can be identified in the region of Twente: acidification, eutrophication and drought. These three problems have a cumulative impact on the Dinkel by acting synergistically on each other (Roos and Vintges, 1991). For example, the groundwater levels in Twente have dropped and therefore acid rain can penetrate deeper into the soil. With a lower groundwater level, the mineralisation of nitrate is slower and therefore more nitrate will enter the stream, causing eutrophication. The main causes of groundwater level reduction are groundwater abstraction for agricultural irrigation and the drinking water industry.

In the 1970s, the Dinkel was channelised in its upstream reach for the purpose of drainage in order to enable urban development in the German town Gronau. This changed the discharge pattern of the Dinkel over time, resulting in low flows in summer ($0.40\,m^3\,s^{-1}$) and high flows in winter (up to $11\,m^3\,s^{-1}$). The water quality was affected by nutrient and organic load from the effluent of sewage plants and intensive cattle and pig farming. As a result of low dilution the water quality is especially poor in summertime. It is obvious that the solution to these problems should be sought in the wider area of Twente.

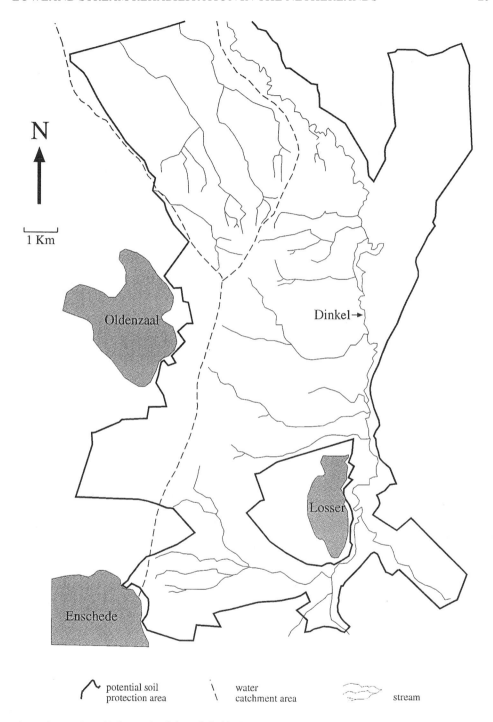

Figure 2.5 The middle reach of the Dinkel in Twente

The provincial government realised the complexity of these problems and in March 1991 decided to develop a new policy, the so-called north-east Twente area directed policy. The main objective was to solve the area-wide environmental problems by 1996. The participants involved in the policy network are shown in Figure 2.6.

Several organisations, including the State Forestry Service, the Overijssel Landscape Foundation and the Overijssel Nature and Environment Foundation, were invited during the development phase of the area directed policy. The invitation of the latter caused some conflict with the Overijssel Agricultural Board as these agencies have different views and objectives on agriculture and the environment, which led to a lack of consensus in the policy network during the development of the area directed policy. This caused a delay of several months before the plan was agreed in March 1993.

The whole plan initially comprised 15 projects and at a later stage three more projects were added. Three of these projects are especially important for the Dinkel; namely, nutrient discharge to surface water, conservation and rehabilitation of the natural stream channels, and retention of water. More research has to be done before the whole policy can be implemented. At present, little can be said on the measures to be taken or even the effects these might have. Proposed measures include research into water volume and retention possibilities in marshes and upper reaches of the system, the location and amounts of nutrient discharge, and the purchase of land adjacent to the streams in order to re-establish meandering and create buffer strips. The final results will not be clear until after the year 2000.

This area directed policy is characterised by multiple projects, many executive and

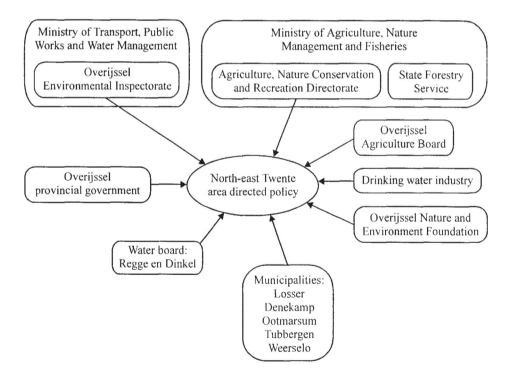

Figure 2.6 Policy network of the area directed policy in north-east Twente

participants, and long-term planning and implementation. Unlike the Tongelreep project, it is impossible to predict whether or not the area directed policy will be successful in restoring the Dinkel system. The choice of this type of policy network is appropriate because the problems the water board is facing are too complex to be solved by this agency alone. At present, most projects are being implemented, but agencies such as the Landinrichtingsdienst have not committed themselves to full implementation. For the implementation of the area directed policy, the provincial government is dependent upon the co-operation of the five Dutch municipalities and the water board Regge en Dinkel (Figure 2.6). There is some concern that the farmers' interest is considered by the municipalities to be of more importance than the collective interest in nature conservation.

DISCUSSION AND CONCLUSIONS

The present-day problems of stream rehabilitation are complex and uncertain and need different kinds of problem solving. A distinction is made between area directed policy and natural lowland stream rehabilitation policy as mechanisms for stimulating rehabilitation projects. Both types of policy are implemented by policy networks and these are considered to be the key to the success of lowland stream rehabilitation as they can improve the problem solving capacity. The problem solving capacity is dependent on the mobilisation of the policy network and the consensus within the policy network. The consequences of a low problem solving capacity are at least a delay of the project and may mean the cancellation of the entire project. However, further research is needed to achieve an understanding of other important factors that can improve problem solving capacity, e.g. the negotiation processes within policy networks.

In the first part of this chapter a theoretical framework was developed in which the main elements were identified as the type of problem, the type of rehabilitation policy, mobilisation of the policy network, and the consensus, all of which have an impact on the problem solving capacity and could possibly delay implementation. Some of these elements were explored in two case studies and they are summarised in Table 2.9.

The aim of the Tongelreep rehabilitation project was to solve a relatively simple problem compared to the complex problems of the Dinkel project, resulting in a different type of rehabilitation policy. The delay in time can be explained by the low to moderate problem solving capacity, leading to the mobilisation of an incomplete policy network and a lack of consensus, although the involved agencies did achieve a compromise. Comparing the participants of the policy networks for the two projects described, some differences become clear (Figures 2.4 and 2.6, Table 2.10). The main difference is the complete absence of private agencies in the Tongelreep rehabilitation project, causing a delay of three years. If farmers' agencies had been involved, this might have been avoided. In the Dinkel rehabilitation project an important agency, the Landinrichtingsdienst, is still absent for reasons explained in earlier sections. The number of participants is different (eight in the Tongelreep project, 12 in the Dinkel project), and there was a lack of consensus in both projects. This lack of consensus was caused mainly by the different views on the objectives of the involved agencies. The involvement of municipalities is important because of their responsibility for the development plans. This plan plays an important role in town and country planning and rehabilitation

Table 2.9 Comparison of two stream rehabilitation projects: the Tongelreep and Dinkel projects

	Tongelreep	Dinkel
Problem	Discharge fluctuations	1. Acidification 2. Eutrophication 3. Drought
Factors	Dimensions	1. Hydrology 2. Water quality
Type of rehabilitation policy	Natural lowland stream rehabilitation policy	Area directed policy
Mobilisation of policy network	Incomplete policy network	Almost complete policy network
Consensus	Compromise between agencies	Compromise, but potential conflicts are still possible
Problem solving capacity	Low to moderate	Moderate to low
Delay	Implementation is delayed for three years	A delay of several months in the planning/development phase

Table 2.10 Comparison of the participants of the policy network of the rehabilitation projects

Participants	Tongelreep	Dinkel
National government agencies	Ministry of Agriculture, Nature and Fisheries Department of Nature, Forest, Landscape and Wildlife management (initiator) State Forestry Service (initiator) Landinrichtingsdienst	Directorate of Agriculture, Nature Conservation and Recreation
Provincial government	Noord-Brabant	Overijssel (initiator)
Water board	De Dommel	Regge en Dinkel
Municipalities	Valkenswaard, Leende, Hamond-Achel	Denekamp, Losser, Ootmarsum, Tubbergen, Weerselo
Private agencies		Overijssel Nature and Environment, Overijssel Agricultural Board, Overijssel Water Company

schemes and projects are implemented only if they correspond with this development plan. In many lowland stream rehabilitation projects, the municipalities are not directly involved in the rehabilitation projects.

Both the water boards, De Dommel and Regge en Dinkel, have responsibilities for both water quantity and water quality.

This chapter has dealt with the theoretical notions relating to the problem solving capacity of policy networks and the success of lowland stream rehabilitation projects. The case studies investigated have supported these notions. The use of policy networks is growing and the environmental projects involved show good results in that a better and quicker accomplishment of the objectives is achieved. It is concluded that the establishment of policy networks will aid in the success of lowland stream rehabilitation projects in the long term.

REFERENCES

Ambting, R. (1992) *Muddling Through*. University of Twente (in Dutch).
Driessen, P.P.J. (1990) *Landinrichting gewogen, De plaats van de milieu-, natuur en landschapsbelangen in het landinrichtingsbeleid*. Thesis, Kerckebosch, Zeist (in Dutch).
Glasbergen, P. (1989) *Beleidsnetwerken rond milieuproblemen*. VUGA, 's Gravenhage (in Dutch).
Glasbergen, P. and Driessen, P.P.J. (1993) *Innovatie in het gebiedsgericht beleid*. SDU Uitgeverij, Den Haag (in Dutch).
Glasbergen, P., Driessen, P.P.J. and Van der Veen, J. (1992) Gebiedsgericht milieubeleid; reflectie op een sturingsconcept. *Milieu en recht*, 393–402 (in Dutch).
Hermens, E.M.P. and Wassink, W.Th. (1992) Natuurtechnisch beekherstel in Nederland. *Landinrichting*, **32**(5), 8–15 (in Dutch).
Higler, L.W.G. (1993) The riparian community of north-west European lowland streams. *Freshwater Biology*, **29**(2), 229–241.
Jasperse, P. (1994) *Beekherstel in Nederland, een farce of fictie?* Universiteit Twente, Vakgroep Civiele Techniek, Enschede (in Dutch).
Maarse, J.A.M. (1991) Hoe valt effectiviteit van beleid te verklaren? Deel I Empirisch onderzoek. *Beleidsevaluatie*, 122–136, Samson H.D. Tjeenk Willink, Alphen aan den Rijn (in Dutch).
Roos, R. and Vintges, V. (1991) *Het milieu van de natuur, Herkennen van verzuring, vermesting en verdroging in de natuur*. Utrecht (in Dutch).
Simons, J. and Boeters, R. (1998) A systematic approach to ecologically sound river bank management. In Waal, L.C. de, Large, A.R.G. and Wade, P.M. (eds) *Rehabilitation of Rivers: Principles and Implementation*. Wiley, Chichester, pp. 57–85.
Tolkamp, H.H. (1980) *Organism–substrate relationships in lowland streams*. Thesis, Wageningen.
Van der Hoek, W. and Higler, B. (1993) *Natuurontwikkeling in beken en beekdalen; verkennende studie naar de mogelijkheden van natuurontwikkeling in beek- en beekdalsystemen in Nederland*. NBP-onderzoeksrapport 3, DLO-Instituut voor Bos en Natuuronderzoek, Wageningen (in Dutch).
Van Heffen, O. (1993) *Beleidsontwerp, resultaat en omgeving, De problematiek van randgroepjongeren in zeven Nederlandse gemeenten*. Febodruk, Enschede (in Dutch).
Verdonschot, P.F.M., Schot, J.A. and Scheffers, M.R. (1993) *Potentiële ecologische ontwikkelingen in het aquatisch deel van het Dinkelsysteem*. Onderdeel van het NBP-project Ecologisch onderzoek Dinkelsysteem, IBN-rapport 004, Instituut voor Bos en Natuuronderzoek, Wageningen (in Dutch).

3 An Approach to Classification of Natural Streams and Floodplains in South-west Germany

ROLF BOSTELMANN,[1] ULRICH BRAUKMANN,[2] ELMAR BRIEM,[3]
THOMAS FLEISCHHACKER,[3] GEORG HUMBORG,[4] INA NADOLNY,[4]
KARL SCHEURLEN[5] and UWE WEIBEL[5]

[1] *Arbeitsgemeinschaft Landschaftsökologie, Karlsruhe, Germany*
[2] *Landesanstalt für Umweltschutz Baden-Württemberg, Karlsruhe, Germany*
[3] *Institut für Geographie und Geoökologie, Universität Karlsruhe, Germany*
[4] *Institut für Wasserbau und Kulturtechnik, Universität Karlsruhe, Germany*
[5] *Institut für Umweltstudien, Heidelberg, Germany*

INTRODUCTION

Stream and river rehabilitation has become an important topic in Germany with the growing concern for ecological issues among the public, legislators and politicians. Although great improvements were made in intensive sewage treatment and water purification, improvement of water quality is often not enough. It became apparent that the lack of natural structures in streams was the limiting factor for ecological recovery of running waters. Rehabilitation projects were often carried out based on subjective ideas about what a stream should look like, with very little basis in experience or scientific research. Due to a fundamental lack of knowledge on streams, the ecological benefits were limited. As a result, the State of Baden-Württemberg (south-west Germany) initiated a programme for stream rehabilitation. In this context the Stream Research Group (SRG, Forschungsgruppe Fliessgewässer) was created. The members of the SRG and the disciplines represented are shown in Figure 3.1.

The *Leitbild* concept (Kern, 1992) provides guidance for ecologically sound stream management. *Leitbild* (or 'vision') is the model image of a stream in its 'natural' condition and can be used to compare the actual state with the desired state. It includes the floodplain as an integral part of the stream ecosystem. On the basis of its *Leitbild*, measures can be adapted to obtain the desired condition taking the individual properties of any stream into account. The *Leitbild* must therefore be stream-type-related and multidisciplinary.

One approach to ecologically sound stream management practice is to carry out individual case studies, though they are generally expensive and time-consuming. Additionally, results are often not applicable to other streams. Consequently, facing limited public budgets, such an approach would benefit only a minority of streams and the

---

Rehabilitation of Rivers: Principles and Implementation. Edited by L. C. de Waal, A. R. G. Large and P. M. Wade.
© 1998 John Wiley & Sons Ltd.

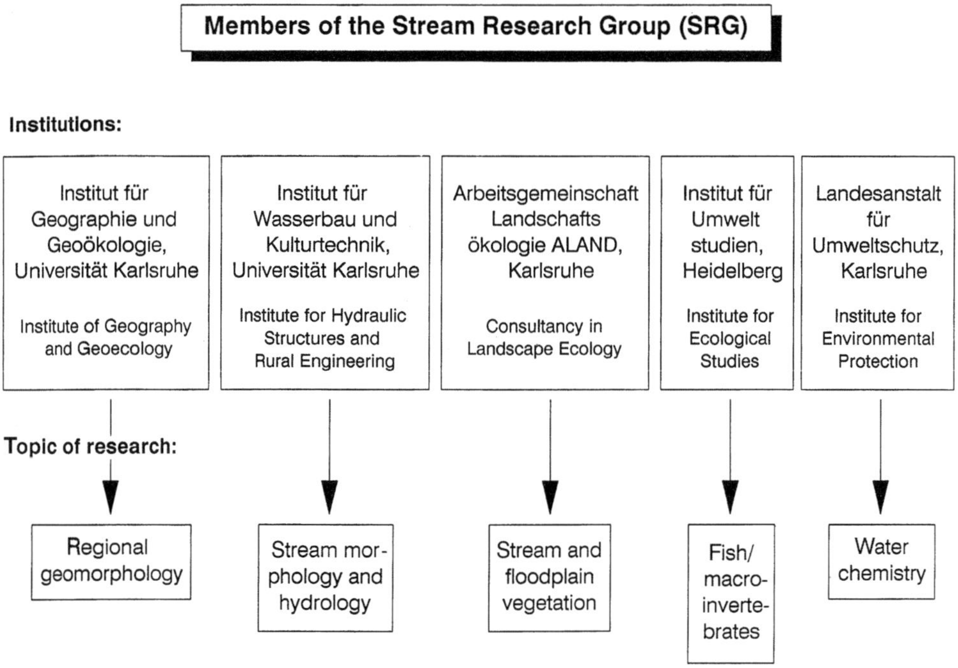

Members of the Stream Research Group (SRG)

Institutions:

Institut für Geographie und Geoökologie, Universität Karlsruhe	Institut für Wasserbau und Kulturtechnik, Universität Karlsruhe	Arbeitsgemeinschaft Landschafts ökologie ALAND, Karlsruhe	Institut für Umwelt studien, Heidelberg	Landesanstalt für Umweltschutz, Karlsruhe
Institute of Geography and Geoecology	Institute for Hydraulic Structures and Rural Engineering	Consultancy in Landscape Ecology	Institute for Ecological Studies	Institute for Environmental Protection

Topic of research:

Regional geomorphology	Stream morphology and hydrology	Stream and floodplain vegetation	Fish/ macro- inverte- brates	Water chemistry

Figure 3.1 Members and disciplines involved in the Stream Research Group (SRG)

majority would be neglected. The aim of the SRG is to develop a stream management tool that can be applied to most small streams in south-west Germany, i.e. a verifiable methodology suitable for use throughout the region.

The SRG studied the complicated task of developing a methodology for a large variety of streams which cannot all be treated in the same way. The grouping of streams with similar properties, i.e. a stream classification, would simplify the work and possibly lead to a stream-type-dependent *Leitbild*. Many different stream classification approaches are available (Otto and Braukmann, 1983; Moog and Wimmer, 1990; Naiman *et al.*, 1992). However they often describe stream types in a rather general fashion, whereas stream management requires detailed as well as generic information. The aim of the project being carried out by the SRG is to overcome the gap between theories of stream classification and the creation of a stream-type-dependent *Leitbild* (Forschungsgruppe Fliessgewässer, 1993; Nadolny, 1994; Humborg, 1995).

METHODOLOGY FOR STREAM CLASSIFICATION AND STREAM TYPES OF BADEN-WÜRTTEMBERG

Two study areas in south-west Germany, the Oberrheinebene (Upper Rhine Graben) and the Odenwald (a Triassic Bunter Sandstone area), were selected to develop methods for stream classification (Figure 3.2). The methodology has been tested over a period of five years.

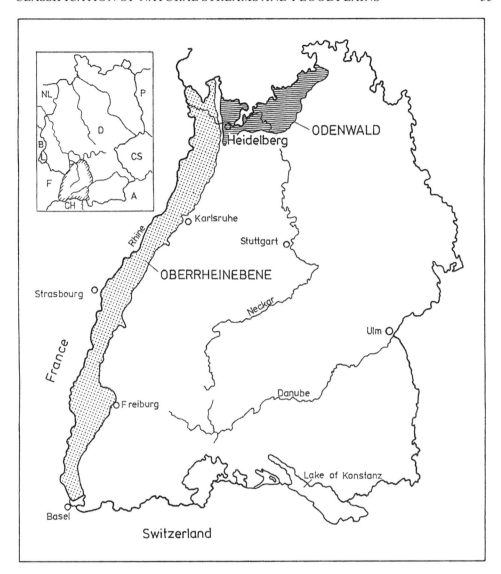

Figure 3.2 Study areas in south-west Germany

The Oberrheinebene is a lowland area whereas the Odenwald represents an upland area. The streams considered in these study areas are small natural streams with a maximum bankfull width of $c.$ 20–25 m; a maximum mean annual discharge of $c.$ 2–3 m^3 s^{-1}; and a maximum catchment area of $c.$ 150 km^2.

'Natural' in this context is defined as being as free of human influence as possible. It is unrealistic to aim for pristine conditions. Therefore it was decided that the *Leitbild* should reflect the natural properties of the stream at the beginning of the Industrial Revolution, thus integrating the cultural history of the landscape into the *Leitbild* concept. At that time the agricultural practices and other human activities were still adapted to local geo-ecological conditions.

The variety of criteria for grouping streams is as wide as the variety of streams itself. Geomorphology is the only discipline with a framework acceptable to all members of the SRG, providing a hierarchical grading of criteria and maintaining consistent interrelationships between biotic and abiotic factors. Geomorphologists have the broadest comparative view on streams and take the long-term genesis into account. Otto and Braukmann (1983) present a classification approach for the former West Germany – the first attempt to combine regional geomorphological patterns and stream patterns in Germany. Lowland, upland and alpine streams are described in general but are not differentiated further. This work provides a good overview of German streams, though the approach is not sophisticated enough for practical local application.

To obtain more detailed regional information with higher spatial resolution, the 66 natural units (Meynen and Schmithüsen, 1962) of the State of Baden-Württemberg were used. They were grouped by the SRG into 11 Main Stream Regions (MSRs). In regard to geology, relief and resulting geomorphological patterns, weathering products and soils, groundwater, soil moisture, surface water, climate and vegetation, the MSRs form homogeneous parts of the landscape. A 1:500 000 scale map showing the MSRs and valley shapes of the streams and rivers was compiled. Figure 3.3 shows a scaled down, black and white image of the original MSR map (Forschungsgruppe Fliessgewässer, 1993).

The assumption was made, and verified for the areas investigated so far, that each MSR can be associated with at least one main stream type. The individual stream types can be further subdivided on a more detailed level of investigation. Special stream types such as bog-streams and lake effluents are not taken into consideration because of their atypical nature. Typical features used to characterise the main stream types of the MSRs are as follows:

- valley and stream density
- water chemistry (carbonate or silicate stream)
- stream planform
- roughness of bed material
- flow characteristics
- suspended load and bed load

The MSR map forms the fundamental basis for a hierarchical stream classification of the State of Baden-Württemberg. In the MSRs under investigation, semi-natural streams were identified for field research. Characteristic stream reaches were selected and surveyed by all members of the SRG. Combining the findings from all disciplines, a stream-type-dependent *Leitbild* can be assembled. The methodology of the SRG to develop a classification system for small streams and their floodplains in south-west Germany is shown in Figure 3.4.

STREAM TYPES IN THE MSRS

Two out of 11 MSRs, a Triassic Bunter Sandstone upland area (Odenwald) and the lowlands of the Oberrheinebene, have been thoroughly investigated. Studies of two more,

CLASSIFICATION OF NATURAL STREAMS AND FLOODPLAINS

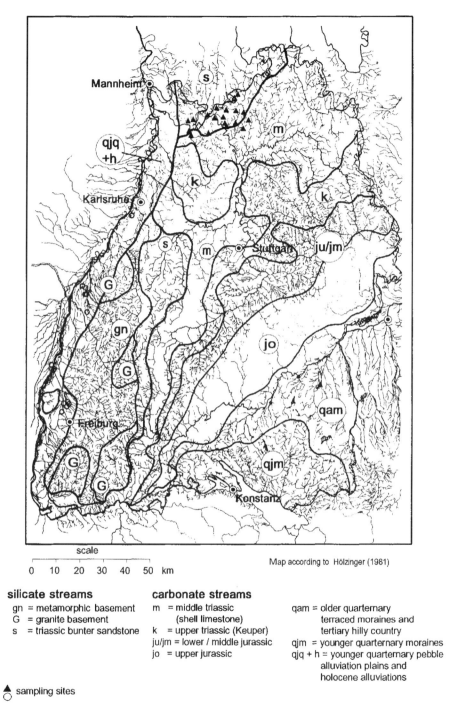

silicate streams
gn = metamorphic basement
G = granite basement
s = triassic bunter sandstone

carbonate streams
m = middle triassic
   (shell limestone)
k = upper triassic (Keuper)
ju/jm = lower / middle jurassic
jo = upper jurassic

qam = older quarternary
   terraced moraines and
   tertiary hilly country
qjm = younger quarternary moraines
qjq + h = younger quarternary pebble
   alluviation plains and
   holocene alluviations

▲ sampling sites
○

Figure 3.3 Main Stream Regions (MSRs) of the State of Baden-Württemberg, south-west Germany

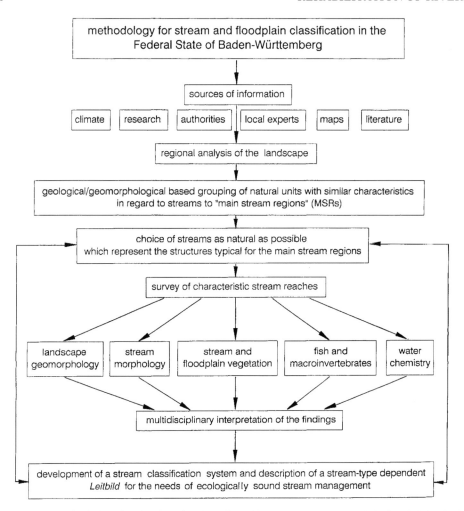

Figure 3.4 Methodology for the classification of small natural streams and their floodplains in the State of Baden-Württemberg, south-west Germany

an Upper Triassic MSR (Keuper) and a metamorphic basement MSR, meanwhile have beem completed (Forschungsgruppe Fließgewässer, 1998).

In all MSRs under investigation, in addition to the main stream type characteristic of the region, it was possible to distinguish several further stream types. In the sandstone area stream type depends on the valley shape and in the Rhine valley on the influences from mountainous upper catchment areas and the geomorphological features of the Rhine itself: terraces, alluvial fans and palaeo-floodplains. Figure 3.5 shows the hierarchical levels of stream-type classification for one MSR, a Triassic Bunter Sandstone area.

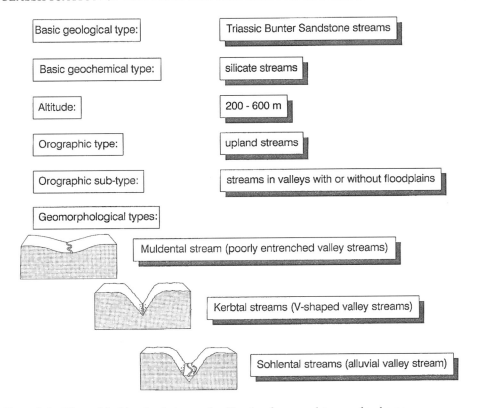

Figure 3.5 Hierarchical levels of stream classification for a sandstone upland area

ELEMENTS OF A *LEITBILD* FOR UPLAND STREAM TYPES IN THE TRIASSIC BUNTER SANDSTONE OF THE ODENWALD

Combining the findings of all disciplines represented in the SRG, a stream-type-dependent *Leitbild* can be formulated. Some features of stream types in the Triassic Bunter Sandstone MSR are given below to illustrate elements of this *Leitbild* (Forschungsgruppe Fliessgewässer, 1993).

GEOMORPHOLOGICAL ASPECTS

In the Triassic Bunter Sandstone area (Bunter) of the State of Baden-Württemberg there are only a small number of rivers and valleys (valley density: 0.7). The characteristic forms are deeply incised valleys with stepped concave longitudinal profiles. All the streams have their origins in flat depressions in the upland area and flow into cuts in the formation of the upper Bunter (so). Further downstream they reach the deeply incised higher middle Bunter (sm2/c2) with a break in the general slope. Then they flow in steep-sided V-shaped valleys with small flat floors (Figure 3.6).

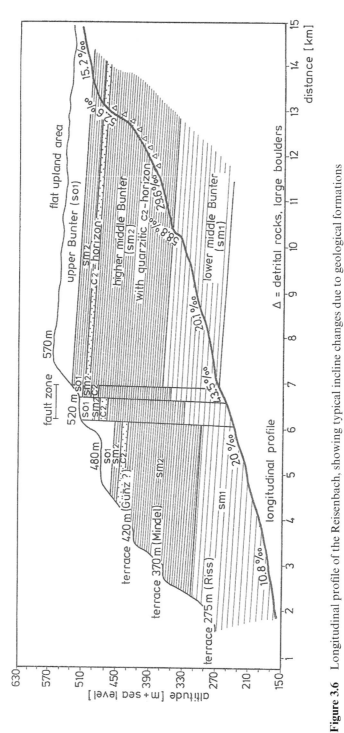

Figure 3.6 Longitudinal profile of the Reisenbach, showing typical incline changes due to geological formations

The weathering products were mainly produced in the youngest glacial period. This material consists of platy congelifracts in a comminuted sandy matrix which covers the solid rock. In the case of the quartzitic bound-down c2-horizon of the middle Bunter, large boulders arise due to weathering which cover slopes, valley bottoms and stream channels. The large effective pore volume of this residual detritus has a large water storage capacity that results in a regular runoff process. From a geomorphological point of view, the natural streams in this sandstone area can be classified into three groups (Figure 3.5):

1. *Muldental streams*: streams flowing in flat depressions in the upland channels, poorly entrenched in loamy substratum with single rock fragments (Figure 3.5).
2. *Kerbtal streams* (V-shaped valley streams): stream channels are embedded in detrital material (congelifractate) with boulders creating pool–riffle and pool–step sequences and cascades; the bed load contains boulders, stones and sand with a lack of gravel; sand is therefore the only material that will be transported at mean discharge (Figure 3.5).
3. *Sohlental streams* (V-shaped valley streams with a flat floor): streams flow over a sedimentary floor accumulated as alluvial fans in the last ice age; platy rubble and sand dominate the bed load (Figure 3.5). Cobbles pave the flat profiles of the stream beds and cause a high roughness; channel migration is therefore typical for these, as are sand accumulations on the floodplain. Sand is washed out and transported even at lower water levels; gravel and coarser bed load is dragged along at higher discharges.

A special form of small V-shaped valleys in the sandstone area of the Odenwald is the *klinge* (ravine). These valleys are characterised by great differences in elevation over short flow distances (Figure 3.7). This special valley type was created during young tectonic uplift and the downcutting of the main river (e.g. gorge of the Neckar). Due to the higher incline and steeper valley sides, the presence of boulders, cascades and log jams are more dominant characteristics in *klingen* than in ordinary V-shaped valleys.

LIMNOCHEMICAL ASPECTS

The ion balance composition and concentration of the streams in the investigated area of the Triassic Bunter Sandstone of the Odenwald show a clear similarity with some silicate streams of the Black Forest. On the other hand, the difference in the ion composition and concentration between silicate and carbonate streams is clearly seen (Figure 3.8).

Geochemically the investigated upland streams in the Odenwald form a relatively homogeneous group of the silicate type. This is represented by the uppermost *muldental* and *kerbtal* valley reaches. The silicate stream type is characterised by low to very low ion concentrations, with electrolytical conductivities of 50–250 $\mu S\ cm^{-1}$, low to very low hydrogen carbonate concentrations (under 0.7 eq l^{-1}) and low to very low calcium concentrations (under 17 mg l^{-1}) (Figure 3.9(a)). In their lower reaches some of the streams – mostly the *sohlental* variety – undergo the transition from a silicate to a carbonate type (FO1 to EL3 in Figure 3.9).

The nutrient content of the investigated streams in this area (Figure 3.9(b)) as measured by the levels of NH_4^+-N, NO_2^--N, NO_3^--N and o-PO_4^{3-}-P which are indicators of anthropogenic pollution, shows a dependence on the size of the stream and the distance from the source. This indicates the influence of nutrient effluents, especially from agricul-

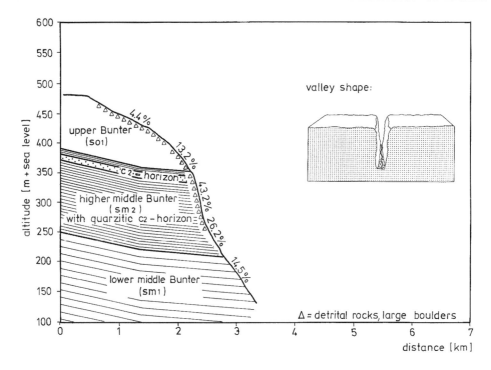

Figure 3.7 Longitudinal profile of a *klinge*

ture. On the other hand, the relatively high nutrient levels might be related to the calcium concentration, as can be observed in many unpolluted streams in rural areas.

Although most of the sampling sites were free from obvious pollution, thus representing 'natural' habitats, some non-point organic pollution and acidification could be observed. Non-point organic pollution was found in some upper reaches of of the *muldental* streams and *kerbtal* streams, leading to a species composition (Illies, 1961) more typical for a community in the lower rhitron areas of *sohlental* streams (for details, see the section on faunal aspects).

Due to their naturally low buffer capacity against the human impact of acidifying precipitations, some of the uppermost stream sections (e.g. TR1 and SA1 in Figure 3.9) can be distinguished hydrochemically as well as hydrobiologically as critically acidified, the consequence being a markedly impoverished benthic flora and fauna, especially in the macroinvertebrate groups of *Mollusca*, *Crustacea* and *Ephemeroptera*. Fish populations, especially brown trout, are also very sensitive to acidification (Gebhardt *et al.*, 1990; Landesanstalt für Umweltschutz Baden-Württemberg, 1991).

HYDRAULIC AND MORPHOLOGICAL FEATURES

Characteristic morphological patterns can be defined in association with the different stream types and linked to the different valley types. The primary identified morphological differences are channel flow capacity, longitudinal profile and stream planform.

CLASSIFICATION OF NATURAL STREAMS AND FLOODPLAINS

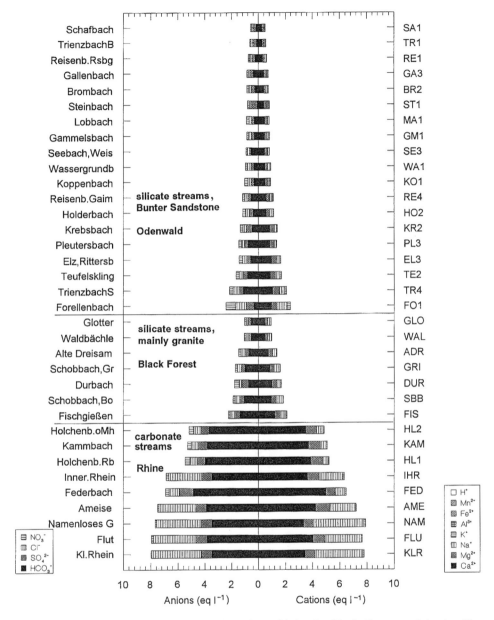

Figure 3.8 Ion balances of streams in the Odenwald, in the Black Forest and in the Oberrheinebene

Stream planform is not described in the following, whereas channel flow capacity is represented by bankfull discharge, and longitudinal profile is indicated by pool–riffle and pool–step sequences.

The bankfull discharge is widely viewed as the parameter responsible for the stream bed shape (e.g. Nixon, 1959; Wolman and Miller, 1960). Water levels at bankfull discharge

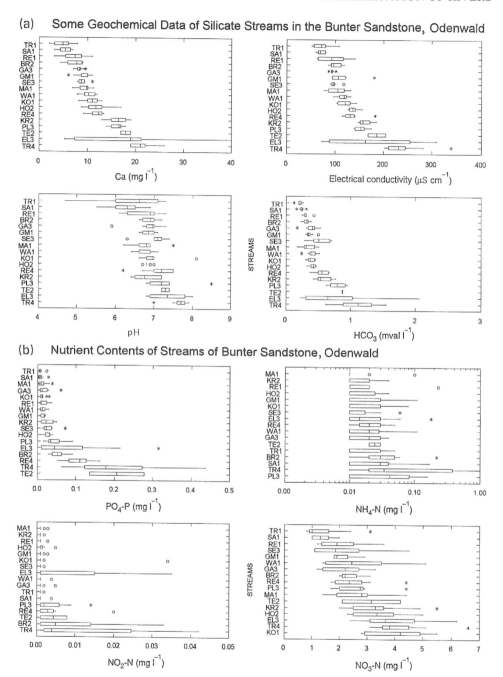

Figure 3.9 (a) Selected geochemical data (streams sorted according to ascending median of calcium content) and (b) selected nutrient data (streams sorted according to ascending median of each parameter) of silicate upland streams in the Odenwald

are used to determine morphologically relevant parameters such as width and depth. Magnitude and recurrence frequency of the bankfull discharge have been the subject of much investigation; it shows a broad scatter of values and demonstrates the impossibility of defining a single or limited range of frequencies (Williams, 1978), although many authors generally assign a return period of one to two years (Dury, 1961; Dury et al., 1963; Morisawa, 1985; Carling, 1987).

The bankfull discharge calculated for the different stream types (Figure 3.10) shows that each stream type covers a range of flow magnitudes and recurrence intervals. The floodplains of *sohlental* streams are flooded frequently (twice a year or at least every second year) whereas *muldental* streams show the widest range of frequencies, i.e. between approximately once a year and once every 50 years. *Kerbtal* streams overtop their banks only during very rare events, with a frequency of 50 to 100 or even more years.

As shown above, stream bed morphology is adapted to a specific discharge capacity. This can be demonstrated by the formation of pool–riffle or pool–step sequences, which are both morphologically important features and one of the most important habitat components in upland streams. In *sohlental* streams pool–riffle sequences were found and their distribution was related to inclination and bankfull width, i.e. the wavelength from pool to pool or riffle to riffle increases with decreasing inclination and increasing width. In contrast, pool–step and cascade patterns are characteristic for steep upland streams, such as *kerbtal* streams, and occur at mean stream bed inclinations greater than 5% (Figure 3.11). Spacing is independent of both stream bed inclination and bankfull width. *Muldental* streams are in an intermediate position showing features of both *kerbtal* streams and *sohlental* streams.

Even based on a limited spectrum of stream pattern, the differences between the individual stream types and their morphological behaviour is confidently demonstrated, justifying the proposed geomorphologically oriented classification.

AQUATIC AND RIPARIAN VEGETATION

The analysis of vegetation was carried out in accordance with the principles of phytosociological methodology (Braun-Blanquet, 1964). The complexes of plant communities (Tüxen, 1978) were investigated according to the method applied by Schwabe (1987) to stream-related complexes in the Black Forest.

Vegetation complexes are spatially associated plant communities within a relatively homogeneous part of the landscape. A study of the sum of plant communities is described as sigma sociology (Σ = Greek letter for sum) (Schwabe, 1989). This method facilitates the study of patterns of the riparian and aquatic plant communities.

Apart from mosses, macrophyte communities cannot generally survive the force of the fast running water in coarse and coarse-bouldery substrate in upland stream beds, where, additionally bed-load transport during floods can cause damage. Mosses dominate the beds of the investigated streams, covering many of the boulders. The distribution of the complexes of aquatic moss communities in the Odenwald (Table 3.1) reflects the longitudinal zonation with increasing nutrient supply and increasing pH values downstream. The *Scapanietum undulatae* complex characterises the upper reaches near the source (a catchment area of about 5 km^2) with oligotrophic conditions sensitive to acidification. The typical complex for the middle and lower reaches is the *Scapanietum*

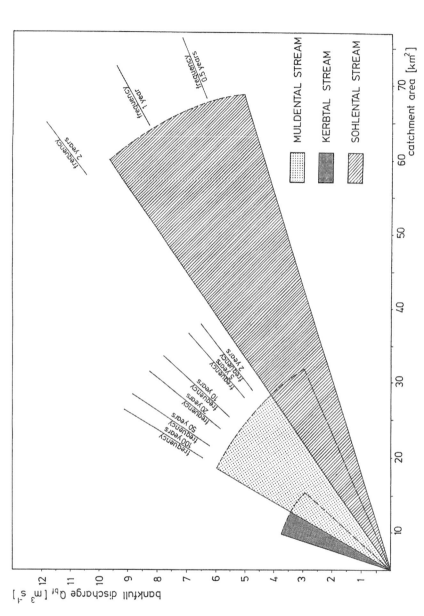

Figure 3.10 Magnitude and frequency of bankfull discharge in accordance with stream types and catchment area in the Odenwald

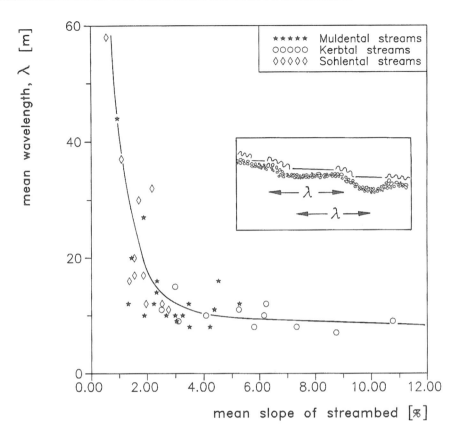

Figure 3.11 Spacing of pool–riffle and pool–step sequences as a function of stream type and stream slope in the Odenwald (pool–step sequences are found in *kerbtal* streams with slopes of stream bed exceeding 5%)

undulatae–Oxyrrhynchietum rusciformis complex. The existence of the *Oxyrrhynchietum rusciformis–Brachytecio–Hygrohypnetum* complex is related to higher nutrient supply, often human-imposed. Under more natural conditions this complex would be very rare in the investigated streams, appearing only in some of the lower reaches. If angiosperms are present, they usually grow towards the sides. Their distribution also follows that of the complexes of the moss communities. Because longitudinal zonation is related to valley form, the distribution of the stream bed complexes is also related to the stream types (in the MSRs) mentioned previously.

The characteristic riparian plant community of the investigated streams is the *Stellario-Alnetum glutinosae*, an alluvial alder forest (Table 3.2). This community is lacking in upper reaches where flow dynamics are inhibitory. Otherwise, this community dominates from the headwaters down to the lower reaches. Frequent floods about twice a year and related scour and fill at the banks are apparently fundamental factors for the presence of the *Stellario–Alnetum glutinosae* community and characterise a genuine alluvial forest. A vegetational complex is named after this *Stellario–Alnetum glutinosae* complex. It can be subdivided according to valley form, local climate and soil type (Table 3.2).

Table 3.1 Complexes of aquatic moss communities in the Odenwald

Group of community complex	1A	1B	1C	2A	2B	3
Number of investigated complexes	3	3	3	6	14	14
Average number of plant communities per complex	3	2	3	8	9	8
1 *Scapanietum undulatae* complex						
Scapanietum undulatae	3	3	3	V	V	I
Sphagnum fallax cushion	3	.	.	.	.	.
2 *Scapanietum undulatae–Oxyrrhynchietum rusciformis* complex						
Brachythecietum plumosi	.	.	3	V	V	IV
Oxyrrhynchietum rusciformis	.	.	.	V	V	V
3 *Oxyrrhynchietum rusciformis–Brachythecio–Hygrohypnetum* complex						
Brachythecio–Hygrohypnetum	.	.	.	.	V	IV
Chiloscyphus polyanthos stock	.	.	.	.	II	IV
Other aquatic moss communities						
Fontinalis antipyretica stock	.	1	.	II	III	III
Fisidentetum pusilli	.	.	.	IV	III	IV
Riccardia chamaedryfolia stock	.	.	.	II	II	+
Hygrohypnetum ochracei	.	.	.	I	.	I
Modothecetum cordaeanae	.	.	.	.	.	+
Aquatic moss community of occasionally flooded boulders						
Thamnobryum alopecurum stock	.	.	.	II	III	V
Angiosperm communities of the stream bed						
Glyceria fluitans community	1	.	1	III	II	II
Stellaria ulginosa cushion	1	.	.	I	I	I
Cardamine amara community	.	.	.	IV	III	II
Veronica beccabunga community	.	.	.	I	II	II
Mentha x *verticillata* community	.	1	.	.	+	+
Ranunculus flammula stock	1	.	.	I	.	.
Juncus bulbosus community	1	.	.	.	.	.
Veronica scutellata–Agrostis canina community	.	1	.	.	.	.
Sium erectum brook reed	.	.	.	.	+	.
Agrostis stolonifera community	.	.	.	.	.	+
Small structures in the stream bed						
Caltha palustris dots	.	.	1	II	III	I
Chrysosplenium oppositifolium cushion	.	.	.	III	IV	I
Epilobium obscurum cushion	.	.	.	II	II	II
Callitriche cf. hamulata stock	.	.	.	II	+	II
Viola palustris cushion	1	.	1	.	.	.

1, *Scapanietum undulatae* complex: A, subcomplex with *Sphagnum* fallax cushions; B, genuine subcomplex; C, subcomplex with *Brachythecietum rivularis*.
2, *Scapanietum undulatae–Oxyrrhynchietum rusciformis* complex: A, genuine subcomplex; B, subcomplex with *Brachythecio–Hygrophypnetum*.
3, *Oxyrrhynchietum rusciformis–Brachythecio–Hygrohypnetum* complex.
Arabic figures give the absolute number of findings (when there are less than five relevés for a group). Roman figures give the frequency in classes: +, 0–10%; I, > 10–20%; II, > 20–40%; III, > 40–60%; IV, > 60–80%; V, > 80–100%.

An example of the distribution of aquatic and riparian plant communities in a cross-section of an Odenwald stream is given in Figure 3.12.

FAUNAL ASPECTS

Samples of macroinvertebrates and fish were taken at 40 sites in the Bunter Sandstone of Odenwald. The samples were subjected to factor analysis, coordinate analysis and cluster analysis.

Macroinvertebrate and fish communities in the Odenwald are typical for silicate upland streams. Species composition and the relative abundance of functional feeding groups show longitudinal zonation patterns. The upper reaches of the streams in the Odenwald area often have an intermittent character. Typically, the invertebrate fauna of these sites has a low constancy and fundamental changes in species composition and biomass occur in the course of a year. Typical for these regions are species commonly found in springs, like the crustacean *Niphargus* spp. Other krenal species are the caddis-flies *Ptilocolepus granulatus* and *Rhyacophila laevis*.

Habitat structure appears to be the most important factor for the composition of the macroinvertebrate community in the middle and lower reaches. Riffle sections are characterised by the occurrence of the planarian *Polycelis felina*, the mayfly species *Baetis alpinus*, *Epeorus sylvicola* and *Rhithrogena* cf. *semicolorata*, and the water beetles *Elmis maugetii/aenea*, *Esolus angustatus* and *Limnius perrisi* (Table 3.3). All species occur in more than 60% of spring and autumn samples. This community is very similar to the lotic communities found in the upland silicate streams in the Eifel and in the mountain streams of the Black Forest (Braukmann, 1987). The lentic pool sections are characterised by the occurrence of the dipteran species *Prodiamesa olivacea*, the alderfly *Sialis fuliginosa*, the water-beetle *Oreodytes rivalis* and freshwater mussels *Pisidium* spp. Variation within the lentic community is much higher than in the lotic community, as a result of substratum stability in the riffle sections of streams in the Odenwald.

The riffle–pool sequence is also of great importance for the fish populations. For example, adult brown trout (*Salmo trutta*) live in the pool section of a stream, whereas juvenile trout settle in the riffle section. Riffles are also the main habitat for the bullhead (*Cottus gobio*).

The input and retention of allochthonous material, especially deciduous leaves, is highest in V-shaped valleys. This results in a relatively high proportion of shredders (invertebrates that feed on coarse particulate organic matter). The invertebrate fauna shows a steady change in the composition of the functional feeding groups along the river continuum. The relative abundance of collectors increases in the downstream regions, whereas the relative importance of shredders decreases. This zonation pattern is typical of streams of temperate regions (Vannote et al., 1980; Hawkins and Sedell, 1981; Statzner and Higler, 1985). The upper reaches of streams are the most sensitive to acidification. In acidic streams the composition of functional feeding groups changes because of the breakdown of crustacean populations, the most important shredder group. Acidification is a major threat to streams with low alkalinity. It leads to a loss of biodiversity, especially in the macroinvertebrate groups of snails, crustaceans and mayflies. Fish populations, especially brown trout, are also very sensitive to acidification (Gebhardt et al., 1990; Landesanstalt für Umweltschutz Baden-Württemberg, 1991).

Table 3.2 Complexes of riparian plant communities in the Odenwald

Group of community complex	1	2	3A	3B	3C	3D	4	5	6
Number of investigated complexes	3	4	3	13	10	4	6	8	2
Average number of plant communities per complex	6	6	5	6	9	7	6	6	6
1 *Sphagnum–Alnus glutinosa* swamp complex									
Sphagnum–Alnus glutinosa swamp (*Alnion glutinosae*)	3	.	.	.	.	.	.	.	.
Salix x multinervis community	2	.	.	+	.	.	.	.	.
Sphagnum palustre community	3	.	.	.	.	.	.	.	.
Sphagnum fallax community	2	.	.	.	.	.	.	.	.
2 *Carici remotae–Fraxinetum* complex									
Carici remotae–Fraxinetum	.	4	.	.	.	.	.	.	.
3 *Stellario–Alnetum glutinosae* complex									
A *Stellario–Alnetum* with *Aconitum napellus*	.	.	3	.	.	.	.	.	.
Phalarido–Petasitetum hybridi	.	.	1	.	+	.	.	.	.
B *Stellario–Alnetum*, typical form	.	.	.	V	V	.	.	.	.
C *Carex brizoides* community	.	.	.	.	V	4	V	.	.
D *Luzulo–Alnetum glutinosae*	.	.	.	.	.	4	I	.	1
4 *Carex brizoides–Alnus glutinosa* alluvial forest complex									
Carex brizoides–Alnus glutinosa alluvial forest	.	.	.	.	.	.	V	.	.
5 *Aceri–Fraxinetum* complex									
Aceri–Fraxinetum	.	.	.	.	.	.	.	II	.
Fraxinus excelsior–Acer pseudoplatanus community	.	.	.	.	.	.	.	V	.
6 *Carpinus betulus* riparian complex									
Carpinus betulus riparian community	.	2	.	II	.	.	I	II	2
Luzulo–Fagetum	.	1	.	.	.	.	.	I	2
Other forest and shrub communities									
Rubus fruticosus community	.	1	1	II	III	1	I	.	.
Rubus idaeus community	.	.	2	.	III	.	I	.	.
Salix cinerea community	.	.	.	I	II	.	.	.	.
Stellario–Carpinetum	.	1	.	.	.	.	.	.	.
Picea abies forest (*Alno–Ulmion/Alnion**)	1*	.	1	II	II	4	.	III	.
Herbal vegetation									
Urtica dioica riparian community	.	2	3	IV	III	1	III	.	.
Athyrium felix-femina riparian community	.	1	1	+	II	1	.	III	.
Scirpetum sylvatici	.	1	.	+	+	1	II	.	.

	1	2	3A	3B	3C	3D	4	5	6
Aruncus dioicus riparian community	.	.	.	.	.	.	.	I	.
Stellaria nemorum riparian community	.	.	.	II	I	.	I	.	.
Filipendula ulmaria community	.	.	2	III	V	.	.	.	.
Impatiens glandulifera community	.	.	.	II	.	.	.	.	.
Phalaridetum arundinaceae	.	.	.	I	I	.	.	.	.
Mentho–Juncetum inflexi (fragm.)	.	.	+	I	I	.	.	.	.
Meadow communities (Molinietalia)	.	.	.	.	II	.	.	.	.
Sphagno–Juncetum acutiflori	1	.	.	.	.	.	.	.	.
Knautia dipsacifolia stock	1	.	.	.	.	.	.	.	.
Polygono–Bidentetum (fragm.)	.	.	1	.	.	.	.	.	.
Ranunculus aconitifolius community	.	.	.	.	.	+	.	.	.
Holcus mollis stock	.	.	.	.	.	+	.	.	.
Glyceria fluitans community (Agropyro–Rumicion)	.	.	.	.	.	+	.	.	.
Polygonum cuspidatum community	.	.	.	.	.	.	.	.	.
Caricion fuscae community	.	.	.	.	.	.	I	.	.
Small structures of the riparian vegetation									
Luzula sylvatica stock	1	.	.	.	I	.	I	II	1
Lysimachia nemorum stock	.	.	+	+	I	.	.	I	.
Isolepis setacea community	1	.	.	.	.	.	.	.	.
Moss and lichen communities									
Pellietum epiphyllae	3	.	III	III	III	3	.	III	1
Diplophylletum albicantis	2	.	II	II	I	2	.	IV	2
Eurhynchietum praelongae	1	1	III	III	+	1	.	IV	1
Orthodicrano–Hypnetum	1	1	II	II	+	.	.	II	1
Isothecietum myosuroidis	1	.	+	.	+	1	III	III	1
Calypogeietum muellerianae	.	1	.	.	+	.	.	I	1
Paraleucobryetum longifolii	1	.	+	+	.	.	.	I	1
Uloletum crispae	.	.	+	+	.	.	.	.	.
Parmelietum furfuraceae	.	.	.	+	.	.	.	.	.

1, *Sphagnum–Alnus glutinosa* swamp complex.
2, *Carici remotae–Fraxinetum* complex.
3, *Stellario–Alnetum glutinosae* complex: A, subcomplex of *Stellario–Alnetum* with *Aconitum napellus*; B, typical subcomplex; C, subcomplex with *Carex brizoides* community; D, subcomplex with *Luzulo–Alnetum*.
4, *Carex brizoides–Alnus glutinosa* alluvial forest complex.
5, *Aceri–Fraxinetum* complex.
6, *Carpinus betulus* riparian complex.
Arabic figures give the absolute number of findings (when there are less than five relevés for a group). Roman figures give the frequency in classes: +, 0–10%; I, >10–20%; II, >20–40%; III, >40–60%; IV, >60–80%; V, >80–100%.

Valley form	Longitudinal zonation	Community of riparian woodland	Dominant tree species	Type
Muldental 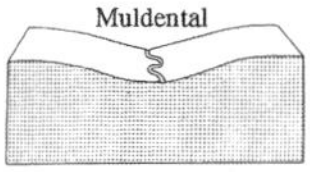	Headwaters	Humid woods of *Alnus glutinosa* dominating; *Stellario–Alnetum* mostly not yet developed	Often exclusively *Alnus glutinosa*; *Fraxinus excelsior* often missing or only single trees	A (C)
	Middle regions	*Stellario–Alnetum* dominating	Mostly in species-rich form with *Fraxinus excelsior*	C
	Special aspects	*Sphagnum–Alnus glutinosa* swamps in the spring regions of the flat upland areas	*Alnus glutinosa* mostly dominating alone; very oligotrophic form with *Betula pubescens*	A
Kerbtal 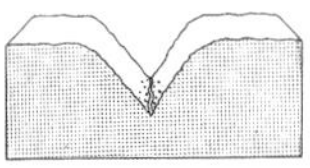	Headwaters	On boulders *Acer pseudoplatanus–Fraxinus excelsior* wood.	*Acer pseudoplatanus*, *Fraxinus excelsior*, *Ulmus glabra*	D
		Also *Stellario-Alnetum*, when the bottom of the valley is slightly trough formed.	*Alnus glutinosa* dominating; *Fraxinus excelsior* mostly only sparing	C
		Small patches of *Carpinus betulus* edges on steep banks	*Carpinus betulus*, *Corylus avellana*, *Fagus sylvatica*	E
	Special aspects	*Carici remotae–Fraxinetum* only in the less mountainous west; missing in the east	*Alnus glutinosa* mostly dominating alone	B
		'Klingen' with *Acer pseudoplatanus–Fraxinus excelsior* wood	*Acer pseudoplatanus*, *Fraxinus excelsior*, *Ulmus glabra*	D
Sohlental 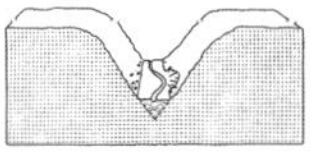	Middle and lower regions	Characterised by *Stellario–Alnetum glutinosae*	Mostly in species-rich form with *Alnus glutinosa*, *Fraxinus excelsior*, *Acer pseudoplatanus*, *Salix fragilis*, etc.	C
	Special aspects	On steep banks stretches of *Carpinus betulus* edges.	*Carpinus betulus*, *Corylus avellana*, *Fagus sylvatica*	E
		Occurrence of *Ulmus effusus* in lower regions near the Neckar River		C

Main types of riparian woodland:

A Alder stocks of the *Sphagnum–Alnus glutinosa* swamps
B Alder–maple–ash of the *Carici remotae–Fraxinetum*
C Alder–maple–ash stocks of the *Stellario–Alnetum*
D Maple–ash–elm stocks of the *Aceri–Fraxinetum*
E Hornbeam stocks on steep banks

Figure 3.12 Riparian woodland communities as a function of longitudinal zonation and valley form in the Odenwald

Table 3.3 Faunal characteristics of the main stream habitats in the Odenwald

	Habitat type	
	Riffle	Pool
General features	Fast flowing habitat	Depositing habitat
	Constantly high oxygen concentration	Varying oxygen concentration
	High abundance of filter-feeders (especially blackflies)	Macroinvertebrate community dominated by detritivores
	Large proportion of highly specialised invertebrates	Invertebrate community dominated by common species
Substratum type	Large stones, mosses	Silt, sand, detritus
Indicator species	*Polycelus felina*	*Sericostoma* cf. *personatum*
	Epeorus sylvicola	*Prodiamesa* cf. *olivacea*
	Rhitrhogena cf. *semicolorata*	*Oreodytes rivalis*
	Perlodes microcephalus	*Habrophlebia lauta*
	Rhyacophila dorsalis	*Sialis fuliginosa*
	Elmis maugetii/aenea	
	Limnius perrisi	
	Cottus gobio	
	Salmo trutta (juvenile)	*Salmo trutta* (adult)

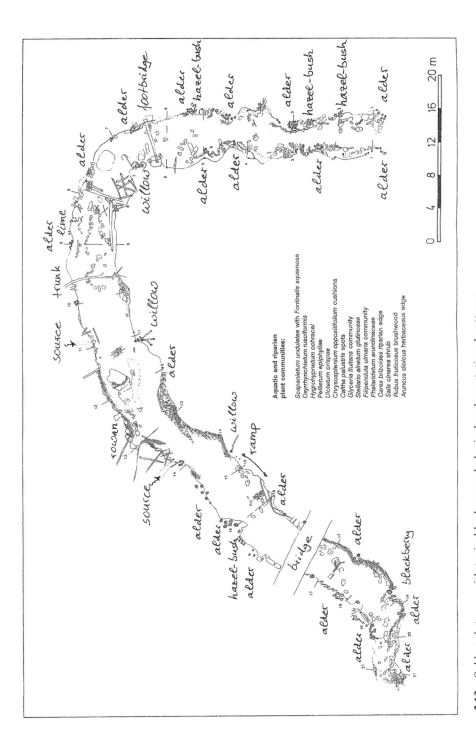

Figure 3.13 *Sohlental* stream with typical hydromorphological and vegetational patterns

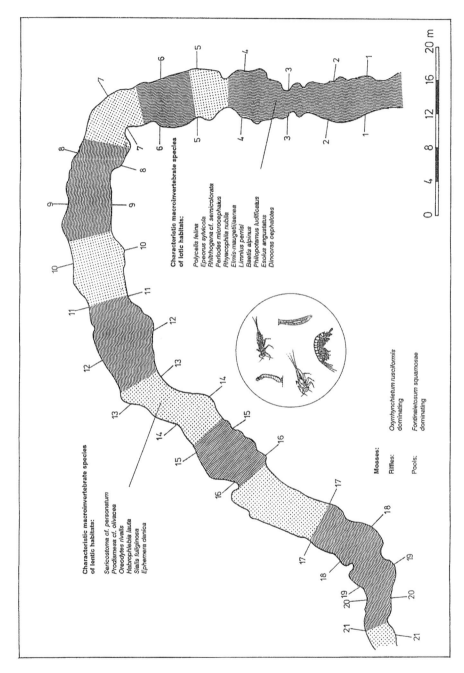

Figure 3.14 In-stream flow characteristics of an Odenwald *sohlental* stream with associated mosses and macroinvertebrate communities

INTERDISCIPLINARY ASPECTS

The results of the different disciplinary studies are incorporated in the *Leitbild*. An example of this holistic view is given in Figures 3.13–3.15. All these figures refer to the same stream, a typical *sohlental* stream in the Odenwald.

Figure 3.13 shows a sketch of a stream reach, including characteristic hydromorphological structures such as cobbles, dead wood and floating tree roots as well as the typical distribution of woody riparian vegetation. The macroinvertebrate fauna and mosses associated with lotic and lentic habitats are given in Figure 3.14. The cross-section in Figure 3.15 represents a part of the stream bed with typical vegetation.

The aim of the development of the *Leitbild* is to establish a holistic view of individual stream types. The focus is to create a clear description of typical stream properties comprehensible to outsiders, i.e. people not directly involved in the process of stream classification. Additionally with this approach the acceptance and justification of stream classification is ensured.

CONCLUSION

The presented multidisciplinary stream classification approach takes into account both lateral geographic and longitudinal stream characteristics. Most significant for a fully

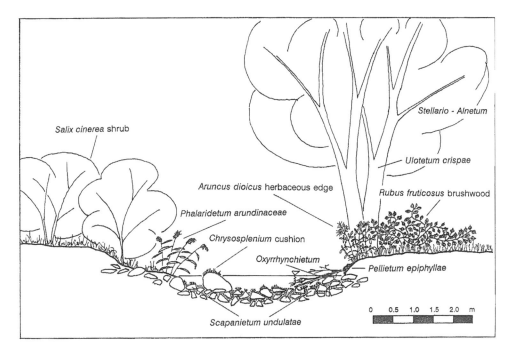

Figure 3.15 A cross-section showing the distribution of aquatic and riparian plant communities in an Odenwald stream of the *sohlental* type

developed stream classification system is its framework character, allowing the integration of an unlimited number of further information levels, based on specifications defined by potential users.

Stream classification can be seen as an indispensable tool for the development of stream-type-adapted management in harmony with the specific regional character of running waters. Typical areas for the application of information derived from stream classification systems include river rehabilitation, environmental assessment studies, maintenance practice, stream conservation, water works and ecological valuation.

A comprehensive summary of the overall stream-type characteristics is the *Leitbild*. The *Leitbild* thus provides a description of the desirable stream properties in accordance with the natural potential. In the context of stream management, it can be used as a guideline in dealing with a variety of problems; however, it needs to be emphasised that users will have to adapt the *Leitbild* to their individual objectives.

REFERENCES

Braukmann, U. (1987) Zoozönologische und saprobiologische Beiträge zu einer allgemeinen regionalen Bachtypologie. *Ergebnisse der Limnologie*, **26**, 1–355.
Braun-Blanquet, J. (1964) *Pflanzensoziologie*, 3rd edition. Springer, Wien, New York.
Carling, P.A. (1987) Bed stability in gravel streams, with reference to stream regulation and ecology. In Richards, K. (ed.) *River Channels*. Basil Blackwell, Oxford, pp. 321–347.
Dury, G.H. (1961) Bankfull discharge – an example of its statistical relationship. *Bulletin of the International Association of Hydrology*, **6**, 48–55.
Dury, G.H., Hailes, J.R. and Robbie, H.B. (1963) Bankfull discharge and magnitude frequency series. *Australian Journal of Science*, **26**, 123–124.
Forschungsgruppe Fliessgewässer (1993) 'Fliessgewässertypologie – Ergebnisse interdisziplinärer Studien an naturnahen Fliessgewässern und Auen in Baden-Württemberg mit Schwerpunkt Buntsandstein-Odenwald und Oberrheinebene. Umweltforschung Baden-Württemberg, Umweltministerium Baden-Württemberg, Ecomed Verlag, Landsberg/Lech.
Forschungsgruppe Fliessgewässer (1998) Regionale Bachtypen in Baden-Württemberg – Arbeitsweisen und exemplarische Ergebnisse an Keuper – und Gneisbächen. *Handbuch Wasser 2*, Band 41, Landesanstalt für Umweltschutz Baden-Württemberg (ed.). Karlsruhe.
Gebhart, H., Linnenbach, M., Marthaler, R., Ness, A., Rapp, N. and Segner, H. (1990) Untersuchungen zur Auswirkung von Gewässerversauerungserscheinungen auf Fische und Amphibien sowie Erarbeitung einschlägiger Bioindikationsverfahren. Unpublished report prepared for Umweltbundesamt Berlin.
Hawkins, C.P. and Sedell, J.R. (1981) Longitudinal and seasonal changes in functional organization of macroinvertebrate communities in four Oregon streams. *Ecology*, **62**(2), 387–397.
Hölzinger, J. (1991) Die Vögel Baden-Württembergs – Gefährdung und Schutz, Grundlagen, Biotopschutz, Landesanstalt für Umweltschutz Baden-Württemberg, Karlsruhe.
Humborg, G. (1995) Typologische und morphologische Untersuchungen an Bergbächen im Buntsandstein-Odenwald, *Mitteilungen Institut für Wasserbau und Kulturtechnik*, Heft 192/1995, Universität Karlsruhe, Karlsruhe.
Illies, J. (1961) Versuch einer allgemeinen biozönotischen Gliederung der Fliessgewässer. *International Review Ges. Hydrobiologie*, **46**(2), 205–213.
Kern, K. (1992) Rehabilitation of streams in Southwest Germany. In Boon, P.J., Calow, P. and Petts, G.E. (eds) *River Conservation and Management*. John Wiley, Chichester, pp. 321–335.
Landesanstalt für Umweltschutz Baden-Württemberg (ed.) (1991) *Die Fischfauna der Bäche des Odenwaldes*. Ökologisches Wirkungskataster Baden-Württemberg: Sonderbericht 1, Karlsruhe.
Meynen, E. and Schmithüsen, J. (1962) Handbuch der Naturräumlichen Gliederung Deutschlands, volumes 1/2. Gemeinschaftsveröffentlichung der Institute für Landeskunde und des Deutschen Instituts für Länderkunde, Selbstverlag Bad Godesberg.

Moog, O. and Wimmer, R. (1990) Grundlagen zur typologischen Charakteristik österreichischer Fliessgewässer. *Wasser und Abwasser*, **34**, 55–211.

Morisawa, M. (1985) *Rivers: Form and Process*. Longman, London and New York.

Nadolny, I. (1994) Morphologie und Hydrologie naturnaher Flachlandbäche unter gewässertypologischen Gesichtspunkten – Gewässermorphologische und hydrologische Grundlagen für naturgemäßen Wasserbau und ökologische Gewässerentwicklung. Mitteilungen Heft 189 / 1994, Institut für Wasserbau und Kulturtechnik, Universität Karlsruhe, Karlsruhe, August 1994.

Naiman, R.J., Lonzarich, D.G., Beechie, T.J. and Ralph, S.C. (1992) General principles of classification and the assessment of conservation potentials in rivers. In Boon, P.J., Calow, P. and Petts, G.E. (eds) *River Conservation and Management*, John Wiley, Chichester, pp. 93–123.

Nixon, M. (1959) A study of the bankfull discharges of rivers in England and Wales. *Proceedings of the Institution of Civil Engineers*, **14** (Paper No. 6322), 157–174.

Otto, A. and Braukmann, U. (1983) Gewässertypologie im ländlichen Raum, Schriftenreihe des Bundesministers für Ernährung, Landwirtschaft und Forsten, Reihe A: Angewandte Wissenschaft Heft 228, Landwirtschaftsverlag GmbH 4400 Münster-Hiltrup.

Schwabe, A. (1987) Fluss- und bachbegleitende Pflanzengesellschaften und Vegetationskomplexe im Schwarzwald. *Diss. Bot.*, **102**, Borntraeger, Berlin, Stuttgart.

Schwabe, A. (1989) Vegetation complexes of flowing water habitats and their importance for the differentiation of landscape units. *Landscape Ecology*, The Hague, **2**, 237–253.

Statzner, B. and Higler, B. (1985) Questions and comments on the River Continuum Concept. *Canadian Journal for Fisheries and Aquatic Sciences*, **42**, 1038–1044.

Tüxen, J. (ed.) (1978) Assoziationskomplexe (Sigmeten) und ihre praktische Anwendung. *Ber. International Symposium Int. Ver. VegKde. Rinteln 1977*, Vaduz.

Vannote, R.L., Minshall, G.W., Cummins, K.W., Sedell, J.R. and Cushing, C.E. (1980) The river continuum concept. *Canadian Journal for Fisheries and Aquatic Sciences*, **37**, 130–137.

Williams, G.P. (1978) Bankfull discharge of rivers. *Water Resources Research*, **14**(6), 1141–1154.

Wolman, M.G. and Miller, J.P. (1960) Magnitude and frequency of forces in geomorphic processes. *Journal of Geology*, **68**, 54–74.

4 A Systematic Approach to Ecologically Sound River Bank Management

JENNIE SIMONS[1] and RENÉ BOETERS[2]
[1]*Centre for Civil Engineering Research, Codes and Specifications, Gouda, The Netherlands*
[2]*Ministry of Transport Public Works and Water Management, Road and Hydraulic Engineering Division, Delft, The Netherlands*

Ecological development along rivers is the rehabilitation of erosion and sedimentation processes and the dynamic spatial and temporal shift of the land–water boundary and the full astonishment watching what happens.
(Klink, 1993)

RIVERS AND RIVER MANAGEMENT PLANS

RIVERS IN THE NETHERLANDS

The Netherlands lie in the delta of the rivers Rhine, Meuse and Scheldt. These large rivers and a large number of smaller rivers form a very characteristic part of the Dutch landscape (Figure 4.1). The sections of the large rivers flowing through the Netherlands are mostly the lower reaches, although, part of the River Meuse can be characterised as a middle reach. In the Netherlands most soils are soft and consist mainly of sand, clay and (in some places) gravel.

Through history the rivers have changed a great deal. In the Middle Ages people started to build dikes along large sections of the Dutch rivers in order to prevent flooding and reduce meandering. These dikes have fixed the river beds to a large extent. Around 1600 the first so-called summer dikes were built, on a small scale, to protect floodplains against flooding in the summer (Douben, 1998). As a result the floodplains could be used for more intensive farming and agriculture. At high water levels sediments settle down on the floodplain, in the area between the main dike and the summer dike, covering old relief patterns such as old river arms, river banks and dunes with clay. Furthermore, many river bends have been straightened and main channels have been narrowed for the benefit of navigation and to improve discharge of water and ice.

Four water types can be distinguished along a river with main dikes and secondary levees (Figure 4.2). Water flows through the main channel (1) and through the side channels (2). In meanders with only one continuous open connection to the main channel (3), the influence of the river current varies considerably. Water bodies without a continuous open connection to the river (4) are characterised by practically stagnant water. Their

Rehabilitation of Rivers: Principles and Implementation. Edited by L. C. de Waal, A. R. G. Large and P. M. Wade.
© 1998 John Wiley & Sons Ltd.

Figure 4.1 View on the River Waal, a branch of the River Rhine, with adjacent water bodies. (Copyright: Meetkundige Dienst Afdeling Grafische Technieken)

water supply comes from seepage and flooding. These different types of water bodies and differences in abiotic factors sustain a wide variety of flora and fauna.

As more than half of the Netherlands lie below sea level, several dams have been built in the north (1932) and the estuary in the south-west (between 1958 and 1987). These dams guarantee protection against flooding by the sea. As a result, the transition from fresh to saline water is abrupt in most of the estuaries and many large freshwater and salt-water lakes have been created. Freshwater tidal areas and brackish water zones, with their characteristic landscapes and ecological communities, have practically disappeared. The construction of dams, locks and weirs and the abrupt changes from saline water to freshwater have caused the near disappearance of fish migration up the rivers.

The transition zone from water to land, the riparian zone, is a dynamic part of the river corridor. This zone, shaped by erosion and sedimentation, creates habitats for specific plant and animal species. Various factors, however, have led to the degradation of many riparian zones. As a result of the construction of summer dikes, a much smaller area of land is in regular contact with the river. Plant and animal species are endangered by grazing cattle, hydraulic loads and substantial erosion in places, as well as soil removal and the fixation and protection of river banks through petrified materials, asphalt and

ECOLOGICALLY SOUND RIVER BANK MANAGEMENT

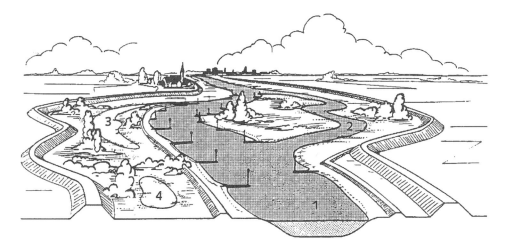

Figure 4.2 Schematic cross-section of a river in the Netherlands (after J. Gorter in Smit & van Urk, 1987)

concrete. Pollution is another major factor of concern for the ecological welfare of Dutch rivers. However, international efforts and consultations on the water quality of the Rhine are beginning to pay off. The water quality of the River Rhine is improving, although the Meuse and the Scheldt are lagging behind. Progress on water quality is largely dependent on the political will to provide adequate funding and back-up.

SCOPE OF THIS CHAPTER

River management includes both the economic and ecological functions of rivers. This chapter mainly deals with the way rivers function ecologically, which has been neglected for a long time. The ecological functioning of rivers requires improvement of the quality of water and soil. It also requires safeguarding and improving habitat and river morphology. For two reasons this chapter will focus only on safeguarding and improving habitats. Firstly, as the water quality improves, safeguards and improvement of habitats have become limiting factors for ecosystems. Secondly, the national government has provided a scheme for integrating the necessary measures regarding river management, to be taken by the various local and regional authorities. In this scheme, the ecological function of rivers is given special attention.

For Dutch rivers, navigation and protection against flooding are major economic issues. These issues determine to a large extent the possible ecological development of rivers. They demand, for instance, a certain depth of the channel, as a result of which there will often be no shallow water, nor small islands, cut claybanks or moving sandbanks. In other words, the ecological opportunities for rehabilitating many main channels are rather limited. River floodplains, however, offer many ecological opportunities. Riparian strips as part of the floodplain play an important role because here, for most of the time, the interaction between land and water takes place. For this reason this chapter focuses on the opportunities of ecological rehabilitation of floodplains in general, with special emphasis on the river banks as a transition zone. A large part of this chapter is based on a Dutch-written manual,

Ecologically Sound Banks (CUR and RWS/DWW, 1994). After introducing rivers in the Netherlands, and the importance of river management plans, river management in the Netherlands and river bank management plans are described. A conceptual approach to create a more ecologically sound river is introduced step by step, providing examples of ecologically sound approaches to river bank management in the Netherlands.

MANAGEMENT DEFINED

This chapter starts from the principle that water management should be responsible for creating and maintaining a certain quality (specified beforehand) of the water system and adjacent banks. In this definition, management includes the organisation of management, ownership, budget, maintenance, monitoring, evaluation and if necessary (re)designing river banks and floodplains. A management plan describes a management strategy for a specified period. Thus, a water management plan is a coherent description of all management activities and implies integration of the present water system, management perceptions, the management goals, and organisation of management and all the management activities such as plans for (re)designing and construction, maintenance, monitoring and evaluating the main channel, banks and floodplains.

THE IMPORTANCE OF RIVER MANAGEMENT PLANS

Developing a coherent management plan takes a lot of time and energy. Yet, in a small and densely populated country such as the Netherlands, land is a vital resource. There is severe pressure on land as a result of many economic activities. The choice for short-term economic benefit instead of the long-term benefits of sustainable water systems is an ever present danger. That is why there is so little natural habitat left in the Netherlands. Integrated efforts are needed to preserve those few natural habitats left and to create possibilities for the development of nature and ecological networks (green corridors) between fragmented habitats. Sophisticated methods are required for formulating management plans for periods of five or ten years with long-term management goals which have been translated into definite objectives. It requires co-operation between a number of local and regional authorities and coherent, three-dimensional thinking in water systems analysis integrating the three abiotic factors soil, water and air. At the same time, it requires teamwork between people of various disciplines, such as biologists and hydraulic and civil engineers, who, by providing their knowledge, techniques and creativity, make systematic planning possible. Although management activities and management information have always been important, the concept of more coherent organisation and planning, as well as better communication, is now more widely appreciated as necessary to aid effective rehabilitation.

Systematic planning has some other advantages (Heidemij, 1991; Oranjewoud, 1991; CUR and RWS/DWW, 1994; Rijkswaterstaat Dienst Weg- en Waterbouwkunde, 1994):

- Different and sometimes conflicting demands and wishes can be dealt with in terms of importance. The input can be balanced and priorities can be set. In the end, this will lead to a coherent output based on well-defined choices and motives.
- The procedure for making a management plan and the communication it implies will help the people involved think and act in the same direction.

- It will provide insight and make possible accurate planning of all management activities within the limits of available staff and budget.
- Changes in objectives and management insights can be implemented consciously and adjusting the input of staff, budget and activities can be done more effectively and efficiently.
- It will make consistent management less vulnerable to changes in staff. Consistency in terms of management is particularly essential to achieve high ecological values.
- It enables monitoring and evaluation of management which can be used to adjust, complete or improve future management.
- Publishing management objectives and activities makes communication with external organisations and the general public easier.

Therefore, a management plan can provide effective, purposeful and cost-effective management tools which link with the increased ecological insights and the demands for a more integrated approach.

RIVER MANAGEMENT IN THE NETHERLANDS

RIVER POLICY PLANS

In traditional river management much attention has always been paid to the regulation of discharges of water and ice, and economic activities such as navigation, fishing and recreation. The ecological function of rivers has been neglected for a long time. In 1985 an important policy document called *Living with water* (Ministry of Transport, Public Works and Water Management, 1985) confirmed a new trend in water management in the Netherlands: 'integrated care for (the condition and use of) water systems, including water, beds, banks and shores, with regard to their physical, chemical and biological composition'. In 1987 the Brundtland report, *Our Common Future*, was published with the message for 'sustainable development'. This means that water systems, as a natural resource, should be exploited in such a way that they can fulfil their functions properly, not only at present but also in the future. Both policy documents led to a change of approach in water management. As a result, the Dutch government adopted the National Policy Document on Water Management (Ministry of Transport, Public Works and Water Management, 1989). In this document the final goals for the most relevant action packages are defined as follows.

The shores of the watersystems offer possibilities for development and sustainable functioning of those systems and for various uses; the shore design is aimed at the functions of the shore, the adjoining water and land. The rivers, including the floodplains are the main axes through the Netherlands for transport, recreation, animal migration and ecology.
(Package 7: Shores and areas outside the dykes)

The morphology, hydraulic design and water regime of the water systems in the Netherlands present good conditions for complete and well-balanced biotic communities and sustained use by man. (Package 8: Restoration of water systems)

Key words in this document are sustainability and multi-functionality. It introduces a more integrated approach to all water systems, economically as well as ecologically (Boeters *et al.*, 1994).

THE RIVER WAAL: AN EXAMPLE OF RIVER MANAGEMENT IN THE NETHERLANDS

The River Waal is one of the busiest rivers in the world. Every year about 160 000 vessels cross the German–Dutch border. Vessels with a cargo capacity of between 250 and 18 000 tonnes carry 140 million tonnes of freight every year. The River Waal is not only an important shipping connection between the port of Rotterdam and Germany, but also for inland shipping (Boeters et al., 1994). Consequently, the river is an important factor in the Dutch economy.

The Dutch government aims at three main points with regard to the River Waal (Douben, 1998):

1. *Sufficient safety against flooding.* The standard chosen is flooding once in every 1250 years. This frequency corresponds with a discharge of 15 000 m^3/s^{-1}.
2. *Improvement of navigability.* By 2010, a minimal river depth of 2.80 m is desired (with the exception of 20 ice-free days at the most) and a minimal width of 170 m. At the moment, a channel depth of 2.50 m and a channel width of 150 m during low water (Q95%) is aimed at.
3. *Rehabilitation of ecological values.* Through the years human activities have had great economic benefits, especially in the shipping and agricultural trade. However, nature had to pay the debts. The River Waal with its wide floodplains is to play an important national as well as international part as an ecological connection. Rehabilitation of ecological values involves the following:
   - designing ecologically sound river banks
   - bringing back natural topography and riverine habitats that belong there, by following the original relief
   - developing floodplain forest
   - increasing hydrodynamics and morphodynamics in the floodplains by removing the summer dikes
   - digging secondary channels

A decrease in agricultural activities in the winter bed of the river will provide more space for the development of nature. The preconditions that originate from the first two objectives will limit the development of nature. The development of nature cannot be completely natural and self-regulatory, but it should be given a positive impetus by means of sophisticated and careful design.

MANAGEMENT PLANS FOR RIVER BANKS

As a result of the physical pressure on river banks and the many functions they serve, looking for opportunities to improve nature conservation on river banks has become a specialist part of river management plans. The action plan described below will aid the design of a management plan for river banks as part of a comprehensive river management plan, and will function as a check-list.

Ideally the description of river bank management includes the following features. The future situation aimed at should be described for as large as possible a part of the river basin or the water system within the control area. This overall description of aims, called

ECOLOGICALLY SOUND RIVER BANK MANAGEMENT

'the target', will be worked up into one or more variants containing concrete objectives which fit in with the possibilities of smaller (partial) river sections. This is part of the management vision, which will have to result in concrete measures in order to achieve the set objectives.

Figure 4.3 shows the interlinking activities making up a river bank management plan. Each phase influences, and is a condition for, the next phase. The order in which the phases are laid out is based on experiences gained in a variety of Dutch projects aimed at integrating the ecology and morphology of water systems. It is therefore advisable to adhere to this order and though it appears complex, it has been tested in a wide range of situations. The local and regional situation, the characteristics and potentials of the water system and its surrounding environment, as well as actions already carried out, should be explored by the manager and will actually determine whether or not the management proposed is ecologically sound.

Figure 4.4 focuses on the first activity, i.e. the drawing up of a river bank management vision.

RIVER BANK MANAGEMENT VISION

PHASE 1: INITIATIVE AND ORGANISATION

The initiative for making a river bank management plan may come from administrative bodies, the organisation responsible for maintenance and control, and society at large.

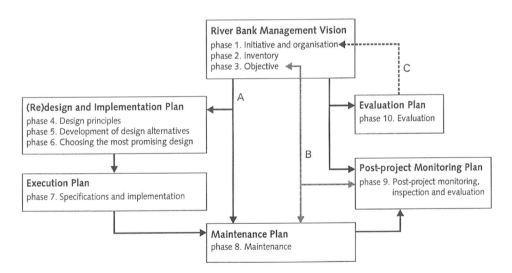

Figure 4.3 The cyclic management process within the various parts of the management plan. A: There is a choice between whether the river bank is redesigned or whether the maintenance plan is adjusted. B: There is a choice between adjusting the maintenance plan or the objectives of the plan before continuing with the remaining phases. C: The organisation which took the river bank management initiative has to decide whether management remains unchanged or whether it will need to be revised. (Copyright: CUR and RWS, 1994)

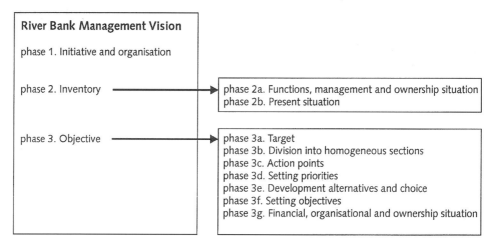

Figure 4.4 The different phases of the action plan for the river bank management vision of ecologically sound river banks. (Copyright: CUR and RWS, 1994)

There can be a number of reasons for such an initiative, e.g. necessary maintenance, excessive erosion or siltation, improvement of the accessibility, and/or the wish to improve ecological development, a holistic approach to water management and multi-functionality. As soon as the decision has been taken to develop a river bank management plan a number of actions should be taken. Investigations should be undertaken to identify parties with whom direct and indirect co-operation is possible and necessary. In addition, in this phase it should become clear whether there is sufficient knowledge, experience and labour available to supervise and execute the entire plan procedure. If not, this should be made available, if necessary by calling in assistance from outside.

Phase 1 is complete when the procedure to be pursued has been described in a so-called 'start document'. Clarity and openness in the decision-making process is important, both to the organisations involved and to third parties. The start document will increase the performance of the planning process, the acceptance of the plans made and the involvement by the participants. All the organisations involved, with their management or general boards in a prominent role, will have to approve the planning procedure. The planning procedure successively describes the following:

1. the necessity of developing a management plan;
2. the outline of the related objectives;
3. supervision of the process (who is responsible for what, and how, when and by whom decisions will be taken);
4. the design of the organisation and how co-operation is arranged;
5. the way external decision-making is arranged;
6. procedures for lodging objections and appeals;
7. the financial organisation for the river bank management plan.

It should be borne in mind that developing a management plan only is not enough. Money should be allocated in order to execute the management plan (design, maintenance, monitoring and evaluation). The budget needed for carrying out the measures

ECOLOGICALLY SOUND RIVER BANK MANAGEMENT 65

described in the management plan can only be assessed when the management vision has been completed (see Phase 3G).

PHASE 2: INVENTORY

In Phase 2, two different inventories should be made. Firstly, the management situation, ownership situation and current and future functions of the river and its banks should be described (Phase 2A). Secondly, the present condition of the river banks should be described within the context of the water system or river basin (Phase 2B). Both elements of this phase are very important. When the analysis is inaccurately made or crucial aspects are overlooked, this may lead to measures and arrangements that do not suit the area to be managed. As a result, the actual bottlenecks and problems are often dealt with inadequately and in the worst cases can even be aggravated.

Phase 2A: Functions, management and ownership situation

The management and ownership situation should be investigated and described accurately in order to make all people involved aware of their responsibilities. Such an inventory is vital and requires identification of the following:

- the boundaries of control and ownership
- the management tasks and the task conceptions of the managers
- the agreements between managers and/or with private sector parties
- the current management (including maintenance)

In case of land leasing, the lease term should be determined. Also, checks should be made as to whether management and ownership boundaries have been synchronised.

It is important to be aware of current and pending policies. If, for instance, there are plans for large-scale operations adjacent to the river bank (e.g. channel improvement, dredging) it is useful to know at what stage these plans are and whether integration of the river bank project is possible. This can be very cost-efficient and can increase the feasibility of the plans. Current and future policies of the authorities and management organisations involved are usually laid down in policy documents, which also mention the current and future functions of the river itself, the river banks and the adjacent land. Some of the river functions imply limiting conditions, which are influences, changes or circumstances that cannot or may not be changed. For large Dutch rivers this usually concerns shipping and the discharge of water. Also the dikes, which offer protection against flooding, are a limiting condition that cannot be tampered with. Besides shipping, water discharge and safety, there can also be autonomous influences or processes on which management organisations have little or no influence. Further limiting factors for management may for example concern water quality, ever increasing shipping or urban development.

The opportunities for the development, improvement and preservation of natural habitats on river banks lie within the limits determined above. Thus limiting conditions must always be described, especially when it is not clear exactly what the consequences will be for the final design of the new river bank. In order to examine the opportunities for nature conservation and development it is important to know how fixed the limiting conditions really are. There might be some latitude, or this can be created, in favour of the development of nature. Only those factors which appear to have no latitude at all should be regarded as definite conditions and the limiting conditions which have some scope for adjustment should be regarded as action items (see Phase 3C).

Phase 2B: Present situation

Proper documentation of the current condition of the river, its banks and adjacent land is an important basis for the next step and in particular for the assessment of opportunities for rehabilitating river banks. In this respect, the autonomous processes that fall beyond the influence of the river bank management plan should be recognised and be taken into account while setting objectives. Also, the recognition of a lack of knowledge can lead to more realistic expectations with regard to the objectives that have to be achieved. This can have the result of leading to additional research.

The following should be documented:
- location
- land use
- hydrology (water level, changes in water level, the possible existence of seepage, undersoil drainage, water supply and water discharge, currents, wave climate)
- geomorphology (soil type and structure, cross-section, longitudinal section, shape of the river bank, erosion and sedimentation, and presence and type of bank protection)
- water quality, quality of the subsoil of the river bed and the subsoil of the river bank
- landscape values and cultural history
- ecology (aquatic, riparian and terrestrial habitats, the degree of replaceability, the 'intactness' and rarity of habitats, the connection with nearby and/or valuable other nature reserves, whether the connections are lengthwise or perpendicular to the bank, the importance for through transport and spreading of plants and animals)
- autonomous processes (such as desalination, drying, acidification, pollution and salination, which are the result of use and management measures)

This phase can involve a lot of time, especially when it is done in an extensive way. For this reason it is especially important to think about to what degree the present situation should be investigated in detail. However, too global a description of the current situation can lead to too global a description of the target image (Phase 3A). This applies particularly to the ecological opportunities of the area concerned (Peters *et al.*, 1991). A global description of the target can lead to objectives which are not very concrete or quantifiable (Phase 3F) and to a management plan that does not fit the bordering water and hinterland.

PHASE 3: OBJECTIVES

In Phase 3 the objectives should be described, i.e. the situation that should be achieved within a set period. Objectives are set for each river bank section and should be based, in a logical and analytical way, on the inventories of Phase 2. The objectives should fit in with the possibilities and conditions of the site and should not be designed 'just because it looks beautiful'. Going through Phases 3A to 3G will help in step by step formulation of good objectives.

Phase 3A: Target

The target describes the desired situation of the river bank. In order to give coherence to this desired situation it can be best described for the maximum possible reach of the control area or the water system. A good description of the target is particularly important to ecological development. A coherent plan will give extra value to ecological rehabilitation, as opposed to possible ecological 'patchwork'. Because the ecological function of river banks has hardly been elaborated on in planning until recently, particular attention should be paid to this while describing the target. In designing and implementing plans for river banks, river bank managers have their own responsibilities and freedoms with respect to the banks. However, there are limits to the freedom in ecological development. The target situation must take into account the original ecological situation that might have occurred on that site, i.e. the ecological reference situation.

Ecological reference situation

The ecological reference situation is described in order to get a feel for what ecosystems would fit the river banks in the managed area for the purpose of rehabilitation. An ecological reference situation describes how the ecosystem or habitats in the management area involved might have looked with regard to the current climatic and bio-geographical situation assuming human activities had not influenced the ecosystem.

In order to describe the ecological reference situation, knowledge of ecology and ecological processes is vital. This ecological reference situation can consist of old descriptions of habitats along the river involved or a description of a comparable (reference) river which has not been much affected by humankind. The non-biotic factors of such a reference river should be comparable with those of the management plan area. It is important to investigate which characteristic floral and faunal groups might exist there and what the determining processes are. For rivers, hydrodynamics and the morphodynamics are particularly important (Rademakers, 1993).

Sometimes it is not possible or not desirable to design ecological reference situations by going back in time. This can, for instance, be the case when part of a river is consciously cut off from the river's influence. In the Netherlands, for example, parts of the Meuse–Rhine estuary were cut off by dams from both the river and the sea and have become large freshwater lakes. In such cases, the circumstances in those areas and also in areas upstream have changed so much that it is more useful in the ecological reference situation to describe what habitats might look like under the present circumstances if they were allowed to develop undisturbed. Ecological theories from the literature may assist here.

Describing ecological reference situations is difficult and time-consuming. Attempts must be made to base description on knowledge that has already been developed inside and outside the region in question. It is, however, essential to adapt the description to the situation that applies to the control area in question.

Dutch rivers have been divided into river sections on the basis of changes in water level, flooding frequency and degree of erosion and sedimentation. Subsequently, the chances for realising ecological development aims can be formulated for each section. The ecological development aims for rivers in the Netherlands are divided into four main categories:

I   almost-natural ecological development
II guided-natural ecological development
III half-natural ecological development
IV multi-functional ecological development

The first aim is the most natural variant, the last one is the least natural (Wolfert, 1992; Janssen & Rademakers, 1993; Rademakers, 1993).

From ecological reference situation to the target

Ecological reference situations are an aid to changing the present situation into a more natural river system; they constitute the 'ideal objective'. Ecological reference situations, however, are seldom feasible. It is up to the river bank manager to investigate how much the target and the ecological reference situation overlap. In formulating the target, the preconditions described in Phase 2 play an important part. These are the conditions that originate from major functions and autonomous influences and processes. The preconditions confine the possibilities for realising the ecological reference situation. Consequently, confronting the ecological reference situation with the preconditions will enable one to choose one or more targets.

In general, a management plan covers a ten-year term. Consequently, a river bank management plan should in any case contain a short-term target for the river banks, i.e. what is feasible within ten years. This short-term target should be realistic. Targets that are too idealistic will reduce motivation, because they are not seen to be feasible.

It is advisable to also describe the target that should be achieved within two or more ten-year terms, i.e. the long-term aim. This can help guarantee continuity and create links between two successive management plans. However, the long-term target should clearly state the length of the term and what measures should be taken in order to achieve the target.

Developing a target does not always imply that the rehabilitation is actually moving in the direction of the ecological reference situation. In some cases the possibility of forming new ecosystems by way of (temporary) development management should be considered. An example of such a dilemma is the partial desalination of the Rhine–Meuse estuary after being cut off from the sea during the period 1965–1987. For such places the target of 'preservation of plants on higher ground that are typical for salt and brackish systems' will be hard to achieve while other possibilities within the area are given too little attention.

Target requirements

The target, whether it is short term or long term, is specific for every control area. It should fit the specific ecological potentials of river banks, within the limits set by the various functions and the autonomous influences and processes. It is 'tailor-made' and designed by the local authorities in the area of the river banks and their surroundings.

Regardless, the target should indicate what types of ecosystems, habitats, succession stages, ecotopes and/or organisms and their biotopes are desired, and in what parts of the water system, the river basin or the drainage area specific elements of the target should be emphasised. A detailed description of the target will offer more guarantees for coherence.

ECOLOGICALLY SOUND RIVER BANK MANAGEMENT

It is important to realise that both along the longitudinal profile in a transverse direction from the main channel, a number of partial ecosystems and several habitats may exist. However, this does not imply that as many as possible ecosystems and habitats should be created in one design. On the contrary, ecosystems and habitats need space in order to be able to develop. In Figure 4.5 an example of a river bank target situation is presented describing the desirable development for different river sections. The river bank targets are the results of a combination of hydrological, morphological and ecological factors (Rijkswaterstaat, Directie Gelderland, 1993).

Phase 3B: Division into homogeneous sections

It will seldom be possible to work out the target in terms of absolute measures for the redesign and maintenance of the river banks in the entire control area, because there are often differences between the various parts of the control area. For this reason the river bank control area must be divided into sections, each with its homogeneous characteristics

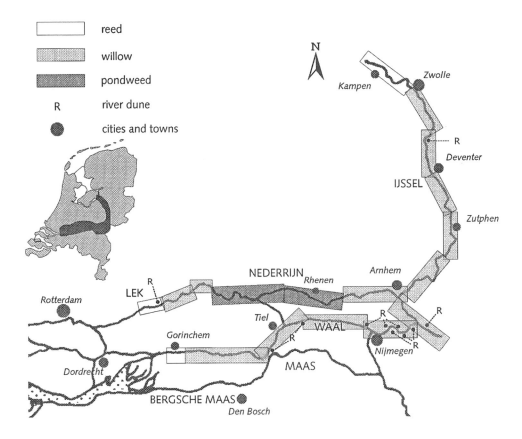

Figure 4.5 The river bank target situation for sections controlled by one of the management departments of the Directorate-General for Public Works and Water Management (Rijkswaterstaat, Directie Gelderland, 1993)

with regard to the functions of the river bank, the water and the land as well as the ecological opportunities. This division will facilitate elaboration and can be used for setting priorities (Phase 3D).

Phase 3C: Action points

When comparing the target with the present situation, the main action points per section can be pointed out. Action points can also be the result of the latitude within the preconditions as they have been formulated in the description of the present situation (Phases 2A and 2B). Action points should be solvable problems. If it is not possible to solve the problems within the term of the plan, they will become preconditions that co-determine the target (Phase 3A). It may thus be necessary to reformulate the target (feedback). Sometimes autonomous developments (e.g. poor water quality) prove to be the preconditions instead of the action points. Unwanted autonomous developments should be mentioned explicitly in the plans and should be discussed with the responsible management and administrative authorities. This certainly applies when the ecological objectives for the river banks are at stake. Solutions for action points within the management plan related to pollution should be examined closely and be realistic, especially when the impact comes from outside.

Phase 3D: Setting priorities

The target should be developed for the various sections. It should be decided which sections should be adjusted first within the term of the river bank management plan. The priority depends on the necessity of taking action, the chances of success and the money available. If necessary, the priority order can be applied to the further elaboration of the management plan. It is preferable, however, to complete the entire Phase 3 for all sections together in order to maintain coherence in the management vision.

Phase 3E: Development alternatives and choice

In this phase, the target decided on in Phase 3A will have to be made concrete for each homogeneous section. On this smaller scale the development, reinforcement or protection of the river bank's nature can be examined in more detail in line with the target. This phase is also the right moment to assess the connection with the ecological value of the surrounding land and water. Sometimes it will be possible to increase the ecological value of both the surrounding area and the river bank when dealing with land, river bank and water at the same time. If the surrounding land has little or no ecological value it may be desirable to reinforce habitats on the river bank in order to establish a connection through the relatively unfavourable surroundings (i.e. a 'green corridor').

The development of various alternatives with different approaches towards achieving the target and/or improving the condition of the surrounding land and water increases the number of options and will give insight into the opportunities of the river bank section involved. The ecosystem, after all, consists of several types of habitats with specific processes and/or function(s) for flora and fauna. Because nature needs time to develop in the desired direction (succession) the term 'nature' or ecological development of the river bank area is generally used. If the area already has a high nature conservation value, how this value can be preserved and enhanced through river bank management should be explored.

Examples of nature development alternatives are as follows:

- an emphasis on types of ecosystems or habitats that are typical for land or water ecosystems or for the specific river bank ecosystems;
- an emphasis on high-dynamic or low-dynamic ecosystems;
- an emphasis on certain types of animal or plant habitats;
- a combination of nature with other interests such as recreation, fishing or water-purifying helophyte zones;
- a choice for an experimental approach in which variations in river bank design and maintenance can be applied if, for example, the development of the sections is hard to assess.

After the various development alternatives have been developed, their values should be described in an easily understandable way. The phase will be completed when a choice has been made between the alternatives. Naturally, an important criterion for making this choice is a certain location's potential for developing a certain habitat or for a specific ecological objective. As well as the chances of success, money and staff availability will be determining factors in the process of choice. A multi-criteria analysis may be of help here. It will facilitate the choice-making process although it should not be forgotten that the outcome is more or less determined by the relative importance given to the criteria. This effect may be examined by carrying out a sensitivity analysis in which criteria are assigned different values.

Phase 3F: Setting objectives

The result of choosing one of the nature development alternatives must be developed in the form of objectives. The objectives are a detailed account of the minimum quality level of the desired nature development, in as testable and quantitative terms as possible. Concrete and detailed objectives will enable the development of plans for (re)design, maintenance, monitoring and evaluation.

Questions to be answered when formulating the objectives are as follows:

- What abiotic demands will the river bank have to comply with?
- What function should the river bank fulfil for what groups of plants and animals and what areas and numbers are involved?
- What natural processes are considered desirable?
- What and how many provisions should be made?

Quantitative and qualitative information on flora and fauna in similar areas which have the desired state of nature, as well as ecological knowledge, can be used to specify objectives in detail. Even when it is doubtful that the objectives can be achieved or when insufficient knowledge is available to develop the objectives in detail, it is still important to present testable and quantitative objectives. This also applies to the experimental approach. In this way adjustments can be made, if necessary. At this stage attempts should be made to indicate in what way, when and with what criteria objectives are to be tested. In this phase the description of testing will be a tool in developing a monitoring plan (Phase 9). Finally, the objectives should be examined to check that they are in line with the target (feedback). If they are not in line, the objectives should be reformulated.

Phase 3G: Financial, organisational and ownership situation

Financial stuation

Once the objectives are set, the method of financing the management plan and its consequent measures should be arranged (see Phases 2A and 3D). The following issues should be given attention:

- budgeting
- the possibility of subsidies for construction and maintenance
- the contributions of various organisations involved
- levies and compensation based on the management plan
- tax facilities
- contracts of compensation for certain actions

This phase influences the possible (re)design plan, the execution and the maintenance of the plan (Figure 4.3). This phase could be carried out while developing Phases 3A to 3F.

Management organisation

During this phase the desired management organisation should be chosen. It is important to map the new management situation. Moreover, in this phase the responsibilities for maintenance, the physical accessibility for inspection and maintenance, and the necessary permissions should be settled. Arrangements should also be made for carrying out maintenance, and issues such as inspection, control and periodic recording of the actual situation and management should be settled. Regulations for compensation due to negligence in maintenance should be set down in this phase.

Ownership

A choice should be made between ownership (one or more owners), purchase of the land or lease (lease term). When land is bought, a management organisation should be appointed, for example the purchasing body, a nature conservation organisation or a private party. New ownership situations need to be clearly defined by mapping.

APPROVAL OF THE MANAGEMENT VISION

By now the management vision, including the objectives, the funds and the management and ownership situation, has been set down. However, the management vision will not have been fully completed until the entire vision has been approved by the management department or the appropriate authority.

After the approval of the management vision the objectives of Phase 3F should be developed in more detail. Changes in the management or ownership situation or employing different maintenance may be sufficient to achieve the objectives (see below). However, it may also be possible to redesign or reconstruct a certain stretch of the river bank in order to give ecological development the best opportunities. In such a case a (re)design plan should be drawn up and in order to implement this plan an implementation plan is needed.

(RE)DESIGN AND IMPLEMENTATION

PHASE 4: FORMULATING DESIGN PRINCIPLES

In a (re)design for the river bank the preconditions and objectives of Phase 3 are translated into design principles. This is achieved by applying knowledge of various disciplines. Ecological design principles must reflect the demands of the habitats, processes and species desired. If, for example, softwood floodplain forest is to be cultivated, its requirements should be known; if a mating place for migratory barbellus fish is projected, the requirements of these fish should be taken into account.

Knowledge about the civil engineering demands upon the river bank is needed in order to provide information about the river bank's necessary strength. This step requires information about the physical loads on the river bank and/or the number and type of ships sailing on the river.

With regard to possible recreational demands upon the design, it is important to assess what types of recreation are involved, what their requirements are and what kind of facilities (fishing places, footpaths, etc.) are needed. The various demands should be ranked on the basis of level of priority, which is generally dependent upon the local conditions. The level of priority should be based on criteria that can be tested and evaluated at a later stage.

Poorly formulated design principles: The Overijsselse Vecht

Along a small precipitation-fed river, rather than placing the river bank protection directly on the river bank itself, the gradient of the slope was decreased by flattening the part above the water level and filling up the part beneath the water level (Figures 4.6 and 4.7). In front of the actual river bank, bank protection was erected. In this way a shallow, wet, 4-6 m wide strip was created behind the bank protection. This type of bank protection is aimed at erosion prevention and improvement of fish stock in the river by increasing the opportunities for mating and growing up. In order to achieve the second objective, marshland and water plants are required (for the hiding and spawning of fish) as well as invertebrates in the wet strip.

As a result of the shallowness of the wet strip (less than 0.6 m), the small amount of water exchange (due to the height of the protective construction and the absence of gaps), river bank sedimentation and river bank erosion (due to the incorrect emplacement of the filter mat in combination with high water levels) the wet strip has closed rapidly. After a favourable start, the function of this wet strip for fish has diminished. On the basis of these experiences the planning and consequent design of similar river bank protections were adjusted and successfully improved.

Figure 4.6 The Overijsselse Vecht, 1987. (Copyright: Jan Koolen)

PHASE 5: DEVELOPMENT OF DESIGN ALTERNATIVES

Various alternative designs can be drawn up, e.g. by using different widths for the river bank, different types of bank protection and different materials. A choice of designs will offer more options and more insight into the opportunities that exist for the river bank involved. In developing design alternatives the composition, expertise and creativity of the design team are very important. The biologists and engineers should work closely together in this phase. The more important ecology is, the more the biologists will function as architects and the engineers as technicians (providing technical solutions).

Before starting the design process, the space available and the potential to purchase land or to zone various activities must be determined. What part of the profile is minimally required for shipping and proper water management must be determined, taking into account current patterns and wave motion. On the basis of the demands formulated in Phase 4, the following need to be checked:

- whether or not different designs are required or desirable for the river bank lengthwise or crosswise;
- whether or not and where structural reinforcements of the cross-section are required and what these should be like (materials and design).

The space necessary for the proper function of habitats and various activities needs to be taken into account in order not to design 'small gardens' and 'flower beds'. Subsequently several alternative designs need to be developed. In some cases it may still be

Figure 4.7 The Overijsselse Vecht, 1990. (Copyright: Geert-Jan Verkade)

possible at this stage that, by making one or more changes in maintenance or through development management, the ecological demands can still be met. All the alternatives should be provided, with guidelines concerning management and maintenance, and should also describe the expected effects and costs. Moreover, the alternatives should as far as possible take into account the following technical demands regarding implementation and maintenance:

- equipment for construction and maintenance should reach the location without damaging the river bank and the surrounding land (e.g. the accessibility of the location by means of a maintenance road);
- it must be possible to inspect the construction;
- in case of damage it must be relatively easy to carry out repair works;
- the way work and future maintenance is carried out should comply with the Occupational Health and Safety Acts pertinent in that region;
- the work should be carried out with minimal interference to other functions (consideration should be given to traffic on the river and on land during works).

Ecological development by fencing out cattle

The objective may be to develop floodplain forest with indigenous riparian plants and animals. If the floodplains and the river bank directly adjacent to the river are intensively grazed, relatively minor intervention may facilitate reaching part or all of the objective. The river current transports the seeds of riparian plants that belong alongside the river, provided the desired plants and trees exist upstream. The cattle, if given the chance, like to eat some of the seedlings and trees; willows especially are favoured. As a result, young trees do not develop into mature trees. This is one of the reasons why floodplains in the Netherlands have changed into grassland. Also trampling, particularly on sandy soils, can be disastrous for vegetation.

Where grazing decreases or even stops, e.g. by putting cattle behind a fence, plants, bushes and trees will rapidly develop, especially along the water's edge (Figure 4.8).

PHASE 6: CHOOSING THE MOST PROMISING DESIGN

During Phase 6 all parties involved will have to assess the designs and make a choice between the various alternatives. In making the best choice the priorities and criteria of Phase 4 and the costs and benefits need to be incorporated.

In a proper cost assessment, both construction costs and the costs of maintenance are included. Costs can be calculated fairly easily, but benefits are much harder to assess. The benefits, including the 'value' of the river bank, should be assessed by the managing authority or the organisation that finances the operation. If it is hard or even impossible to make a choice, multi-criteria analyses or simplified versions of these may offer some help. They quantify the assessment criteria or aspects of evaluation, including costs.

Once a design has been chosen, it is important to lay down the motivation for this choice. The choice may also dictate adjustment of the objective formulated earlier, if it becomes clear that the aim cannot be achieved, neither through (re)design nor through maintenance. This phase is followed by the necessary procedures for acquiring permissions.

Figure 4.8 Development of floodplain forest after fencing out cattle. (Copyright: Ingeborg van Splunder)

PHASE 7: SPECIFICATIONS AND IMPLEMENTATION

When the choice for a design has been made and (if necessary) the choice has been approved, the specifications will need to be drawn up and the work put out for tender, after which work can start. Both in the specifications and during the implementation, details on the role of ecology in the design require special attention. It may be somewhat pretentious to speak of an implementation plan, but these details should indeed be thought of and laid down.

The way in which the work will be carried out is described in specifications clearly outlining every detail of the work (what, where, when, how and how much). In addition, it mentions the quality demands the work has to meet, the conditions within which the work has to be carried out and sometimes the way in which it should be carried out. It also contains the terms of payment. These specifications consequently function as a contract for the parties involved as well as the basis for price-setting.

Creating ecologically sound river banks involves a number of entirely different (technical) demands and conditions compared with the civil-engineered river banks built until recently. A detailed description of these demands forms part of the specifications. In this respect the specifications are a technical guideline for carrying out the work and also a legal instrument. Proper and relevant information must be given to constructors, those responsible for future maintenance and to the people living in the neighbourhood. Such public relations activities have a positive effect on the quality of the work as well as on its acceptance by residents.

Constructing ecologically sound river banks is a specialist activity which requires checks at crucial moments, during construction and after completion. The consequences of the use of inferior materials or the inadequate storage or deployment of living materials will usually only show after some time. An ecologically sound river bank, after all, will need some time to develop after completion. Do not forget to record the situation after completion to aid in post-project monitoring and appraisal of the success or otherwise of the scheme.

A good start

For developing or reinforcing the value of habitats on river banks a sound starting point is important. This can be achieved by

- clearly indicating boundaries of the site;
- indicating areas where people are not, or only under certain conditions, allowed to enter;
- determining in advance the suitable period for carrying out the construction activities;
- clearly indicating suitable location(s) for depots;
- laying down conditions for carrying or possibly *not* carrying out certain activities;
- restricting the release of nutrients and toxic substances when moving earth, preventing condensation of the soil and establishing sufficient micro-relief (Pruisen and Schippers, 1991);
- preferably working from the channel outwards;
- using equipment that disrupts the ground underneath as little as possible;
- preventing damage to living material, especially by paying special attention to roots.

MAINTENANCE PLAN

PHASE 8: MAINTENANCE

The objectives, the chosen design or the design that already has been realised, require appropriate and specific maintenance. Maintenance implies carrying out activities periodically with regard to the management of the river bank. It often involves brief and local activities which do not substantially affect the construction, the profile or the design. For maintenance, clear, feasible and concrete objectives are important. For example, as nesting birds pose different maintenance demands different to those of resident or migrant birds, the priorities and motivation for these priorities should be clearly presented.

A PLAN FOR MAINTENANCE

A maintenance plan encompasses all the activities that are relevant to the entire management plan and that should be carried out during the period planned. Such activities include both the periodical maintenance that is carried out regularly and the occasional and/or the more fundamental maintenance.

At first sight, making a maintenance plan may seem time-consuming and unnecessary, however, the advantages of such a plan are as follows (after Rijkswaterstaat DWW, 1994):

- Ecologically sound river banks often need to develop for some time before fulfilling their function optimally. Sometimes it is necessary to guide the development of the river bank in a more ecologically sound direction through maintenance.
- Achieving the desired ecologically sound river bank (the main objective) poses demands to the methods applied, the equipment, and the timing and frequency of maintenance.
- Continuity and stability in maintenance and management are very important to the natural environment of river banks. The art of maintenance is directed at maintaining or developing the desired long-term natural environment. Interruptions to maintenance should be avoided as much as possible. This implies that the necessary means (including money and labour) should be secured in the long term by the authorities.
- Planning can achieve optimal input of personnel, material and financial resources. This information can also be used in planning work, supervision and costs.
- Planning maintenance activities will make it possible, when testing the objectives for the river bank, to evaluate the effectiveness of maintenance and management and, if necessary, to make timely adjustments.

Contents of a maintenance plan

Making a maintenance plan requires co-operation on policy at executive level and between people of various disciplines. The plan should be approved by the relevant authorities and should contain the following elements:

1. River bank management objectives.
2. Maintenance activities and the planning of these activities, including descriptions of methods, frequency, location, time, costs and organisation.

3. Description of what to do in exceptional cases (e.g. disease, plagues, species invasion, changes and calamities) and if repairs are needed. In the case of repairs the cause of the damage should be assessed. If the damage is incidental, the river bank can be repaired according to the original design. If the damage is structural, it may be necessary to change the construction or even possibly the objectives of the design.
4. Annual records of maintenance activities that have actually been carried out, costs and the labour used. Locations, times, circumstances and specific field conditions should be specified on maintenance forms.

PRINCIPLES OF MAINTENANCE

Above all, maintenance must be practical. During the early years following construction of the bank, the structure of the ecosystem will change. During this period maintenance should be adapted according to the desired ecological development. If the desired quality level is reached it is advisable to choose a continuous form of maintenance for a number of years since this favours ecological development.

When choosing maintenance measures, take into account as much as possible (1) the annual cycle of the desired (or undesirable) plants and animals and (2) the stage of development of the natural environment to ensure efficient maintenance. While developing a maintenance plan use methods and equipment that will minimise damage to the development of habitats on river banks. Try to opt for low and least harmful maintenance frequency and spread maintenance measures over time and/or space.

The difference between minor repairs, major repairs and restoration is related to the desired quality level as described in the river bank's target (see Figure 4.9). Carrying out frequent major repairs is bad for the ecosystem on the river bank due to the scale of activities required. It stimulates fast-growing, often common species. An ecosystem, after all, needs some time to develop (plants and animals need some time to establish into a habitat that has a higher nature conservation value compared to pioneering or disturbed situations). The significance of such a river bank for plants and animals belonging to more 'mature' ecosystems will increase through the years.

Flooding can have a naturally regulating effect on the succession of the river bank. As a result of flooding the river bank can regress to an earlier succession stage. If such regulatory flooding occurs frequently, it may be desirable not to carry out any maintenance activities or as little maintenance as possible, perhaps only on parts of the river bank. This natural situation can increase the ecological value of the river bank. It may, however, also cause a reduction in the ecological value and protective capacity of habitats on the river bank. A temporary decrease may be characteristic of a transition into a different succession stage, but it should be ensured that the river bank does not deteriorate permanently. Repairing damage to ecologically sound river banks may take a lot of time and cost a lot of money. The positive or negative evaluation depends on the situation and the aim for the river bank.

Maintenance costs depend on the method chosen, the frequency, the accessibility, the special provisions required, the degree of complexity and the use, processing and transport of materials and environmental change.

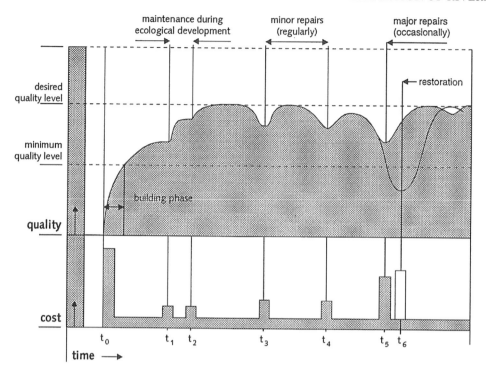

Figure 4.9 Interaction between desired quality level objective, maintenance and maintenance costs. (Copyright: CUR and RWS, 1994)

POST-PROJECT MONITORING PLAN

PHASE 9: POST-PROJECT MONITORING, INSPECTION AND EVALUATION

After completion of the river bank construction or after adjusting the maintenance, the development of the river bank should be monitored. Monitoring makes use of a number of measurable parameters and consists of at least two successive measurements carried out in one location, using the same method each time. Monitoring should be well documented in order to evaluate the development of the river bank against the objectives set. To ensure proper monitoring and evaluation it is necessary that these objectives are quantifiable and described in as much detail as possible.

Evaluation or, in other words, checking whether the river bank functions successfully, offers the opportunity to adjust the development, if necessary. In this way, future damage or quality loss can be avoided or limited. The river bank should also be inspected regularly to assess any potential damage or further maintenance requirements. In some cases extensive monitoring is required for there is still some lack of knowledge about ecologically sound river banks. Thus monitoring can offer an increase in understanding, as well as the opportunity to evaluate the project.

ECOLOGICALLY SOUND RIVER BANK MANAGEMENT 81

Monitoring and evaluation, however, involve time, money and specialist expertise. Consequently, it is important to assess whether monitoring and evaluation are the most suitable ways of providing this information (Duijn and Huys, 1992). For river bank inspection and for monitoring and appraisal, a monitoring plan is strongly recommended.

CONTENTS OF A MONITORING PLAN

When making a monitoring plan the following six aspects are important (after Vos et al., 1990).

Objectives

The objectives formulated for the river bank and the period in which these objectives should be achieved to a sufficient degree, must be listed. Moreover, the aim of monitoring should be described. Monitoring ecologically sound river banks is often directed towards testing the objectives formulated for the bank, including its structure. This is monitoring with a control function. Observing and predicting developments are additional functions (Udo de Haes, 1985). When choosing objectives, it should also be stated who, when and for what purpose the information provided will be used.

Evaluation

For each objective there must be a description of the information required and its quality for the purpose of judging the objective of monitoring. This implies that information should be checked for the following:

- usefulness,
- reliability,
- testability and
- processability.

Describing the criteria for measuring objectives properly will make the choice of parameters much easier.

Choice of parameters

The choice of parameters means deciding what should be measured and largely depends on type of location, objectives and criteria. In addition, they depend on the following:

- desired accuracy of the data and whether the data can be reproduced;
- methods of analysis (an indication for certain relations, interpretability, mutability);
- dimensions and accessibility of the test area or sampling points;
- resources available;
- expertise and skills available.

Methods, techniques and capacity

The different parameters can be measured by various methods. In choosing the right method for measuring and sampling, the following questions are important:

- With what frequency should sampling and measuring take place and how often should they be repeated?
- At what moment should measuring and sampling take place, how much time should the observation take and how precise should the indication of time be?
- Where should measuring and sampling take place?
- What is the desired scale (details and differentiation)?
- How much time, labour and money are available?
- Is experience for the method chosen available?

A method that is simple and requires little labour and at the same time gives a good evaluation of the target of ecologically sound river banks should be preferred.

Processing method

A monitoring system can only function properly if a suitable database is available for the data collected in the field (computer, photographs, etc.). It is important that third parties also have easy access to these data. When designing the database three aspects should be taken into account (Vos *et al.*, 1990):

- the time series of the collected data involving constantly growing data files;
- the requirement that different analyses should be possible for different selections of data;
- the monitoring system may need to be changed or extended.

In addition, the question of how to analyse the collected data should be dealt with here.

Organisation, planning and reporting

This includes a description of what tasks have to be carried out by whom, in what order and in what period. It is often possible to measure more than one parameter at a time, which could save a lot of time. Time and frequency of measurement should take into account maintenance and management activities. Reporting and presentation of data and conclusions are of major importance to the proper functioning of the monitoring system. The report is the basis for the actual assessment of the completed river bank construction. A good report not only states data and conclusions, but also uncertainties and assumptions.

RECOMMENDED PARAMETERS

The parameters listed in Table 4.1 are recommended for measuring, if relevant to the river bank involved (after Leemans, 1993). The parameters marked with an asterisk should only be measured if they are mentioned in the river bank objective. The reason for not recommending that all parameters are measured is that the data are hard to collect and hard to interpret, and that methods for collecting are very expensive. It is often better to have specific measurements carried out by experts. It is important that measuring is carried out thoroughly, so that adequate conclusions can be drawn and actions taken.

Table 4.1 Recommended parameters

- General data about location and reasons for monitoring
- Protective construction and other civil engineering structures
- Human and animal influence
- Maintenance
- River bank profile
- Hydrology
- Soil*
- Hydraulics*
- Ecology
  - Vegetation
  - Invertebrates
  - Fish*
  - Amphibians and reptiles*
  - Birds*
  - Mammals*
  - Micro-flora and fauna*

* These parameters should only be measured if they are mentioned in the management plan objective.

EVALUATION OF THE ENTIRE RIVER BANK MANAGEMENT PLAN

PHASE 10: EVALUATION

Monitoring and evaluation are not the end of the process. The overall management should occasionally, for instance at the end of each plan term (usually 10 years), be assessed with regard to the objectives mentioned in the management plan. The objectives themselves should be assessed and adjusted if necessitated by social developments. Such assessments may have consequences for the design and maintenance of the river bank(s) involved. The precise nature of these consequences can be determined by using the action plan. Even feedback of this kind, therefore, benefits from a systematic and planned approach.

SUMMARY

In this chapter a systematic approach to river bank management has been presented which will enable the development of ecologically sound river banks. The approach is based on a set of related plans, which cover all phases and activities within these phases that river bank management has to deal with. These activities include the setting up of an organisation when starting a river bank management project, making an inventory of the present situation, drawing up design alternatives, executing the chosen design, maintenance, post-project monitoring and evaluation of the project. The most important element of the approach is a clear and complete description of the objective, i.e. the situation that should be achieved within a set period. The objective provides the framework for the design alternatives, prescribes the method of execution and directs post-project monitoring. Finally, a comprehensive description of the objective makes a meaningful evaluation of river bank management possible.

Applying the approach may seem a time-consuming operation. There are, however, many advantages to be gained. The approach offers the possibility of integrating the many functions of river banks and in particular ensures that their ecological function is properly dealt with. It gives directions for the way authorities and other organisations concerned should co-operate during the entire management period. It also provides a clear view of how all disciplines involved, such as ecology and civil engineering, can be combined and geared to one another to achieve the best result possible.

ACKNOWLEDGEMENTS

We would like to thank Olav Beugels for translating the Dutch text and Rob van der Laag for transposing the figures. Also we would like to thank Cor Beekmans (Directorate Oost-Nederland), Klaas-Jan Douben, Marita Cals and Renske Postma (all of the Institute for Inland Water Management and Waste Water Treatment (RIZA)) for their valuable suggestions and comments.

REFERENCES

Boeters, R.E.A.M., Meesters, H.J.N., Schiereck, G.J., Simons, H.E.J., Stuip, J., Swanenberg, A.T.P., Verheij, H.J., and Verkade, G.J. (1994) Waterways with room for nature. 28th International Navigation Congress, Permanent International Association of Navigation Congresses (PIANC), June 1994, Sevilla. Section I, Inland waterways and ports (for commercial and pleasure navigation); Subject 4, Compatability between nature preservation and waterways, pp. 77–97.
CUR and RWS/DWW (1994) Natuurvriendelijke oevers. Eindredactie Simons, H.E.J., Koolen, J.L. and Verkade, G.J. CUR Report 168, Gouda.
Douben, K-J. (1998) Fairway improvement combined with ecological rehabilitation: an Utopical plan for the Dutch Rhine? Ministry of Public Works and Water Management, Rijkswaterstaat, Institute for Inland Water Management and Waste Water Treatment (RIZA). Congress 16–20 May 1994 on East–West, North–South encounter on the state-of-the-art in river engineering methods and design philosophy. State Hydrological Institute, Petersburg, Russia.
Duijn, P.P. and Huys, P.J.J.W. (1992) Richtlijnen voor de monitoring ter evaluatie van milieuvriendelijke oevers. Rijkswaterstaat, Dienst Weg- en Waterbouwkunde (DWW). MI-91-52.
Heidemij (1991) CUR-milieuvriendelijke oevers; aandachtspunten voor het opstellen van beheersplannen voor oevers. Heidemij adviesbureau.
Janssen, S.R.J. and Rademakers, J.G.M. (1993) Nature development along rivers. *Landscape*, **10**(3), 49–68 (in Dutch).
Leemans, J.A.A.M. (1993) *Ecologisch Meetnet Rivieroevers; Methode voor Monitoring van het Rivieroevermilieu*. Stichting voor Toegepaste Landschapsecologie, Nijmegen in opdracht van Rijkswaterstaat (RIZA), STL Report 93-3.
Klink, A. (1993) Natuurontwikkeling in het rivierengebied: omgaan met onzekerheden. Hydrologisch Adviesburo Klink b.v. Wageningen; rapporten en mededelingen 45.
Ministry of Transport, Public Works and Water Management (1985) *Living with Water*.
Ministry of Transport, Public Works and Water Management (1989) *National Policy Document on Water Management*.
Oranjewoud (1991) Milieuvriendelijke Oevers: Systematische aanpak voor de inrichting van oeverzones. Oranjewoud, Oosterhout.

Peters, J.S., van den Hark, M.H.C. and Bakker, C. (1991) Ecologische advisering natuurvriendelijke oevers: een methodische leidraad. Rijkswaterstaat, Rijksinstituut voor Integraal Zoetwaterbeheer en Afvalwaterbehandeling (RIZA), Note 91.086.

Pruisen, H. and Schippers, W. (1991) Natuurbouw en uitvoeringstechnieken. *Landinrichting*, **31**(3).

Rademakers, J.G.M., (1993) Natuurontwikkeling uiterwaarden & ecologisch onderzoek; een verkennende studie. Deelprogramma Natuurontwikkeling. Grontmij en DLO-IBN, Wageningen.

Rijkswaterstaat Dienst Weg- en Waterbouwkunde (1994) Handleiding Beheer Groenvoorzieningen: Opstellen van en werken met beheersplannen beheer en onderhoud van groenvoorzieningen, P-DWW-93-726.

Rijkswaterstaat, Directie Gelderland (1993) Oeverture. Inrichtingsplan oevers Rijntakken. Report no. GLD 93/05-01.

Smit, H. and van Urk, G. (1987) Het herstel van de ecologische waarden van de Rijn: over de zalm en ecologische doelstellingen. H_2O, **20**, 427–430.

Udo de Haes, H.A. (1985) Milieunetten. Inventarisatie, analyse en perspectief. RMNO No. 14, Rijswijk.

Vos, P., Orleans, A.B.M., Janssen, M.P.J.M., Meelis, E., ter Keurs, W.J. (1990) Natuur en milieumeetnetten voor het beleid. Deel 1: Hoofdrapport. Het ontwerpen van meetnetten. Milieubiologie en Instituut voor Theoretische Biologie, R.U. Leiden.

Wolfert, H.P. (1992) Geomorphological differences between river stretches: differences in nature development potentials. Summaries of the contributions to the European workshop 'Ecological Rehabilitaion of Floodplains', Arnhem, the Netherlands, 22–24 September 1992. Rijkswaterstaat RIZA Lelystad; CHR/KHR Lelystad; Vituki Budapest, pp. 137–144.

5 The Influence of Riparian Ecotones on the Dynamics of Riverine Fish Communities

MACIEJ ZALEWSKI and PIOTR FRANKIEWICZ
Department of Applied Ecology, University of Łódź, Poland

INTRODUCTION

The most promising and realistic strategy toward conservation of river ecosystems and their fish communities is the creation or recreation of land–water ecotones which play a key role in integrating catchment and stream channel. The understanding of the functioning of land–water ecotones is now sufficient to use the concept for the conservation and rehabilitation of this narrow zone. One of the most interesting aspects in this field has been the relationship between riparian vegetation and fish communities. Fish, as the most vulnerable element of the riverine ecosystem, are strongly influenced by changes in channel structure and water quality. They are often the most important socio-economic tool in environment conservation, being an important protein supply and the basis for recreation in the form of fishing.

In this chapter, an attempt is made to demonstrate the complexity and extent of the effects of land–water ecotones on fish communities and the role of ecotones in the rehabilitation of riverine habitats for fish.

THE ROLE OF RIPARIAN ECOTONES IN STRUCTURING FISH HABITATS

The riparian ecotones of rivers modify living conditions for fish by increasing habitat diversity. Undercut banks with tree roots, overhanging branches and logs influence not only depth, current velocity and substrate coarseness, but also serve to increase invertebrate biomass and create shelter for fish. As availability of food and habitat diversity are the two most important factors determining species separation in fish assemblages (Moyle and Senanayake, 1984; Welcomme, 1985; Ross, 1986), the above modifications will improve spatial segregation and food availability and thus reduce intra- and interspecific relations such as competition and predation.

Riparian vegetation exerts control over the occurrence of in-stream macrophytes by influencing the light regime of the channel. Managed correctly, riparian vegetation can further enhance habitat diversity. Modified or re-established riparian vegetation should be characterised by high heterogeneity, e.g. a diversity of trees and bushes, but with

Rehabilitation of Rivers: Principles and Implementation. Edited by L. C. de Waal, A. R. G. Large and P. M. Wade.
© 1998 John Wiley & Sons Ltd.

medium structural complexity to maintain light access to the channel. The diverse plant communities on the stream banks are sequentially supplying organic matter to decomposer organisms in the stream, the trophic base for fish. A diversity of species ensures differences in decomposition rate. Such a structure creates an intermediate complexity in riparian ecotones (Zalewski at al., 1994). Such ecotone types not only enhance stream trophy but also reduce the variability and unpredictability of abiotic factors, e.g. velocity and temperature, thus stimulating an increase in the richness and biodiversity of fish species.

Different fish species and/or ontogenetic stages show a wide range of habitat preferences (e.g. depth, current, substrate) (Ross et al., 1987; Nowak and Zalewski, 1991) which reflect their ecomorphological adaptation (Webb, 1984) and enable them to avoid severe competition and to partition resources successfully. Simplification of river channel structures may result in the reduction of fish diversity (O'Hara, 1986), mainly through elimination of those fish species which require, even for a short period of their ontogenetic development, access to different types of habitats (Portt et al., 1986; Schiemer and Spindler, 1989; Zalewski et al., 1998). For drift-feeding fish, the optimal current speed that maximises the ratio between energy gain and expenditure, depends on fish size (Webb et al., 1984). In river sections with a slow current speed, despite low energy expenditure by fish, the supply of drifting organisms for fish feeding in the water column is low. Thus to survive they must migrate to more heterogeneous sections of the stream channel, if such sections exist in a given river. On the other hand, the uniform, high water velocity characteristics of channelised rivers with a poorly developed riparian ecotone force many fish to expend more energy than they gain (Bachman, 1982) (Figure 5.1). As a result, a decline of small fish and specimens with low energy reserves, which are unable to tolerate such conditions, may be observed (Cunjak and Power, 1986). Such a reduction in fish numbers and size ranges in both fish populations and communities increases the chance of species extinction and in consequence leads to a decline in biodiversity.

THE ROLE OF RIPARIAN ECOTONES IN IMPROVING TROPHIC CONDITIONS

Riparian ecotones influence directly and indirectly the amount of food provided to fish. The quantity of terrestrial prey (primarily invertebrates) that enters the river depends directly on the ecotone complexity, while the aquatic food resources in the river depend indirectly on it through both allochthonous organic matter supply and the control of primary production through the light regime. Terrestrial insects may provide the major part of the fish food in low order streams (Hunt, 1975; Zalewski et al., 1985). However, it has been suggested many times that fish biomass and production in shaded streams is lower than in open streams (Murphy and Hall, 1981; Hawkins et al., 1983; Mann and Penczak, 1986; Bilby and Bisson, 1987). This indicates the importance of the autotrophic food chain for fish populations even where allochthonous organic matter is provided to the stream (Bilby and Bisson, 1992). This is because in shaded streams low temperatures and low levels of primary production inhibit biological processing activity, and as a result the amount of energy transferred through food webs and thus available to fish is low. These relationships were clearly shown in the

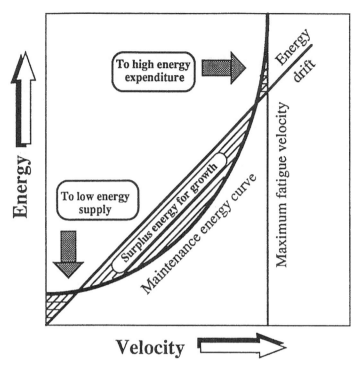

Figure 5.1 The energy expenditure–energy gain relationship for drift-feeding fish as an example of necessity to maintain diverse flow in the stream channel (from Zalewski et al., 1998)

case of the upland Lubrzanka River in central Poland where the biomass of both benthic invertebrates and fish was significantly lower in shaded sections of the stream with low primary production, than in neighbouring sections with better light regimes (Figure 5.2).

The quality, quantity and timing of allochthonous organic matter supply from the riparian ecotones depends on the composition of plant species. Various plants provide leaf biomass at different seasons (e.g. deciduous and coniferous trees), and due to differences in the decomposition rate of leaves, the timing of energy availability for invertebrates and fish can differ greatly (e.g. Kaushik and Hynes, 1971; Petersen and Cummins, 1974). In addition, the intensity of agricultural use of the catchment may influence palatability of leaves for invertebrates. For example, Irons et al. (1988) found that leaf consumption was positively correlated with nitrogen content and negatively with tannin concentration. Thus, the nutrient status of riparian vegetation, which depends on the nutrient supply from the catchment, may strongly influence energy transfer through a stream's food web. According to Cummins (1992), the primary and secondary succession of stream/river biotas is mostly determined by the pattern of organic matter supply from riparian/floodplain ecotones.

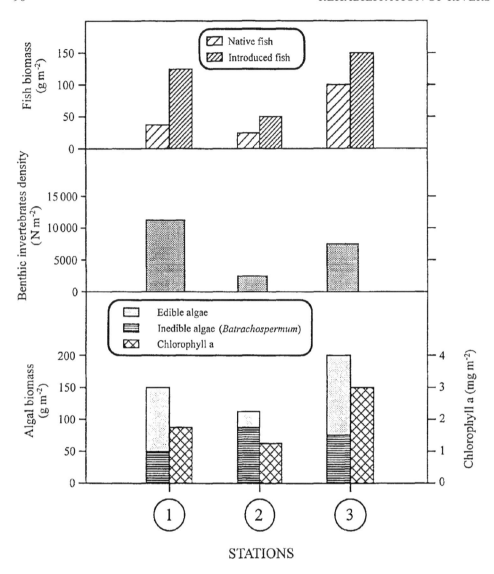

Figure 5.2 The effect of light reduction by riparian ecotone on the structure and biomass of benthic algal communities, the biomass of benthic invertebrates, and the biomass of native and introduced fish in upland streams (central Poland). Station 1: stream flowing through mixed forest; Station 2: stream heavily shaded by deciduous forest; Station 3: stream flowing through meadows with a low density strip of alder on its banks. (Data from Zalewski *et al.*, 1985, 1994)

THE ROLE OF RIPARIAN ECOTONES IN WHOLE SYSTEM PROCESSES

Riparian ecotones function as filters for nutrients, pollutants and catchment erosion. Their filtering efficiency may reach values even above 80% (Gilliam et al., 1986). When banks are unprotected, soil erosion increases turbidity and siltation through rapid runoff from agricultural areas (Schlosser and Karr, 1981). Also pesticides are transferred directly from fields to the river and can be accumulated in food webs. A properly structured land–water buffer zone allows not only the retention of nutrients and pesticides transported to a river, but also it may accelerate nutrient spiralling through optimisation of in-stream primary productivity (*sensu* Ellwood et al., 1983) and thus increase the self-purification ability of the river (Figure 5.3). The filtering capacity of riparian ecotones, together with their ability to increase fish production by acceleration of nutrient and energy transfer through food chains, reduces the amount of organic matter transported to reservoirs or estuaries, thus slowing down their eutrophication rate.

By creating land–water ecotones of intermediate complexity in small rivers (Figure 5.4) we can expect on the one hand the improvement of habitat diversity, organic matter supply and retention, and on the other, the acceleration of nutrient spiralling and energy transfer to the top trophic levels. All these processes will result in an increase in fish productivity, improvement of self-purification, and consequently the reduction of downstream eutrophication. Expanding the view to the whole river system, Schiemer and Zalewski (1992) concluded:

"the network of microhabitats required during critical life stages and the inter-linking of them under conditions of strong water level fluctuations and floods will determine recruitment and year class strengths. The input of organic material from floodplains and lentic backwaters, the interconnectivity between water bodies and light access are responsible for the trophic conditions and the carrying capacity of the fish fauna".

HOW TO RESTORE RIVERINE SYSTEMS?

The data presented above clearly indicate the optimal strategy for riverine ecosystem restoration leading to enhancement of biotic structure and fish communities. According to the concept of fish community regulation by a continuum of abiotic–biotic factors (Zalewski and Naiman, 1985), and more generally the idea of habitat templet (Southwood, 1977), the abiotic factors such as flow velocity, temperature, and habitat structure are of primary importance in shaping biotic communities in rivers. That is why the restoration strategy should be focused on two fundamental elements: first, a reduction in the strength of abiotic/stochastic factors, which at extreme values eliminate some fish populations mainly by forcing them to expend too much energy; secondly, improvement of trophic conditions for fish, as a faster growth rate and an increase in stored reserves of energy-rich substances may enhance fish survival when abiotic factors become unfavourable for fish.

Both aims can be achieved by restoration of riparian/floodplain ecotones integrating nature conservation, cultural aesthetic values, and recreation. The proposed design for the rehabilitation of riverine habitats for fish communities is summarised in Table 5.1. The general assumption is that restoration techniques should be adjusted to the type of the

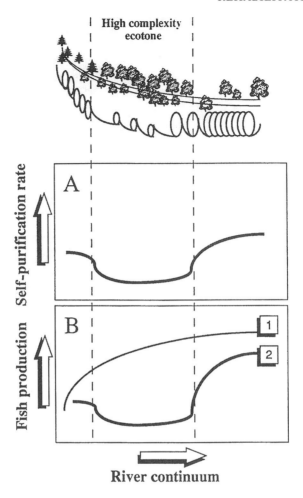

Figure 5.3 The effect of light reduction in a high complexity riparian ecotone on (a) river self-purification rate and (b) fish biomass distribution. 1, Fish biomass distribution along the river continuum in temperate rivers (data from Zalewski *et al.*, 1985). 2, Modification of fish biomass distribution due to light access reduction (data from Zalewski *et al.*, 1985)

stream (*sensu* Huet, 1959). The two main parameters are stream gradient and width. The recommended riparian area expressed as surface of insolated channel should be between 40 and 90%. The lowest value refers to upland trout and grayling streams, where not only light intensity but also water temperature may be critical factors for fish populations. (Care should be taken in removal of the riparian canopy as this may elevate temperature beyond the threshold of salmonids thermal tolerance.) Large rivers should have better insolation than smaller upland rivers, as both a good light regime and a high water temperature stimulate biological processes in the river channel resulting in enhancement of the self-purification rate. In upland and mountain trout and grayling streams, the presence of such structures as undercut banks with tree roots is of great importance, enabling territorial fish such as trout to coexist. In smaller streams dominated by trout a higher proportion of riffles than pools is recommended. Riffles with high primary productivity are

THE INFLUENCE OF RIPARIAN ECOTONES

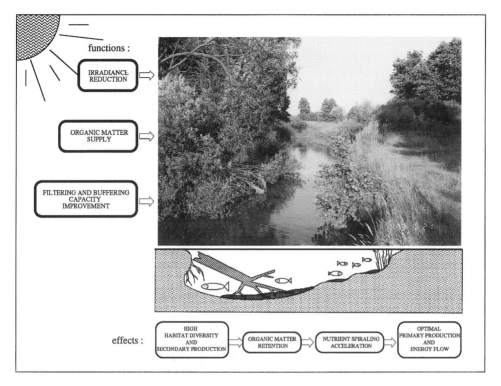

Figure 5.4 The role of riparian ecotones as a factor modifying riverine fish communities

a source of drifting invertebrates; however, to enable large and small trout to segregate spatially, pools with a minimal depth of 0.4 m are required (Milner, 1982; Nowak, 1995). On the other hand, in rivers dominated by grayling the proportion of pool areas should be higher. In such rivers shallow side pools are necessary to provide rearing areas for juvenile fish (P. Gaudin, pers. comm.). In large rivers where bream and barbel dominate, river connections with wetlands and oxbows, as well as gravel beaches for juvenile reophylous fish, are important in maintaining a diverse fish community (Penczak and Zalewski, 1974; Schiemer and Waidbacher, 1992). According to the flood pulse concept (Junk et al., 1989), the floodplain connections are also important for the river purification rate, as most of organic matter is transported downstream during the peak-period short floods.

In a completely channelised river, two types of restoration techniques may be recommended. Firstly, there are semi-intensive techniques: this includes planting trees and bushes on the river banks. After several years, the bank's structure will gradually become diversified by tree roots, floating branches and fallen trees. This will generate dams of organic debris which are of great importance for river channel processes, for example, increasing the retention of organic matter (Gregory, 1992). Secondly, there are intensive techniques: these achieve results more rapidly but involve more expensive enterprises, such as side-slope reduction and/or replacement of meanders (Petersen et al., 1992). Although river conservation techniques are beginning to incorporate whole catchment processes (e.g. Newson, 1992), from the point of view of the restoration of the fish community the most important are those at the channel and floodplain level.

Table 5.1 The rehabilitation of riverine habitat for fish by restoration and management of riparian ecotones

Type of river	Gradient (%)/ mean width (m)	Recommended riparian area expressed as surface of insolated channel	Bank structures: undercut banks with tree roots	Recommended riffle/pool proportion (%) and pool depth (m)	Technique of restoration
Trout streams	Upland: > 4.5/10 Mountain: > 6.0/15	Upland: 40%* Mountain: 60%	Recommended: 80% Minimal: 50%	60/40 0.4	Side-slope reduction, planting trees and riverine bushes, sequential cutting of trees on the southern banks
Grayling streams	Upland: > 1.0/25 Mountain: > 2.0/30	Upland: 50%* Mountain: 70%	Recommended: 80% Minimal: 60% Shallow side pools as rearing areas for juvenile fish	40/60 0.6	Side-slope reduction, planting trees and riverine bushes, sequential cutting of trees on the southern banks
Barbel rivers	0.3–3.0/60	> 80%	Riparian zone development, connections with wetlands and oxbows in the valley, gravel beaches as habitats for juvenile reophylous fish		Hydro-engineering works (elimination of levees, recreation of meanders, etc.)
Bream rivers	0.1–1.0/100	> 90%	Riparian zone development, connections with wetlands and oxbows in the valley		Large-scale hydro-engineering works (elimination of levees, floodplain restoration)

*If necessary should be adjusted according to local summer temperature conditions.

ACKNOWLEDGEMENTS

During the preparation of this paper our investigations on processes in land–water ecotones were supported by British Council grant no. WAR/992/025 and State Committee of Scientific Research grant no. PB 1614/6/91.

REFERENCES

Bachman, R.A. (1982) A growth model for drift feeding salmonids: a selective pressure or migration. In Brannon, E.L. and Salo, E.O. (eds) *Salmon and Trout Migratory Behaviour Symposium.* University of Washington, Seattle, WA, pp. 128–134.

Bilby R.E. and Bisson, P.A. (1987) Emigration and production of hatchery coho salmon (*Oncorhynchus kisutch*) stocked in streams draining an old growth and a clear-cut watershed. *Canadian Journal of Fisheries and Aquatic Science,* **44**, 1397–1407.

Bilby, R.E. and Bisson, P.A. (1992) Allochthonous versus autochthonous organic matter contributions to the trophic support of fish populations in clear-cut and old-growth forested streams. *Canadian Journal of Fisheries and Aquatic Science,* **49**, 540–551.

Cummins, K.W. (1992) Catchment characteristics and river ecosystems. In Boon, P.J., Calow, P. and Petts, G.E. (eds) *River Conservation and Management.* John Wiley, Chichester, pp. 125–135.

Cunjak, R.A. and Power G. (1986) Winter habitat utilization by stream resident brook trout (*Salvelinus fontinalis*) and brown trout (*Salmo trutta*). *Canadian Journal of Fisheries and Aquatic Science,* **43**, 1970–1981.

Ellwood, J.W., Newbold, J.D., O'Neill, R.V., Stark, R.W. and Singley, P.T. (1983) The role of microbes associated with organic and inorganic substrates in phosphorus spiralling in a woodland stream. *Verhandlungen der Internationalen Vereinigung für theoretische und angewandte Limnologie,* **21**, 850–856.

Gilliam, J.W., Skaggs, R.W. and Jacobs, T.C. (1986) Controlled agricultural drainage: an alternative to riparian vegetation. In Correll, D.L. (ed.) *Watershed Research Perspectives.* Smithsonian Institute Press, Washington, DC.

Gregory, K.J. (1992) Vegetation and river channel process interactions. In Boon, P.J., Calow, P. and Petts, G.E. (eds) *River Conservation and Management.* John Wiley, Chichester, pp. 255–269.

Hawkins, C.P., Murphy, M.L., Anderson, N.H. and Wilzbach, M.A. (1983) Riparian canopy and substrate composition interact to influence the abundance of salmonids, sculpins and salamanders in streams. *Canadian Journal of Fisheries and Aquatic Science,* **40**, 1173–1185.

Huet, M. (1959) Profiles and biology of Western European streams as related to fish management. *Transactions of the American Fisheries Society,* **88**, 153–163.

Hunt, R.L. (1975) Use of terrestrial invertebrates as food for salmonids. In Hasler, A.D. (ed.) *Coupling of Land and Water Systems.* Springer Verlag, New York, pp. 137–151.

Irons, I.G., Oswood, M.W. and Bryaant, J.P. (1988) Consumption of leaf detritus by a stream shredders: influence of tree species and nutrient status. *Hydrobiologia,* **160**, 53–61.

Junk, W.J., Bayley, P.B. and Sparks, R.E. (1989) The flood pulse concept in river–floodplain systems. In Dodge, D.P. (ed.) *Proceedings of the International Large River Symposium. Canadian Special Publication on.* Fisheries and Aquatic Science, **106**, 110–127.

Kaushik, N.K. and Hynes, H.B.N. (1971) The fate of the dead leaves that fall into streams. *Archiv für Hydrobiologie,* **68**, 465–515.

Mann, R.H.K. and Penczak, T. (1986) Fish production in rivers: a review. *Polskie Archiwum Hydrobiologii,* **33**, 233–247.

Milner, N.J. (1982) Habitat evaluation in salmonid streams. In *Proceedings of 13th Annual Study Course.* Institute of Fisheries Management, College of Librarianship, Wales, pp. 47–72.

Moyle, P.B. and Senanayake, S. (1984) Resource partitioning amongst the fishes of the rainforest streams of Sri Lanka. *Journal of Zoology, London,* **202**, 195–223.

Murphy, M.L. and Hall, J.D. (1981) Varied effects of clear-cut logging on predators and their habitat in small streams of the Cascade Mountains, Oregon. *Canadian Journal of Fisheries and Aquatic Science*, **38**, 137–145.

Newson, M.D. (1992) River conservation and catchment management: a UK perspective. In Boon P.J., Calow, P. and Petts, G.E. (eds) *River Conservation and Management*. John Wiley, Chichester, pp. 385–396.

Nowak, M. (1995) The heterogeneity of the river channel and riparian ecotones complexity as a factor determining diversity and biomass of fish community. PhD thesis, University of Łódz.

Nowak, M. and Zalewski, M. (1991) The fish distribution in habitats of lowland river Grabia. *Acta Universitatis Lodziensis, Folia Limnologica*, **5**, 153–165.

O'Hara, K. (1986) Fish behaviour and the management of freshwater fisheries. In Pitcher, T.J. (ed.) *The Behaviour of Teleost Fishes*. Croom Helm, London, pp. 496–521.

Penczak, T. and Zalewski, M. (1974) Distribution of fish numbers and biomass in barbel region of the river and the adjoining old river-beds. Ekologia Polska, **22**, 107–119.

Petersen, R.C. and Cummins, K.W. (1974) Leaf processing in a woodland stream. *Freshwater Biology*, **4**, 343–368.

Petersen, R.C., Petersen, L.B.-M. and Lacoursiere, J. (1992) A building-block model for stream restoration. In Boon, P.J., Calow, P. and Petts, G.E. (eds) *River Conservation and Management*. John Wiley, Chichester, pp. 293–309.

Portt, C.B., Balon, E.K. and Noakes, D.L.G. (1986) Biomass and production of fishes in natural and channelized streams. *Canadian Journal of Fisheries and Aquatic Science*, **43**, 1926–1934.

Ross, S.T. (1986) Resource partitioning in fish assemblages: a review of field studies. *Copeia*, **1986**, 352–388.

Ross, S.T., Baker, J.H. and Clark, K.E. (1987) Microhabitat partitioning of southern stream fishes: temporal and spatial predictability. In Matthews, W.J. and Heins, D.C. (eds) *Community and Evolutionary Ecology of North American Stream Fishes*. University of Oklahoma Press, Norman, London, pp. 41–51.

Schiemer, F. and Spindler, T. (1989) Endangered fish species of the Danube River in Austria. *Regulated Rivers: Research and Management*, **4**, 397–407.

Schiemer, F. and Waidbacher, H. (1992) Strategies for conservation of a Danubian fish fauna. In Boon, P.J., Calow, P. and Petts, G.E. (eds) *River Conservation and Management*. John Wiley, Chichester, pp. 363–382.

Schiemer, F. and Zalewski, M. (1992) The importance of riparian ecotones for diversity and productivity of riverine fish communities. *Netherlands Journal of Zoology*, **42**, 323–336.

Schlosser, I.J. and Karr, J.J. (1981) Riparian vegetation and channel morphology impact on spatial patterns of water quality in agricultural watersheds. *Environmental Management*, **4**, 397–407.

Southwood, T.R.E. (1977) Habitat, the templet for ecological strategies? *Journal of Animal Ecology*, **46**, 337–365.

Webb, P.W. (1984) Form and function in fish swimming. *Scientific American*, **251**, 58–68.

Webb, P.W., Kostecki, P.T. and Stevens, E.D. (1984) The effect of size and swimming speed on locomotor kinematics of rainbow trout. *Journal of Experimental Zoology*, **109**, 77–95.

Welcomme, R.L. (1985) River fisheries. *FAO Fish Biology Technical Paper no. 262*, pp. 1–330.

Zalewski, M. and Naiman, R.J. (1985) The regulation of riverine fish communities by a continuum of abiotic–biotic factors. In Alabaster, J.S. (ed.) *Habitat Modifications and Freshwater Fisheries*. Butterworth, London, pp. 3–9.

Zalewski, M., Frankiewicz, P. and Brewinska, B. (1985) The factors limiting growth and survival of brown trout, *Salmo trutta* m. *fario* L., introduced to different types of streams. *Journal of Fish Biology*, **27** (Supplement A), 59–73.

Zalewski, M., Frankiewicz, P., Przybylski, M., Banbura, J. and Nowak, M. (1990) Structure and dynamics of fish communities in temperate rivers in relation to the abiotic–biotic regulatory continuum concept. *Polskie Archiwum Hydrobiologie*, **37**, 151–176.

Zalewski, M., Puchalski, W., Frankiewicz, P. and Bis, B. (1994) Riparian ecotones and fish communities in rivers – intermediate complexity hypothesis. In Cowx, I.G. (ed.) *Rehabilitation of Freshwater Fisheries*. Fishing News Books, Blackwell Scientific Publications, pp. 152–160.

Zalewski, M., Bis, B., Łapinska, M., Frankiewicz, P. and Puchalski, W. (1998) The Importance of the riparian ecotone and river hydraulics for sustainable basin-scale restoration scenarios. *Aquatic Conservation: Marine and Freshwater Ecosystems*, **8**, 287–307.

6 Rehabilitation of Rivers by Using Wet Meadows as Nutrient Filters

ANN FUGLSANG
Funen County Council, Department of Nature and Water Environment, Odense, Denmark

INTRODUCTION

Increasing eutrophication problems in the coastal areas off Denmark during the 1970s and in the beginning of the 1980s were caused by increasing discharges of nutrients from the major sources during the same period of time. During the period 1976–1987 non-point sources (primarily agriculture) accounted for an average 82% of the nitrogen runoff and 22% of the phosphorus runoff to the coastal waters surrounding Funen, Denmark. The volume of the nitrogen runoff from the land has been shown to be closely correlated with the intensity of agricultural activities in the catchment area. Approximately 90% of the nitrogen runoff in the watercourses is nitrate. Extensive application of artificial fertilisers and an inexpedient use of manure with a very poor degree of nitrogen utilisation are the primary causes of the large nitrogen runoff from farmlands (Funen County Council, 1990).

On Funen, the total use of nitrogen fertilisers has doubled since the mid-1950s causing eutrophication of surface waters and pollution of groundwater. In addition, with the installation of drain tiles, drainage from farmlands has increased during this period causing a lack of nutrient transformation in wetlands and riparian zones. The installation of drain tiles made it possible to cultivate wet areas along rivers and lakes. Furthermore, many large areas have been drained by pumping away water from the wetlands.

The number of natural or grassed areas in the agricultural landscape of Funen has been halved since the 1950s, from 10% to 5% (Denmark's Central Statistical Office, 1957, 1992) which is equivalent to 14 000 ha (the total area of Funen is 356 500 ha). The natural areas also include non-rotational fallow land. The grassed areas include non-rotational grasslands only.

Funen County Council is trying to show the scope for re-establishment of wet meadows along rivers in relation to reducing the contamination of surface and coastal waters and the re-introduction of meanders in rivers. The objective of this countryside rehabilitation project (at a catchment scale) is to re-establish the hydrological, chemical and biological interactions between the river and the floodplain.

Prior to this catchment-scale project, Funen County Council carried out investigations of nutrient transformation in a wet meadow at Storå on Funen. Other Danish investigations (e.g. Rabis Stream in Jutland) are similar to the investigations at Storå on Funen.

Rehabilitation of Rivers: Principles and Implementation. Edited by L. C. de Waal, A. R. G. Large and P. M. Wade.
© 1998 John Wiley & Sons Ltd.

Both investigations showed that the lack of wetlands and other wet areas give a lack of nitrate transformation (denitrification and plant uptake) of approximately 400 kg NO_3-N ha^{-1} $year^{-1}$ (Brüsch, 1991; Fuglsang, in press).

THE FUNEN COUNTRYSIDE REHABILITATION PROJECT

OBJECTIVES

The objective of the Funen Countryside Rehabilitation Project is to set up practical initiatives with the aim of reducing the diffuse leaching of nutrients from the agricultural land into Funen's waterways and the sea. A further objective is to demonstrate possible elements of sustainable farming for the future, whereby the interests of the water recipients and nature conservation are comprehensively catered for.

MOTIVATION

The intensification of farming has led amongst other things to previously uncultivated areas being brought into rotation. This has resulted in an increased leaching of nutrients from the land, a more uniform cultivated landscape and a paucity of natural habitats for flora and fauna.

The intention is for this trend to be reversed through countryside rehabilitation with a view to meeting the goals set by the national environmental policies in the National Environmental Action Plan and the Plan of Action for sustainable agricultural growth, including the goal of halving the leaching of nitrates into waterways, such leaching being principally due to the waste of nitrates from farming.

In order to recreate an agricultural landscape which would have the appearance of a mosaic of cultivated and uncultivated acreage, giving free scope to nature, it is necessary to show the way by means of the example set by practical initiatives. The setting aside of uncultivated areas alongside waterways offers the opportunity for interplay between both riparian areas and the waterway, especially if maintenance of the waterways (weed cutting and dredging for flood prevention) is avoided. The uncultivated areas should be set aside as wet meadows for nitrate transformation. Nitrate transformation can occur by irrigating the meadows with drainage water or water from the rivers. This would contribute to a reduction of the diffuse leaching of nutrients from farmland, a more varied cultivated landscape, and the establishment of an environment with a larger number of habitats and corridors for the natural animal and plant life.

Maintenance is an unnatural intrusion in the waterways, often disturbing the natural flora and fauna. In order to establish a more natural balance in the waterways as little maintenance as possible is currently carried out, within statutory limits. These maintenance provisions have been made to allow for the drainage requirements of farming and do not further the interaction between the waterways and larger areas surrounding the waterways.

Today substantial resources are used for the maintenance of the waterways of Funen County. Every year the Funen County Council spends 5.2 million kroner maintaining the 600 km of county watercourses, representing 8.70 kroner per metre. Some of these resources could be diverted to the setting aside of uncultivated areas alongside waterways in return for the payment of compensation to the landowner, an arrangement that would be advantageous to the landowners, nature and the water environment.

PRACTICAL INITIATIVES

The countryside rehabilitation project has to be a catchment-scale project where many aspects are taken into account. The adoption of more extensive cultivation methods and the laying fallow of areas bordering waterways (meadows) are important elements in this project.

The surface leaching will thereby be limited to the appropriate waterways, and the use of fertilisers and pesticides close to waterways will be avoided. Flooding of meadows with drainage water from adjacent cultivated areas, thus restoring the function of these meadows as filters for nitrates, is another important element. Water quality will be improved and the overburdening of the waterways, and consequently the sea, with nitrates will thus be reduced.

A third element is to manage the flooded meadows, where appropriate. Varied types of habitats, including areas under grass advantageous to wild flora and fauna, could thereby be provided. A fourth important element is to avoid the maintenance of waterways, thereby improving and protecting the interaction between waterways and the neighbouring environment and improving the quality of the waterways.

FOLLOW-UP STUDIES

To demonstrate that the changes in land use along waterways have an effect on the diffuse leaching of nutrients from agriculture to the waterways, Funen County Council has chosen a watercourse system (a catchment area of approximately 80 km^2) with an existing water quality monitoring station for this countryside rehabilitation project. Stations for monitoring water quality in the watercourses on Funen have existed since 1967, but most of the monitoring stations were established in 1976. Funen County Council has long series of water quality data from about 30 of the largest watercourses on Funen. Measurements in future years can be compared with measurements from the past and thereby it can be demonstrated whether the nutrient removal objectives for the project are reached or not.

COLLABORATION WITH FARMING ORGANISATIONS AND LANDOWNERS

Funen County Council is collaborating with farming organisations in order to establish contact with landowners in the areas bordering the selected watercourse. The farming organisations are participating in a working group as part of the project, together with Funen County Council. An information meeting for the landowners along the selected watercourse has shown that many of the landowners (around thirty) are interested in participating in the project. Close collaboration is going to be set up with these landowners.

PROPOSALS FOR MANAGEMENT

Management proposals will be individually assessed for each separate area along the watercourse, but according to the following overall guidelines wherever possible. Uncultivated areas have to remain uncultivated, except where management can protect specific botanical and ornithological interests. Grasslands have to remain grasslands wherever possible. The grasslands should be used for grazing, or at least one crop of hay (late summer/autumn) should be taken annually. Cultivated areas have to be sown with grass and used for grazing wherever possible, or at least one crop of hay (late summer/autumn) has to be taken annually.

REHABILITATION PROJECTS IN DENMARK

In 1982 the Danish legislation concerning waterways, the Watercourse Act, was revised. The new legislation gives equal priority to drainage and environmental quality, and not just to drainage of cultivated fields as had previously been the case.

The new legislation has resulted in the new and ecologically more gentle maintenance practice of waterways by the watercourse authorities in Denmark. In addition, the watercourse authorities, primarily the county councils, have started a substantial amount of restoration work in the watercourses. Most of the restoration projects are carried out to improve the population of fish, such as the establishment of spawning grounds and making any obstructions passable for fish and invertebrates. A smaller number of restoration projects which attend to more general improvements of nature, such as recovery of sinuosity, have also been carried out.

Many county councils have started to make the obstructions in the watercourses passable for fish, e.g. are Southern Jutland, Funen, Viborg, Ribe, Northern Jutland and Vejle. At the end of 1993 about 17 larger restoration projects for recovery of sinuosity in watercourses in Denmark were carried out, and of these, five were on Funen (Ministry of the Environment, Denmark, 1994).

In addition, Funen County Council initiated the above-mentioned countryside rehabilitation project in 1994, which attends to more general improvements of nature by establishing wet meadows alongside waterways. Prior to this project, Funen County Council carried out investigations of the nutrient transformation in a wet meadow along Storå on Funen. Results from these investigations have formed the basis of the countryside rehabilitation project in relation to improvement of the water quality.

INVESTIGATIONS OF NUTRIENT TRANSFORMATION IN A WET MEADOW ON FUNEN IN DENMARK

STUDY AREA

In 1989, Funen County Council established a study area to investigate the nutrient transformation in a wet meadow. The study area is situated along the Storå watercourse at Brenderup on the island of Funen in Denmark (Figure 6.1). The investigations will be ongoing until the year 1999.

WET MEADOWS AS NUTRIENT FILTERS

The study area (0.8 ha) is predominantly clay soil with a secondary content of organic matter and a low hydraulic conductivity. The land is relatively flat with a 1% slope down to the Storå watercourse. The study area was cultivated and sown with barley until the summer of 1989. In the autumn of 1989 the study area was equipped with sheet piling, sampling stations, etc., and flooded with drainage water. The study area gradually became established with grasses and herbs, and looked like an uncultivated area in the spring of 1990.

As water began to flow over the area the directions of the main surface water currents were recorded. The path of water flow through the flooded area was subsequently confirmed in March 1991 using NaCl as a tracer to determine the movement of the

Figure 6.1 Location map. The arrow shows where the study area is situated (on the north-western part of Funen).

surface water and residence time (Figure 6.2). The residence time averaged 28 hours and ranged from about 9 hours to 3 days.

Figure 6.2 shows measurements of the specific conductance in the outlet water during the period of the tracer test. NaCl (18 kg) was added to the inlet water and the movement of the NaCl across the study area was found by measuring the specific conductance in the surface water running across the study area. About 7 hours later, 8 kg of NaCl was added to the surface water on the study area at the front of the NaCl wave, because the added amount of NaCl was diluted by water on the study area. The added amount of NaCl at the front of the NaCl wave gives the curve a bulge shape. The increase in the specific conductance started about 9 hours after the 18 kg of NaCl had been added, which was the shortest residence time. The tracer test showed that about 50% of the added amount of NaCl was carried across the study area in about 28 hours. At that time the decrease in specific conductance became linear. Most of the added NaCl was carried across the study area in about 3 days. The registration of specific conductance in the outlet water stopped in about 38 hours. Measurements of specific conductance across the study area have revealed no groundwater seepage entering the study area prior to flooding with drainage water, nor at the time of the NaCl tracer test (Brüsch, 1991).

EXPERIMENTAL DESIGN

The study area was surrounded with sheet piling (to a depth of 1 m) to facilitate monitoring of surface runoff and determination of the water balance. Five outlet pipes were mounted on the sheet piling and connected to a single pipe to continuously monitor surface runoff from the field using an electromagnetic flowmeter. In order to ensure that drainage water flows and seeps evenly into as much of the study area as possible, a ditch was constructed along the upslope part of the study area. Inflow of drainage water to the study area was also measured continuously by an electromagnetic flowmeter. Across the study area 18 sampling stations were placed (Figure 6.3).

SAMPLING AND ANALYSIS

Water samples were collected at the inlet and the outlet from the study area twice a month throughout the whole period when flood water was present (normally from August/September to May). Water samples (surface and shallow subsurface water) were collected at each sampling location in the study area twice in May, August, November and December in 1990, in March, May, August and October in 1991 and in February, April, May and December in 1992. In general, these samples were collected four or five times during the period with water flowing through the study area.

Surface and shallow subsurface water (0–10 cm depth) was collected from 18 stations in the study area during each sampling period. In addition, water samples were collected from a depth of 10–20 cm and 90–100 cm at four stations in deep well piezometers. Surface and shallow subsurface samples were collected in sampling tubes. A cross-section of the study area and installed sampling units is presented in Figure 6.3.

Prior to sampling, the stagnant water in the sampling tubes in the study area was collected using a vacuum pump. New water (approximately 500 ml, depending on season and analysis programme) moving into the sampling tubes was vacuum-pumped and

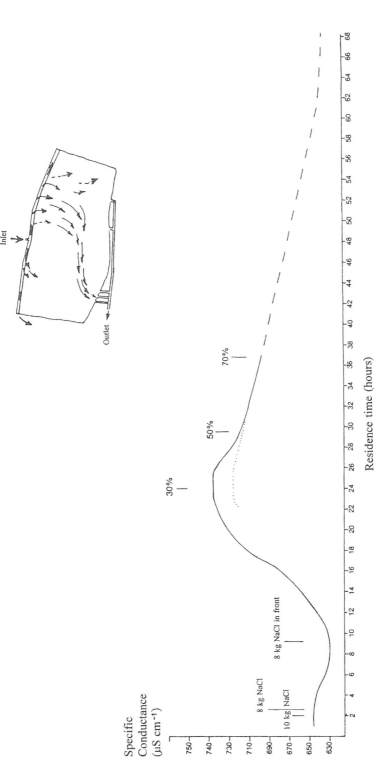

Figure 6.2 Tracer test in March 1991. The specific conductance in the outlet from the test field as a function of time (hours) after adding NaCl. Amount of added NaCl and % transported NaCl are shown. The directions of the main surface water currents measured at the tracer test are shown at the small sketch

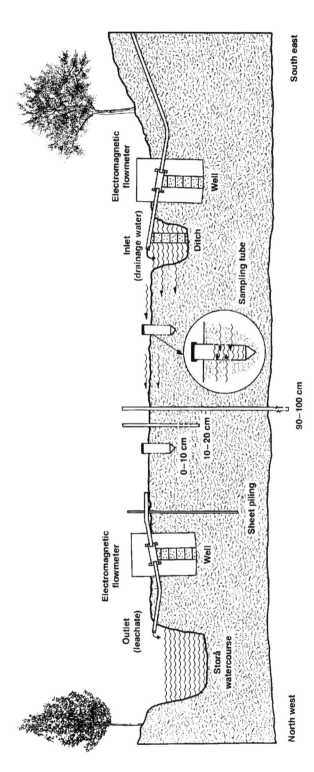

Figure 6.3 Sketch of experimental design showing the movement of drainage water, surface water and soil water plus different types of sample stations

analyses. In the field, water used for nitrate analyses (up to 100 ml) was filtered and preserved by adding thymol ($C_{10}H_{14}O$) to prevent transformation before analysis.

All samples were analysed for nitrogen fractions (total-N, $NH_3 + NH_4^+$-N, $NO_2^- + NO_3^-$-N), phosphorus fractions (total-P, PO_4^{2-}-P), Cl^-, SO_4^{2-} and pH. Danish Standard methods were used for analyses (Fuglsang, in press). Temperature and specific conductance of samples were measured in the field. Twice a year (in May and December) all the samples were also analysed for macro ions (K^+, Na^+, Mg^{2+}, total-Ca, total-Fe, Fe^{2+} and Mn^{2+}).

WATER BALANCE

A water balance was calculated for the period from March 1991 to January 1993 (Figure 6.4. Each month is represented by two columns).

Potential evapotranspiration (as an average of the County of Funen) was calculated by the Danish Institute of Plant and Soil Science using the Makkink equation (Mikkelsen and Olesen, 1991). Air temperature and global radiation were used for these calculations. The difference between the two columns in Figure 6.4 must correspond to the infiltration. On average, the estimated infiltration is 1500 mm year^{-1} in the study area, which is predominantly clay soil with a secondary content of organic matter.

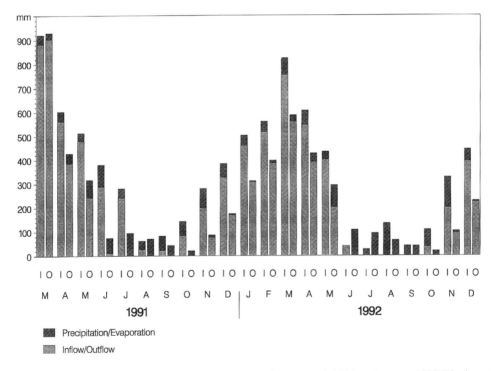

Figure 6.4 Water balance calculated for the period from March 1991 to January 1993. The input to the study area is shown in the first column as the inflow of water and the precipitation and the output is shown in the second column as the outflow of water and the evapotranspiration from the whole study area

Other studies in flooded fields at Syvs Stream in Zealand, Denmark reported much higher infiltration: up to 1500 mm month^{-1} (Hoffmann, 1991) at a study area composed of peat soil.

NITRATE TRANSFORMATION

Nitrate transformation ((NO_2^- + NO_3^-)-N) has been calculated on the assumption that one-third of the area (c. 0.3 ha) converts nitrate effectively (Brüsch, 1991). The size of the effective area was estimated on the basis of the tracer test that also defined the movement of surface and subsurface water as shown in the upper part of Figure 6.2.

The lowest nitrate transformation rates have been found during periods of high runoff and low temperatures and when nitrate inputs are low, typical in the summer. The highest nitrate transformation rates have been found during spring time. In the period from 1991 to 1992, where the waterflow was measured continuously, the lowest nitrate transformation corresponded to 20% of the nitrate inflow (in March 1991). In the period from March 1991 to January 1993 the amount of transformed nitrate varied from 6 to 100 kg N ha^{-1} month^{-1}, when there was a nitrate input. This corresponds to 100% and 80% of the nitrate inflow, respectively (Figure 6.5). In the winter period (from November to March) the nitrate transformation was 50 kg N ha^{-1} month^{-1} as an average corresponding to 60% of the nitrate inflow (both low temperature and low light intensity).

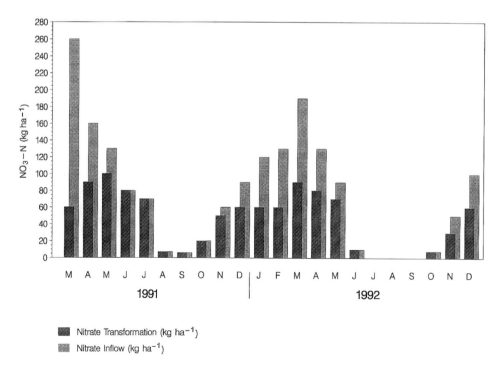

Figure 6.5 Nitrate transformation (kg ha^{-1}) and nitrate inflow (kg ha^{-1}) to the study area in the period from March 1991 to January 1993

The annual rate of nitrate transformation has varied from approximately 470 to 690 kg N per ha^{-1} year^{-1}, with the highest rate in the beginning of the investigation period, where the plant uptake has been largest, because the new vegetation (grass and herbs) had to establish. The infiltration of nitrate is approximately 3% of the total nitrate transformation and it corresponds to the aerial deposition of nitrate on the study area. The deposition is not added to the amount of nitrate in the inlet, so these two amounts neutralise each other.

The ranges in nitrate transformation reported here are substantially greater than the ranges in nitrate transformation calculated for an originally uncultivated area at Rabis Stream in Jutland, Denmark (Brüsch and Nilsson, 1990). Substantially lower values in the spring and higher values in the autumn are observed on Funen. Such differences in the range of nitrate transformation can be attributed to the fact that the geological characteristics of the two areas are different. The area at Rabis Stream in Jutland is composed of organic sediments with underlying meltwater sand, while the area at Storå on Funen primarily is composed of clay. The differences in the range of nitrate transformation may also reflect the fact that the system at Rabis Stream is more stable and consists of different plant species to the system at Storå, because it has not been cultivated. The study area at Storå on Funen is at present an unstable system. This instability is caused by the change from being drained and intensively cultivated to being an uncultivated field.

Studies from a riparian forest in the USA estimated the amount of transformed nitrate at 47.2 kg N ha^{-1} year^{-1} (89% of the initial concentration), or an average 4 kg N ha^{-1} month^{-1} (Peterjohn and Correll, 1984). The nitrate transformation in the uncultivated and non-forested Danish areas along watercourses is substantially higher. A recent summary of some of the available literature on nitrate transformation gives a range of 68% to 100% reduction in nutrient concentration, depending on the initial concentration and factors such as soil type and width of uncultivated areas along watercourses (Petersen et al., 1992).

How much of the transformed nitrate is sequestered and utilised by plants and how much is lost due to denitrification has not yet been determined. Brüsch and Nilsson (1991) reported plant uptake at an originally uncultivated area at Rabis Stream in Jutland, Denmark, to be 5–7% of total nitrate transformation, which amounted to approximately 400 kg $(NO_2^- + NO_3^-)$-N ha^{-1} year^{-1}. Plant uptake at the study area on Funen has probably been greater than 5–7% of the total nitrate transformation. In 1990 in particular, the vegetation grew very high on this area. The study area on Funen was previously cultivated and is undergoing succession by a new plant community.

PERSPECTIVES

ON FUNEN IN DENMARK

Results from the investigations on Funen show that riparian zones along watercourses, if properly selected and established, can be used as an important element, among others, in reducing nitrogen leaching from intensively cultivated farm land and thereby improving the water quality of the waterways.

Riparian zones, corresponding to an area of just 2% of the farm land on Funen, have a potential to reduce the nitrogen runoff from farm land to the Funen watercourses by 25%. This calculation is made from knowledge about the nitrogen runoff from the cultivated farm land to the watercourses on Funen, which is about 6 630 000 kg nitrogen year^{-1}. This amount of nitrogen is calculated as an average for the period 1976–1987 on the basis of measurements in 20 watercourses. Approximately 90% of the nitrogen runoff in the watercourses is nitrate, corresponding to 5 967 000 kg nitrate year^{-1} from agriculture to the watercourses on Funen (Funen County Council, 1994). Funen County Council has chosen the goal to be a 25% reduction of this amount of nitrate discharged to the watercourses from agriculture by using wet meadows as nitrate filters; 25% of 5 967 000 kg nitrate year^{-1} corresponds to 1 491 750 kg nitrate year^{-1}. To transform this amount of nitrate per year by establishing wet meadows as nitrate filters would require approximately 4000 ha wet meadows in the light of the fact that riparian zones consisting of wet meadows have the capacity to transform about 400 kg N ha^{-1} year^{-1} (investigations presented above); 4000 ha corresponds to about 2% of the farm land on Funen. Two per cent of the farm land on Funen corresponds to less than half of the natural and extensively cultivated areas that have been brought under cultivation since the 1950s. To re-establish riparian zones as wet meadows is very important in reducing the nitrogen leaching to the surface water, but it does not solve the problem of nitrate leaching to the groundwater. Such problems can only be solved by measures on the cultivated fields.

In Denmark a lot of money is spent on river maintenance (5.2 million kroner per year in the County of Funen). By choosing to re-establish the riparian zones as wet meadows, maintenance of the watercourses can be stopped and landowners can be compensated. In that way, nutrient transformation and natural watercourses are achieved. In addition, the landowners are satisfied because they get economic compensation for the impeded drainage of their fields. At the same time the hydrological, chemical and biological interactions between the river and the floodplain can be re-established. Funen County Council has therefore initiated a rehabilitation project incorporating these aspects along a section of watercourse approximately 10 km long, as described above.

IN DENMARK

In Denmark about 200 000 ha of meadow have the right characteristics to act as nitrate filters with more or less nitrate transformation (Brüsch, 1987). The geological and other physical conditions are very different in different regions of the country, so it is difficult to calculate how much of the nitrate leaching from agriculture in Denmark wet meadows can be transformed. This 200 000 ha of meadow corresponds to about 7% of the agricultural land in Denmark.

SUMMARY AND CONCLUSIONS

This chapter describes a demonstration project in which countryside is rehabilitated by establishing wet meadows alongside waterways on Funen in Denmark. Results from investigations of the capacity of uncultivated riparian zones along watercourses to reduce nutrient leaching from cultivated farm land on Funen have formed the basis of the countryside rehabilitation project in relation to improvement of the water quality. Both the demonstration project and the investigations were carried out by Funen County

Council. In addition, the potential for using wet meadows as nutrient filters more generally on Funen are described.

Funen County Council has planned and initiated a countryside rehabilitation project. The objectives of the project are to set up practical initiatives with the aim of reducing the diffuse leaching of nutrients from the agricultural land into Funen's waterways and the sea. In addition, the project will demonstrate possible elements of sustainable farming in the future, whereby the interests of the water recipients and nature conservation are comprehensively catered for.

The practical initiatives will be to adopt more extensive cultivation methods and laying fallow areas bordering waterways (meadows). Surface leaching will thereby be limited to the appropriate waterways, and the use of fertilisers and pesticides close to waterways will be avoided. The meadows will be flooded with drainage water entering from adjacent cultivated areas, thus restoring the function of these meadows as filters for nitrates. The overburdening of the waterways, and consequently the sea, with nitrate will thus be reduced. The flooded meadows will be managed where appropriate. Varied types of habitats, including areas under grass advantageous to wild flora and fauna, could thereby be provided. The maintenance of waterways will be avoided where appropriate. Thereby the interaction between waterways and the neighbouring environment is improved and protected and the quality of the waterways is improved. The farming organisations are participating in a working group and together with Funen County Council the farming organisations establish contact with landowners along the selected watercourse system. Close collaboration with landowners interested in participating in the project is ongoing.

The study, which started in 1989, was designed to investigate the capacity of uncultivated riparian zones along watercourses to reduce nutrient leaching from cultivated farm land. The area consists of a 0.8 ha area of formerly cultivated land along the Storå watercourse that was left uncultivated and flooded with drainage water, so that the whole area could act as a natural nutrient filter. Measurements of nutrient concentrations (nitrogen and phosphorus) were made in drainage water (inlet), surface water, soil water and in water leaving the field (outlet) four or five times each year during the period 1990–1993. In addition, measurements of nutrient concentrations in the water from the inlet and outlet of the experimental site have been made every two weeks since October 1991.

Since March 1991 the inflow and outflow of water from the study area has been measured continuously to estimate the water balance and the nutrient transformation rate on a monthly and annual basis.

In the period from March 1991 to January 1993 the amount of transformed nitrate ((NO_2^- + NO_3^-)-N) varied from 6 to 100 kg N ha^{-1} $month^{-1}$, corresponding to a variation of 20% to 100% of the nitrate inflow, when there was a nitrate input. The transformation includes both denitrification and plant uptake.

In the winter period (from November to March) the nitrate transformation was 50 kg N ha^{-1} $month^{-1}$, on average corresponding to 60% of the nitrate inflow. The annual rate of nitrate transformation varied from 470 to 690 kg ha^{-1} $year^{-1}$, with the highest rate being at the beginning of the investigation period, where the plant uptake has been largest.

On Funen, the scope for re-establishing riparian zones as wet meadows is great. Riparian zones, corresponding to an area of just 2% of the farm land on Funen, have the potential to reduce the nitrate runoff from farm land to the Funen watercourses by 25%.

However, re-establishment of riparian zones does not prevent nitrate leaching to groundwater resources. This can only be tackled at source, i.e. the agricultural fields.

If the riparian zones are re-established as wet meadows along a whole watercourse section, the maintenance of that watercourse section can be stopped. There is no need for weed cutting, dredging, etc., in the watercourse to ensure that drainage water can leave the cultivated fields near the watercourse, if these fields are laying fallow or are wet meadows. At the same time, water from the watercourse can flood the wet meadows in the winter period. The landowners must be compensated for impeded drainage and exacerbated flooding of their fields. In this way, the resulting nutrient transformation and thereby reduction of the leaching of nitrate to the watercourse can occur, leading to a natural watercourse interacting with the surroundings, and satisfied landowners who can contribute to fulfil the national goal of halving the leaching of nitrates into waterways from agriculture.

ACKNOWLEDGEMENTS

There have been many contributors to the investigations described in this chapter. The author wishes to express her thanks and appreciation to Jørgen Dan Petersen and Verner Hastrup Petersen who made it possible to carry out investigations on a scientific level within the scope of Funen County Council; Stig Eggert Pedersen who contributed with technical knowledge; Jette Gelsbjerg and Jan Mørch Sørensen who ran the sampling programme; Hans Brendstrup who calculated the water balance and performed the calculations of nutrient transformation; Inge Møllegaard who produced the graphs; and Birgitte Skjøtt who accomplished different practical tasks. All these contributors are employed by Funen County Council.

Walter Brüsch contributed with technical knowledge and Torben Jensen ran parts of the sampling programme, both of whom are employed by the National Geological Research Institute. Carl Chr. Hoffmann of the National Environmental Research Institute contributed with technical knowledge.

REFERENCES

Brüsch, W. (1987) Groundwater Chemistry in Selected Meadow Areas. Marginal land and Environment Interests. Ministry of the Environment, Denmark. National Agency of Environmental Protection, Technician Report No. 20 (in Danish).

Brüsch, W. (1991) Determination of the residence time and the movement pattern of the surface water across a re-established meadow along the Storå, Funen County. The National Geological Research Institute, Report (In Danish).

Brüsch, W. and Nilsson, B. (1990) Nitrate transformation and water movement in a wetland area. *NPO Research from Ministry of the Environment*, Report No. **C15**, pp. 1–52 (in Danish).

Brüsch, W. and Nilsson, B. (1991) Nitrate transformation and water movement in a wetland area. *NPO Research from Ministry of the Environment, C-abstracts, Nitrogen and Phosphorus in Fresh and Marine Waters*, Report No. **C15**, pp. 241–258

Denmark's Central Statistical Office (1957) *Statistics on Agriculture, Gardening and Forestry, 1955.* The Statistical Department, Volume 4/167/1.

Denmark's Central Statistical Office (1992). *Agricultural Statistics, 1991.* The Statistical Department, Volume 56.

Fuglsang, A. (in press) Nutrient transformation in a reestablished riparian zone along the Storå River, Denmark.

Funen County Council (1990) Eutrophication of coastal waters. Coastal water quality management in the County of Funen, Denmark, 1976–1990. Funen County Council, Department of Technology and Environment, Report.

Funen County Council (1994) Water environment monitoring, watercourses 1993. Funen County Council, Department of Nature and Water Environment, Report (in Danish).

Hoffmann, C.C. (1991) Water and nutrient balances for a flooded riparian wetland. *NPO Research from Ministry of the Environment, C-abstracts, Nitrogen and Phosphorus in Fresh and Marine Waters*, Report No. **C13(b)**, pp. 203–220.

Mikkelsen, H.E. and Olesen, J.E. (1991) *Correlation between Methods to Determine Potential Evapotranspiration.* Danisk Institute of Plant and Soil Science. Report no. S2157 (in Danish).

Ministry of the Environment, Denmark (1994) The watercourses – ten years with the new Watercourse Act. Ministry of the Environment, Denmark. National Agency of Environmental Protection, *Environment News*, **10** (in Danish).

Peterjohn, W.T. and Correll, D.L. (1984) Nutrient dynamics in an agricultural watershed: observations on the role of a riparian forest. *Ecology*, **65**, 1466–1475.

Petersen, R.C., Petersen, L.B.M. and Lacoursiére, J. (1992) A building-block model for stream restoration. In Boon, P.J., Calow, P. and Petts, G.E. (eds) *River Conservation and Management.* John Wiley, Chichester, pp. 293–309.

7 Practical Approaches for Nature Development: Let Nature Do Its Own Thing Again

JAN P. M. VAN RIJEN

Ministry of Agriculture, Nature Management and Fisheries, Regional Policy Department South, Eindhoven, The Netherlands

INTRODUCTION

A dominant feature of the Dutch landscape is the delta of the large rivers Rhine, Meuse and Scheldt. The western and northern parts of the country are low lying and flat with large valuable wetland areas. In contrast to this, the eastern and southern parts of the country are characterised by higher ground mostly covered by sandy soils. In these areas small rivers and brooks are common and possess their own special features, forming an important part of the ecological structure of the Netherlands, not only from a national but also from an international perspective. The upper reaches of the catchment areas of these rivers and brooks lie in bordering countries, Germany to the east and Belgium to the south.

The province of Noord-Brabant is situated in the southern part of the Netherlands with most of the brooks and small rivers draining the sandy soils, in a complex geological and hydrological situation (Figure 7.1). The lowland stream systems in this province flow mainly in a northerly direction, discharging water into the River Meuse which flows into the North Sea. A large number of the streams originate in Belgium, which indicates that European policy is needed to deal with the problems occurring in these catchments. The Dommel, the largest and most important system, is located in the central part of Noord-Brabant.

The first part of this chapter concerns the experiences with nature development for two lowland streams, both part of the Dommel stream system. The second part presents a practical method that can be used for decision-making and to assess goals and perspectives for specific projects. It also gives a possible strategy for nature development in balance with the present degraded environment, referred to as 'autonomous nature development'.

NATIONAL NATURE POLICY PLAN

The Ministry of Agriculture, Nature Management and Fisheries is the governmental organisation responsible for the execution of the Nature Policy Plan of the Netherlands

Rehabilitation of Rivers: Principles and Implementation. Edited by L. C. de Waal, A. R. G. Large and P. M. Wade.
© 1998 John Wiley & Sons Ltd.

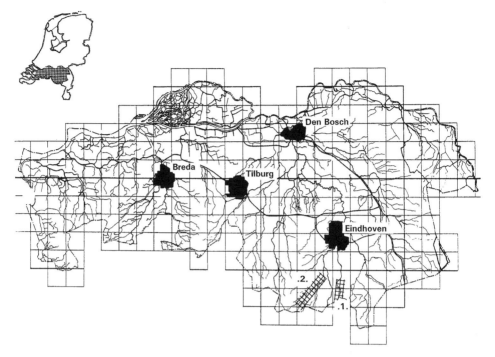

Figure 7.1 The location of Noord-Brabant in the Netherlands and the Achelse Kluis (1) and Keersop (2) nature development projects

(Ministry of Agriculture, Nature Management and Fisheries, 1990). This plan's aim is to set goals for governmental nature and landscape policy for a period of approximately 20 years. The plan's most important goal is to support ecological sustainability. At present, sustainability is limited, largely because areas of high ecological value are small and patchy, and are very sensitive to external influences. The plan aims to reduce or eliminate the weak points by enlarging existing nature conservation areas, by ensuring that such areas are not isolated and by intensifying measures to safeguard such areas from negative external influences. As part of this policy a large variety of projects and action issues will be implemented, with the most important issue being the realisation of the National Ecological Network.

The National Ecological Network is a coherent network of areas that forms a sustainable basis for the ecosystems and species considered important in a national and international context. The network consists of core areas, nature development areas and ecological corridors. Their sustainable development is supported by a buffer policy aimed at removing or minimising negative external influences.

The core areas are mainly existing nature conservation areas but also include estates, woodlands, large water bodies and some valuable agricultural landscapes. These are areas of existing national and/or international ecological value. Nature development is a relatively new concept in Dutch nature policy. This concept is based on the development, by human action, of abiotic conditions that will provide an adequate basis for natural processes to take over (Baerselman and Vera, 1989). Since the 'creation' of a new nature conservation area called the Oostvaarderplassen (Vera, 1989), the development of 'artificial' nature on a large scale is considered a successful tool in environmental policy-making. In the Nature Policy Plan the focus is no longer just on the conservation and management of nature, but also on the opportunities to develop new nature areas or new features of high ecological value. Those areas that offer realistic potential for such developments have been incorporated in the National Ecological Network. The large areas of agricultural land lying between the present nature reserves and woodlands can be considered, from an ecological point of view, as missing links. By buying this land, changing its use and drawing up nature development plans, the already existing woods, nature reserves and wetlands can be linked into the National Ecological Network. Ecological corridors are areas that connect core areas and nature development areas. They are important for improvement of the distribution of plant and animal species, by restoring the opportunities for migration between core areas and nature development areas.

The Ecological Network in the Noord-Brabant Province (Figure 7.2) is mainly based on stream corridors. The areas included are not only the present nature reserves but also large nature development areas and ecological corridors, e.g. through the towns and cities and connections with Belgium. When analysing this network, two specific issues can be identified. The first issue is related to the fact that the network incorporates brooks, lowland streams and small rivers running through urban environments, e.g. Breda, Eindhoven and Den Bosch. These cities are not only located at the confluence of brooks and small rivers, but also at important junctions in the ecological network. The second

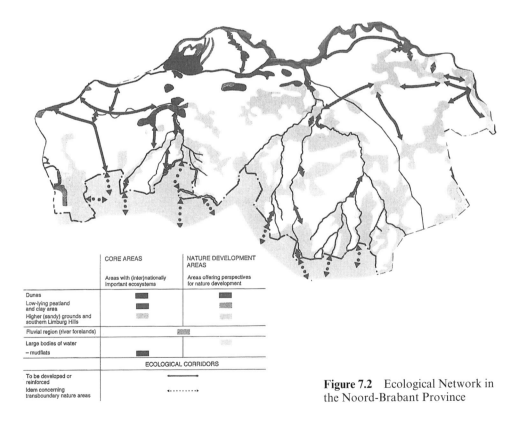

Figure 7.2 Ecological Network in the Noord-Brabant Province

issue relates to the development of the network in conjunction with Belgium. Environmental policy planning and the structure of governmental organisations in Belgium are not yet compatible with the Dutch situation.

The nature development plans for the two projects located in the Dommel stream system (Figure 7.1) are completed and were implemented in 1995. The Achelse Kluis project (1) is situated adjacent to the Tongelreep stream and the Keersop project (2) is a stretch of the Keersop stream.

THE ACHELSE KLUIS PROJECT

In 1989, the Dutch Ministry of Agriculture, Nature Management and Fisheries was given the opportunity to buy an area of 90 ha of agricultural land and 17 ha of existing woodland from the Achelse Kluis monastery (NBLF/LNV, 1989). The nature development project that was set up for this area was named Achelse Kluis after the former landowner. The project area incorporates a 2 km long stretch of the Tongelreep Brook and is surrounded by extensive woodland areas and nature reserves (Figure 7.3).

PRESENT CHARACTERISTICS

The total catchment area of the Tongelreep is 13 425 ha, of which 8950 ha are situated in Belgium. Since 1900, the Tongelreep Brook has been channelised and within the project area the brook was regulated by means of a weir with a fall of around 1.5 m. The brook was contained by an embankment and planted with now-mature poplar trees (*Populus* sp., Figure 7.4). At present, the average discharge is approximately $1 \text{ m}^3 \text{ s}^{-1}$, with a maximum discharge of $3.5 \text{ m}^3 \text{ s}^{-1}$. Land drainage schemes and urban developments have changed both the discharge pattern of the Tongelreep and the water quality compared with the original situation. For example, the municipality of Achel, Belgium, discharges untreated sewage effluent into the brook on a regular basis. In 1996, the building of a waste-water treatment plant in Achel was started; however, it does not include a phosphate removal installation.

The total Achelse Kluis area has been severely influenced by human activities during recent centuries. The original relief has been levelled, the soil structure has been changed and drainage systems have been constructed to optimise agricultural production of potatoes, sugar beet and corn. The soils are considered to be phosphate-saturated and polluted with cadmium and zinc (Oranjewoud, 1990).

OBJECTIVES AND TARGETS

The main objective of the Achelse Kluis nature development project was to accomplish an optimal abiotic starting point for the spontaneous natural development of a new nature reserve, including a meandering stream with the associated processes and habitats, ecologically integrated with the adjacent woodlands and nature conservation areas in both Belgium and the Netherlands. This main objective was translated into several targets, including the following:

1. nature development by means of spontaneous natural processes, such as erosion and sedimentation processes, inundation, seepage, and natural succession of vegetation;

PRACTICAL APPROACHES FOR NATURE DEVELOPMENT

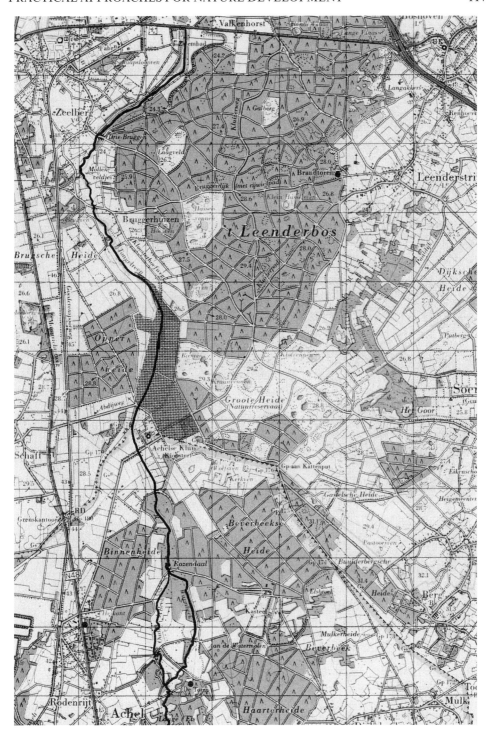

Figure 7.3 Location of the Achelse Kluis project

Figure 7.4 Poplar trees on the embankments along the Tongelreep Brook

2. minimum human influence, i.e. no planting of shrubs and trees, no intensive management system and no re-introduction of species;
3. aiming to achieve a quick result, mainly for policy reasons, in order to use the project as a demonstration site for other, especially lowland stream, nature development projects;
4. implementation of the theory of nature development on former agricultural land by means of a trial and error process.

These targets should result in a nature reserve with a meandering lowland stream, with pool–riffle sequences, and natural erosion and sedimentation processes. The floodplain will be inundated three to four times a year. Marshland, swamps and woodland will develop within the brook corridor benefiting from the inundation process. In the transitional zone, between the stream corridor and the existing nature reserve on the higher ground, a mosaic landscape will develop, influenced by extensive grazing (NBLF/LNV, 1989).

WATER MANAGEMENT PLAN

The most important part of the water management plan (Heidemij, 1991; Oranjewoud, 1992) was the construction of a meandering channel, 4.3 km in length, along the channelised reach of the Tongelreep. The original channel as well as most of the existing ditches in the Achelse Kluis area were infilled. The design of this meandering stretch was based on calculations using various computer simulation models and three different types of cross-section. These models are based on inundation of a floodplain area of approximately

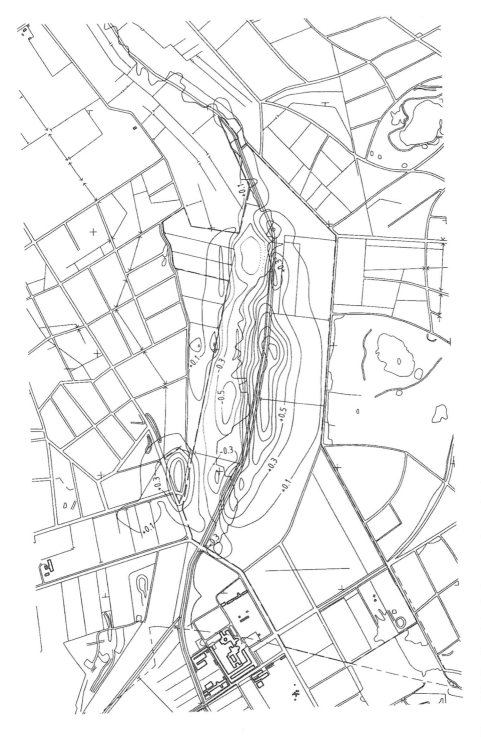

Figure 7.5 Predicted rise in the groundwater levels

10 ha adjacent to the meandering stretch during 20 to 30 days per year. In order to prevent inundation of the floodplain with presently nutrient-rich water, a diversion channel was constructed parallel to the meandering section of the brook by the construction of a fixed weir. When the water quality has improved, this weir will get covered with sand and gravel and the diversion channel will develop as a swamp, again through natural processes.

The water management plan includes a pump to guarantee sufficient drainage for the surrounding agricultural area, part of which is located in Belgium. Sand traps were constructed to prevent siltation of the meandering brook, and weirs to enable fish and other macrofauna to pass by. In several locations excavations took place in order to allow the development of marshland, swamps and wetlands to provide habitat for a range of fauna, such as amphibians.

GROUNDWATER

The project has increased in-stream water levels and has caused changes in groundwater levels, thereby influencing the natural succession of vegetation in a much larger area than just adjacent to the brook. A significant rise in the groundwater level was predicted (Figure 7.5) and calculated on the basis of average winter groundwater levels (LD/LNV, 1992; Oranjewoud, 1992). In the bordering heathland area the rise was predicted to be approximately 0.1 m and in the area adjacent to the brook around 1 m. This rise in groundwater level is considered to be one of the key processes for the desired natural development and therefore one of the most valuable effects of the creation of the meandering stretch in the Tongelreep.

SITUATION IN JANUARY 1995

In 1991, the Achelse Kluis area was under extensive agricultural use which impoverished the soil. This period was followed by leaving the cropped area fallow to allow natural regeneration. By 1995, the area was covered by shrubs of up to 3 m in height, mainly *Salix aurita* along with *Alnus glutinosa*, *Betula pubescens* and some *Quercus robur*. Extensive grazing by cattle from neighbouring farms creates a heterogeneous structure in the developing vegetation. This illustrates that through the absence of active human management, i.e. by doing nothing, the opportunity is created for natural processes to take over, and that nature is extremely capable of 'doing its own thing', even on former agricultural land.

The plans for the Achelse Kluis project were executed in 1995 and have already had some spin-offs. For example, the Belgian authorities have purchased 40 ha of agricultural land in the Tongelreep valley which borders the Dutch project area. It has enabled them to commence a nature development project alongside the Dutch project.

THE KEERSOP PROJECT

The Keersop project is a nature development project with a different approach to the Achelse Kluis project. The total catchment area of the Keersop is approximately 8500 ha, of which 1600 ha is situated in Belgium. Most of the land use in the catchment area is intensive agriculture, mainly cattle farming. The drainage system and the *ruilverkaveling*

PRACTICAL APPROACHES FOR NATURE DEVELOPMENT 121

(the redistribution of agricultural fields in order to optimise the agricultural operations) were optimally adjusted to this land use in the 1970s. Adjacent to the brook some small woodlands and nature reserves exist and remnants of the original meandering channel are found on the Keersop Brook floodplain. In several locations natural gravel beds are found within the stream channel, but unfortunately they are silted up. The sewage systems of two local towns, Bergeyck and Valkenswaard, regularly overflow into the Keersop Brook and the generally reasonably good water quality is temporarily disturbed. As a result, oxygen levels can drop to concentrations lethal for aquatic species, such as fish.

INITIATIVE

In 1988, a local angling group with a special interest in salmonids proposed a relatively simple plan to improve environmental conditions for salmonids, brown trout (*Salmo trutta fario*) and Grayling (*Thymallus thymallus*). For several years, the anglers had been stocking fry of these two species but they observed that the overall conditions for these salmonids could be improved. They suggested a change in the management practice from machine mowing to cutting by hand, and an improvement in the reproduction conditions for salmonids by creating more gravel beds. They also asked the local waterboard to draw up a plan in order to limit silting up of existing gravel beds, and to take measures to prevent the sewage systems from overflowing into the Keersop.

OBJECTIVES AND TARGETS

After this initiative by a local angling group, the project was extended to incorporate the waterboard and local municipalities. The main objective of the project was to improve the habitat conditions for natural populations of salmonids. It also aimed to improve the aquatic environment for the biota such as macrofauna and macrophytes, and to make an overall contribution to the ecological value of the landscape.

The Keersop project was divided into three sub-projects:

- management and maintenance of the Keersop Brook;
- monitoring of the fish population, the macrofauna and the vegetation present in and alongside the brook;
- producing a plan for the development of riparian zones in locations where this could be realised within a few years.

The project has resulted in changing the present weirs in order to make up- and downstream migration of species, especially fish, possible again. In 1989, the waterboard agreed to a change in management practice, from machine mowing to manual cutting, for the total reach of the Keersop Brook. A plan was also prepared for the reconstruction of riparian zones in three different locations.

THREE EXPERIMENTAL STRIPS

The plan to develop riparian zones in three different locations was set up in co-operation with the waterboard and local municipalities (Kindt and de Baaij, 1991). The plan proposed the development of riparian vegetation over a total length of 1.8 km of the brook and a width of 10 m on either side (Figure 7.6). In this stretch an old meander was

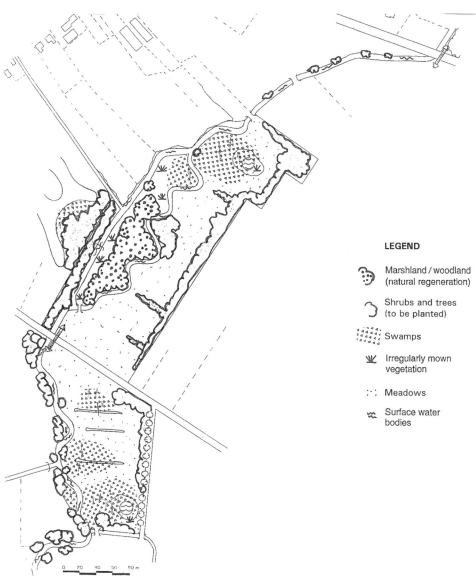

Figure 7.6 Proposal for riparian vegetation (experimental strip no. 3) as part of the Keersop development project

reinstated, marshland will develop, and trees and shrubs were planted alongside the brook. The land area necessary for development of these riparian zones was made available by the waterboard, two municipalities, and a local brewery. The development of these three zones has an important policy implication such that it is used as a demonstration project for other nature development initiatives.

Similar to the Achelse Kluis project, this project has also had some spin-offs. In 1994, the waterboard started a project downstream of the Keersop project area, where me-

PRACTICAL APPROACHES FOR NATURE DEVELOPMENT

ander remnants were reconnected to the brook to regain their function within the water system again. Furthermore, one of the local municipalities produced a plan to improve the landscape structure and the ecological values in the whole catchment area of the Keersop.

COMPARISON OF THE TWO PROJECTS

The two projects outlined above can be used for comparison, especially to illustrate their main differences and to achieve a better insight in nature development for lowland streams systems.

LAND-USE CHANGE

The Achelse Kluis project has changed the function of 90 ha of agricultural land and 17 ha of existing woodland into a nature reserve. The Keersop project is on a much smaller scale and has only changed 3.6 ha of agricultural land into riparian zones. The available surface area is an important issue when setting the objectives and targets for a nature development project. A larger area obviously offers a greater variety of opportunities. It is important to adjust the aims and objectives to the land surface area available as well as to other aspects of the present situation, such as the position in the hydrological system. For example, in the Achelse Kluis project it would have been inappropriate to aim at certain species or species groups, since the space available and the possibility to stimulate certain processes were plentiful. Within the Keersop project it was not possible to influence the processes; however, the environmental conditions for certain species of fauna and flora could be improved.

FINANCES

The purchase of 107 ha of land for the Achelse Kluis project cost approximately £1.5 million and the execution another £530 000. The grand total of the project is around £2.2 million, which works out at a total cost of £20 560 per ha of land. Execution of the Achelse Kluis project, i.e. excluding purchase cost, is estimated at only £5000 per ha. The Keersop project is much cheaper, the purchase of land costing £60 000 and the execution £85 000. However, calculated per hectare of land, the costs are higher than for the Achelse Kluis project. The Keersop project cost £40 277 per ha, including the purchase of the agricultural land, and the execution costs alone were £23 611 per ha. This shows that from a financial point of view it is more efficient to spend tax-payers' money on larger nature development projects than on smaller ones.

OBJECTIVES AND TARGETS

The objectives and targets of the two projects are also different. Achelse Kluis aims at the processes, limiting human influence and the construction of a meandering stream channel. The Keersop project aims at the establishment of suitable habitat conditions for two species of fish, the overall improvement of the aquatic system and landscape aspects. Both projects, however, do contribute to the Ecological Network as proposed in the National Nature Policy Plan, but on a different aspiration level.

ORGANISATION

The initiative for the Achelse Kluis project was taken by the Ministry for Agriculture, Nature Management and Fisheries, as well as the project management. The local waterboard and municipalities have participated in several committees, but did not contribute to the project in term of finances or the provision of land. A systematic approach was adopted from the start and led to a project that was properly taken through all the procedures. The project started in November 1989; by March 1993 the plans were ready to be implemented and all the necessary permits and licences were provided. At this time, two local farmers objected to the project and suspension of the permits followed. Despite the regular informative meetings with the local farmers' organisation and the thorough and detailed site models designed, the farmers could not accept that the project would not have any negative impacts on their farming land. After a delay of two years, the implementation of the project was started in March 1995.

The initiative for the Keersop project was taken by a local angling group. For each part of this project a special committee was formed to design the project's plans in conjunction with the local water authority and municipalities. These local organisations gave their support, not only in terms of money but also through the provision of land. Although these organisations are predominantly interested in agriculture and politics, they did agree to the terms of a nature development project and gave it their full support. This so-called bottom-up approach, in combination with starting the project on a small scale, has certainly had positive effects in an agriculturally dominated society. Even so, it took five years of consultation and meetings with all interested parties. However, the end result is considered to be worth all the effort. In 1993, a declaration of intent for the whole project was signed by all the contract partners, along with a contract for implementation of the three riparian zones. The contract for the construction of weirs was signed in 1995.

A PRACTICAL APPROACH

Based on the experience gained from these two projects, together with other practical approaches (van der Hoek and Higler, 1993; Mulders, 1993), a practical methodology can be produced for decision-making in nature development projects, illustrating the possibilities and aspiration levels (Figure 7.7).

The horizontal scale in Figure 7.7 indicates the increasing hectarage of land that is necessary and the amount of money that can be spent. The available land area and finances determine the method that may be adopted, e.g. from small adjustments in management practice, the technical reconstruction of weirs and the reinstatement of meanders up to the rehabilitation of larger areas or even parts of the total catchment. The vertical scale (Figure 7.7) indicates the increasing nature conservation value and aspiration level: from migration zones, to the rehabilitation of certain habitats, up to nature development by natural processes. It also indicates a decreasing level of human or cultural influence.

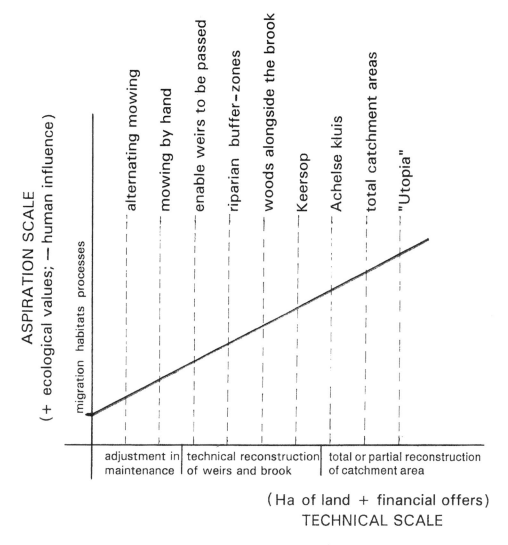

Figure 7.7 Systematic approach for nature development projects

The pre-project conditions can generally be placed at a level above zero, indicating that even in the present situation most sites are of some nature conservation value. However, by implementing rehabilitation measures, further nature conservation value can be added. Minor measures, such as alternating machine mowing (i.e. one year one river bank, next year the other) or manual cutting, can offer certain species improved habitat conditions within that specific area. The demolition of weirs enables fish and other fauna to migrate up- and downstream, with the distribution of these species over a larger area. The development of riparian zones or even floodplain forests improves not only the in-stream habitat for macroinvertebrates and fish, but also the terrestrial habitat for birds. The Keersop project combines several of these relatively minor rehabilitation measures and can be classed as the first step towards a systematic approach. In the

Achelse Kluis project, the development of a meandering stream channel has strong positive impacts on the groundwater level in the surrounding area and therefore improves natural processes and species diversity in a larger terrestrial area. The diagonal line in Figure 7.7 ends where total catchment area can be rehabilitated for nature conservation purposes and nature development. Unfortunately, for most countries this situation is a Utopian one and compromises have to be made, e.g. the Netherlands where 15 million people share a small country with, amongst others, 15 million pigs.

For both the Achelse Kluis and the Keersop projects there is the potential to move up along the scale of aspiration. For example, buying more land would enlarge the Keersop project which would make a larger area available for terrestrial habitats. Through implementation of the project for the downstream stretch of the Tongelreep Brook and by co-ordinating the plans for a Belgian partnership the Achelse Kluis project can include a considerable part of the total catchment area. This practical example illustrates that it is important to have a realistic approach, based on the present situation and opportunities combined with a more idealistic view on what should be achieved in the future.

By using this systematic approach, nature development projects of various types and at different aspiration levels can be classified, to enable a logical and coherent planning. This method is not a scientific approach for the rehabilitation of rivers, but is based on practical experience. It is especially valuable for strategic and policy planning with regards to regions or catchment areas. This approach creates a coherent structure for the realisation of a diverse ecological network of streams and brooks in the Noord-Brabant province. At present, the aims and objectives of most projects in the Netherlands are focused on the restoration and rehabilitation of nutrient-poor (hay) meadows (van der Hoek and Higler, 1993; Mulders, 1993). It is important to prevent single focus goals and targets for all the lowland streams, especially when biodiversity is a main objective. The methodology presented here can be used to achieve a variety of goals and targets, and thus aim at diversity in ecological and landscape values.

AUTONOMOUS NATURE DEVELOPMENT (LETTING NATURE DO ITS OWN THING)

At present, with regards to Dutch nature conservation policy and planning, most of the nature development projects are aimed at a more classical target, i.e. the restoration of nature values that once existed, especially at the end of last century. This type of nature requires the restoration of the environmental conditions of this pre-disturbance state. This could be referred to as environmental conditions that have *once been*. Nature development projects should aim to restore natural processes in balance with the present-day conditions. The natural processes can be regarded as the catalyst for natural development, and therefore as a more sustainable method of nature development.

Environmental problems, such as eutrophication, drought and deposition of acid rain, cannot be solved easily or quickly. Degraded systems and processes, especially within the geo-hydrological system, cannot be restored within the next few decades. It will take more time and even then it is uncertain whether or not it is possible to achieve all the aims of environmental policies and pay for these. Meanwhile, time and money are invested in nature which can only be achieved once the major environmental problems are dealt with. Furthermore, this forced approach is combined with a traditional concept of nature development concentrating on nature which is dependant on human management. A model is designed to illustrate the possible strategies that could be adopted to achieve natural development of lowland stream corridors (DHV-Water BV and Adviesbureau HNS, 1991). Most of our ecosystems, especially lowland stream systems, have changed completely and are severely degraded. In Figure 7.8, the degradation history is shown from the Middle Ages up to the present time. Ecosystems have increasingly been influenced by human activities such as the cultivation of land, especially since the introduction of fertilisers. This change can be recognised as the solid line moves from the nutrient-poor (clean) to the nutrient-rich (polluted) compartment in Figure 7.8, a change that has been further intensified during recent decades. Human activities have led to major environmental problems such as eutrophication, drought, over-fertilisation and the deposition of acid rain. In small nature reserves, that still exist, human influence in terms of management has led to further degradation of natural processes and structures, since the natural processes are suppressed and guided into certain directions. The available area is generally too small for natural processes (abiotic processes such as inundation as well as biotic processes such as predation) to dominate. Nature's ability for self-regulation is generally ignored as environmental policy and management aim to restore a previous state of biodiversity.

The important issue is the overall aim of nature development strategies. Using historical information, one possibility is to restore nature by reinstating traditional agricultural management practices and restoring the overall environmental quality necessary to sustain the restored nature. Such an approach aims to reinstate the biodiversity of the past, which generally means high costs for the community, not only as a result of the need

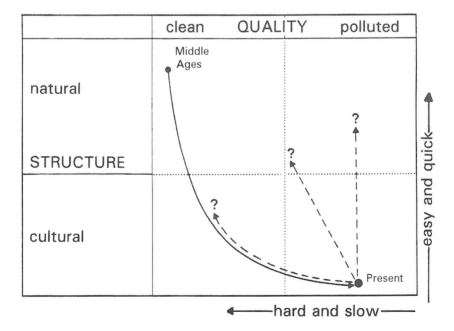

Figure 7.8 Degradation history and possible directions for nature development

for continuous management but also through the need to restore the overall environmental quality.

The time has come to start nature development projects in which the present environmental condition is accepted as it is, including eutrophication and drought, and where human influence is limited. An alternative strategy that aims to restore natural processes, which are sustainable even with the present nutrient-rich and dry conditions, would be to let nature do its own thing again. Let the present-day environmental quality dictate the kind of nature that will develop, without any after-care or management – the kind of nature that can cope with a certain level of environmental degradation and should not cost too much. In the Netherlands, the Oostvaardersplassen are an example of how nature is very able to do its own thing, leading to high ecological values, even from an international perspective (Vera, 1989).

Large parts of the natural, lowland aquatic system, especially the more downstream parts, can be considered as eutrophic. The water quality of most of the small rivers and streams can be classed as hypertrophic. It is evident that the overall water quality needs to be improved, from this hypertrophic state to a more eutrophic state, before large-scale rehabilitation projects can be launched. There is some scientific evidence to suggest that restoring the natural structures of ecosystems on macro- and micro-scales is more critical, especially for lowland streams, than further soil or water clean up (Voigts, 1976; van der Veen and van Wieren, 1980; Bink and van der Made, 1986; Weller, 1987; Vera, 1989).

Important abiotic and biotic processes in the cover-sand areas and the lowland stream systems are determined mainly by hydrological processes. The available space for natural processes and the absence of human influence are key factors for regulation by nature itself. An alternative strategy should also be based upon the question of whether or not the environmental quality of the past, necessary to restore the nature of the past, is environmentally and financially feasible, and socially acceptable. As a basic principle, priority should be given to naturalness and self-regulation by nature without setting strictly defined goals and targets, such as certain species or landscapes. This strategy, called *autonomous nature development* (Figure 7.9), has no aims or objectives other than

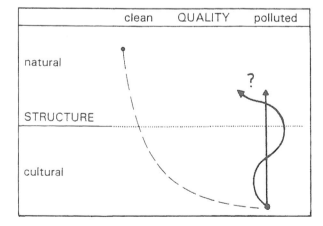

Figure 7.9 Potential future for nature development projects without restoration of the overall environmental quality

self-regulation itself. Autonomous nature development concentrates on the abiotic and biotic processes within an ecosystem without human interference, such as management. This will lead to diversity in natural structures and habitats based on the present environmental quality.

In order to kick-start the main processes operating within the hydrological system, additional measures can be proposed, especially with regards to blocking off watercourses. This kind of temporary human interference should be restricted to one single operation at the beginning of the project. For example, in order to improve (bio)diversity and create variety in terms of ecological structure, large herbivores can be introduced. After such an introduction, the ecosystem has to become self-regulating again. In case these herbivores are not able to survive, e.g. due to a lack of nutrition or the ill effects of certain pollutants, the situation should also be accepted as a part of self-regulation by nature.

The outcome of autonomous nature development is to a large extent uncertain. For example, vegetation succession on phosphate-saturated soils without human management should also be considered as a natural process, although the long-term results are yet unknown. Important abiotic processes such as erosion and sedimentation, which create variety in structure, and biotic processes, such as spontaneous vegetational succession, herbivory, predation, and certain activities such as digging by badgers or wild boar, need to be accepted. Overall, catastrophes and calamities, plagues and diseases are to be accepted as part of the process of self-regulation. People will need to have faith in these processes which will create a diverse landscape with an improved ecological value for the future.

Nature development based on the present environmental quality is the only option if society is unwilling to pay for large-scale sanitation of diffuse contamination or for programmes aiming to reduce acid rain and/or drought. Autonomous nature development is an alternative strategy which is financially feasible because it is inexpensive. At the same time, the opportunity is created to develop structural diversity initiated by natural processes. This will possibly lead to a different type of nature than people are used to at present, but it will be one that can cope with the present environmental quality and one that does not cost a lot.

REFERENCES

Baerselman, F. and Vera, F. (1989) *Nature Development, A Ministry Report*. Ministry of Agriculture, Nature Management and Fisheries, SDU Uitgeverij, The Hague.
Bink F.A. and van der Made, J.G. (1986) Dagvlinders en Herbivoren. *Levende Natuur*, 87(5/6).
DHV-Water BV and Adviesbureau HNS (1991) Raamplan voor een Gebiedsgerichte benadering van het stroomgebied van de Beerze en de Reusel, Utrecht.
Heidemij Adviesbureau BV (1991) Onderzoek en planvorming Beekdal Tongelreep, s'Hertogenbosch.
Kindt and de Baaij BV (1991); Inrichtingsplan Keersop voor 3 proefstroken, Nijmegen.
LD/LNV, Landinrichtingsdienst, afdeling Onderzoek Tilburg (1992) Hydrologie van de Achelse Kluis, internal report, Tilburg.
Ministry of Agriculture, Nature Management and Fisheries (LNV) (1990) *Nature Policy Plan of the Netherlands*, The Hague.
Mulders, J. (ed.) (1993) *De Toekomst van Beekdalen; Besturen van Stromen*. Stichting Natuur en Milieu, Utrecht.
NBLF/LNV, Department for Nature, Environment and Wildlife-Management of the Ministry for

Agriculture, Nature Management and Fisheries (1989) Nature development in the project Achelse Kluis, unpublished report.
Oranjewoud BV (1990) Grondonderzoek Achelse Kluis. Oosterhout.
Oranjewoud BV (1992) Natuurontwikkeling Achelse Kluis. Deel I Onderzoeksrapport, Deel II Inrichtingsplan, Oosterhout.
Van de Veen H.E. and van Wieren, S.E. (1980) Van grote grazers, kieskeurige fijnproevers en opportunistische gelegenheidsvreters. Over het gebruik van grote herbivoren bij de ontwikkeling en duurzame instandhouding van natuurwaarden. Instituut voor Milieuvraagstukken Vrije Universiteit, Amsterdam.
Van der Hoek, W. and Higler, B. (1993) Natuurontwikkeling in Beken en beekdalen: verkennende studie naar de mogelijkheden voor natuurontwikkeling in beeken beekdalsystemen in Nederland. IBN-DLO, Wageningen.
Vera, F.W.M. (1989) *De Oostvaardersplassen*. IVN, Amsterdam.
Voigts, D.K. (1976) Aquatic invertebrates abundance in relation to changing marsh vegetation. *American Midlands Naturalist*, **95**.
Weller, M.W. (1987) *Fresh Water Wetlands: Ecological Processes and Management Potential*. Academic Press, New York.

Part Two

IMPLEMENTATION: STRATEGIC APPROACHES

8 The River Restoration Project and Its Demonstration Sites

NIGEL T. H. HOLMES
River Restoration Centre, Silsoe Campus, Silsoe, Bedfordshire, UK

INTRODUCTION

For centuries the ecological interest of British rivers, as elsewhere, has declined as a result of engineering works to improve flood protection and enable more intensive agricultural production on floodplains. Accelerated losses occurred between 1940 and 1980. For instance, the Countryside Commission (1987) reported that a relatively rural catchment, the River Exe in south-west England, had a 1300% increase in urbanisation between 1940 and 1980. Williams and Bowers (1987) state that very little river-dependent wetland remains, perhaps as little as 1% in East Anglia where over a third of the rivers have been channelised (Brookes *et al.*, 1983). The last decade has seen a remarkable reduction in damage and a revolution that has brought a reversal of some past damage.

Considerable benefits to people have been achieved through improved drainage for agriculture and better protection from flooding for urban communities but this has had its toll on river corridor environments. Infrastructure developments, e.g. new roads and service distributions, large residential and industrial developments, and mineral winnings, have led to the deepening, straightening or diverting of rivers, all of which normally contribute towards degradation of river naturalness and loss of floodplain habitats.

The same mechanism that led to the destruction has been involved in many of the rehabilitation measures being taken. Much of the remedial work is undertaken by drainage authorities when executing 'maintenance' works or planning capital flood alleviation schemes. However, there is a limit to which they can be expected to rehabilitate rivers and increasingly landowners and local authorities are initiating improvements of their own. This can be a daunting task without professional assistance from those who have been involved in similar work in the past.

Despite many encouraging changes in the past ten years, restoring rivers to the more semi-natural state they were centuries ago has been slow and very few rivers have had more than very short lengths improved (Holmes, 1992; River Restoration Project, 1993a,b). This is understandable because many of the environmental improvements have been achieved on the back of management works being carried out primarily to alleviate flooding. Attention has also focused on improving in-channel interests rather than

Rehabilitation of Rivers: Principles and Implementation. Edited by L. C. de Waal, A. R. G. Large and P. M. Wade.
© 1998 John Wiley & Sons Ltd.

including a more holistic approach which restores some floodplain interests and takes account of the contribution that the emerging science of fluvial geomorphology can play (Sear, 1994). Lack of pre-work environmental appraisal also means that the benefits of such efforts cannot be quantified in later years through post-project monitoring.

There are many initiatives and organisations in the UK which raise hopes for the spectre of river rehabilitation gaining public, political and financial support in the years to come. The formation of the River Restoration Project (RRP), now the River Restoration Centre (RRC), is just one example of the growing awareness and desire to turn degraded river corridors back to features of wildlife and landscape value which are both appreciated by people and utilised as part of the country's economic infrastructure. One of the aims of the organisation is to bring to the fore in the UK experiences learnt from some major restoration projects in mainland Europe.

Some especially pertinent examples come from Germany (e.g. Kammbach, Enz and Sandbach – see Kern, 1992), Austria (e.g. Leitha – see Rojacz, 1992), Switzerland (e.g. Emme – see Jaeggi, 1992), the Netherlands (e.g. River IJssel and floodplain restoration – Gerritsen, 1992; Havinga, 1992; Litjens, 1992) and Denmark (e.g. Gelsa, Breda and others – see Madsen, 1995; Nielsen, 1996a,b; Nielsen et al., 1990).

FORMATION OF THE RIVER RESTORATION PROJECT

At the 1990 international conference on river conservation in York (Boon et al., 1992) a group of delegates discussed the need for a national catalyst to encourage, and help facilitate, the restoration of rivers. After several meetings, and drafting of aims, the RRP was formed in 1991. It was formed as a non-profit-making, independent organisation established to promote the restoration of rivers to benefit wildlife, landscape and recreation within the rural and urban framework through which rivers flow today.

The RRP has a Board, its members working with a small group of technical specialists who are advised by a Steering Group. Expertise within the Board and Technical Group encompasses river ecology, landscape ecology, public perception, planning, fisheries, biology, water quality, fluvial-geomorphology and river engineering. The Steering Group is represented by organisations which have a great influence over, or interest in, floodplains and river management. Members include the Environment Agency, the Ministry of Agriculture, Fisheries and Food, the Department of Agriculture Northern Ireland, the Association of Drainage Authorities, statutory conservation and landscape agencies and large organisations with land-holding or other impinging interests (e.g. the National Trust (NT), the Royal Society for the Protection of Birds, the Country Landowners Association and the Confederation of British Industry (Minerals)).

AIMS OF THE RIVER RESTORATION PROJECT

From the outset the principal aims of the RRP were as follows:

(i) to establish international demonstration projects which would show how state-of-the-art restoration techniques can be used to re-establish more natural ecosystems in damaged river corridors;

(ii) to improve understanding about the effect of restoration work on nature conservation value, water quality, landscape and recreation, and to determine the benefits of this;
(iii) to serve as a catalyst or focal point to encourage and facilitate others to restore streams and rivers by dissemination of knowledge;
(iv) to develop methods of establishing partnerships for structured collaborations for river restoration between institutions and interested land-holders with differing powers, resources and responsibilities but who share a common aim of improving rivers;
(v) to disseminate knowledge about effective river restoration methods.

WHY THE NEED FOR A RIVER RESTORATION PROJECT?

Whilst a considerable amount of river rehabilitation has recently been undertaken by organisations such as the National Rivers Authority (now the Environment Agency) in England and Wales, as well as by Groundwork Trusts and others, much of this has been piecemeal. Apart from work to restore fisheries, river rehabilitation only began in Ireland in 1995/6, and no major schemes have been reported for Scotland. Even in England and Wales restoration has rarely encompassed both river and floodplain rehabilitation or been adequately monitored to determine the effects. Also major restoration activities have often been very small-scale and not executed over long lengths of rivers. Many very good examples of such rehabilitation works which have improved river and bank habitats are given in *The New Rivers and Wildlife Handbook* (RSPB/NRA/WT, 1994) and in RRP (1993a).

Equally important in the field of river rehabilitation is the fact that no single organisation in the UK has comprehensive responsibility for river management or restoration. For example, responsibility for flood defence rests with the Environment Agency in England and Wales, the Department of Agriculture in Northern Ireland and the Regional Councils in Scotland. Statutory responsibility for nature conservation and landscape is vested with English Nature (EN) and the Countryside Commission (CC) in England, the Countryside Council for Wales in Wales (CCW), Scottish Natural Heritage (SNH) in Scotland and the Department of the Environment (DoENI) in Northern Ireland. A similar disparate situation is apparent for agricultural departments.

With its formation in 1991, the RRP thus became the only UK organisation with a single and dedicated aim of promoting the restoration of degraded rivers. RRP's unique features include its independence and its ability to operate across political and operational boundaries. Both have been crucial in winning wide support from individuals and organisations and extending the opportunities to promote restoration beyond the narrow limits of individual organisations' jurisdiction.

RIVER RESTORATION PROJECT'S ACTIVITIES TO DATE

In 1993 a Business Plan was produced (RRP, 1993c) which gave details of RRP's ethos, information on phases of activities it proposed over the five year period to 1998, and the manner in which this would be managed and financed. The four phases proposed were as follows:

Phase 1: research, undertake and demonstrate river restoration
Phase 2: promote partnerships for restoration
Phase 3: develop a network of information on restored sites
Phase 4: promote restoration through education, training and publications

For clarity of expressing objectives, four distinct phases were identified, but all four are interlinked and develop from each other. For instance, it is not possible to undertake model demonstration restoration projects (Phase 1) if the basic groundwork for effective development of model partnerships (Phase 2) has not been carried out. For simplicity, the aims of each, and the progress made so far, is summarised below.

PHASE 1: DEMONSTRATION RESTORATION PROJECTS

A key priority of RRP at its inception was to promote restoration on a number of sites to show how collaborative partnerships between landowners and institutions with common goals could be developed, and then to design, construct and monitor an extensive variety of restored features so that their effects on landscape, ecology, land use, flood risk, public recreation and perception could be identified. Equally important was the need to show what problems arise from the first ideas to the completed scheme so that others can avoid costly mistakes in the future.

Stage 1 of Phase 1 was to undertake a feasibility study (RRP, 1993a,b) to determine the extent, types and new innovative approaches to restoration being undertaken in the UK and elsewhere. This study was also charged with assessing the scope, and need, for restoration of rivers in the UK. The main report (RRP, 1993a) provided descriptions of a broad range of techniques employed for restoration together with assessments of benefits and dis-benefits of restoration measures. The report also gave some case examples of selective restoration/rehabilitation schemes in the UK, Europe and the USA. A short summary of this report has also been prepared (RRP, 1993b).

To enable restoration works to be undertaken on selected degraded reaches of river, and hence fulfil Stage 2 objectives of Phase 1, potential sites had to be identified and funding for the work found. In 1993 the EU 'Life' programme announced it would grant funds for the establishment of major river restoration demonstration sites in Denmark and the UK. The application had been jointly applied for by RRP and the South Jutland Council. This led to RRP's priorities being diverted to ensure adequate matching funds were secured. In tandem with this the process of site selection, developing partnerships, designing schemes and undertaking a plethora of key pre-scheme surveys went on apace. The complexity of these, and the valuable insight for others, is the reason why more details of the two demonstration sites are given later.

PHASE 2: DEVELOPING MODEL PARTNERSHIPS FOR RESTORATION

Phase 2 has been progressed in tandem with Phase 1, taking two distinct forms. The key first step for RRP, as it should be for a similar organisation in any country, was to identify which organisations have statutory responsibilities, consenting powers, interests and budgets to promote river restoration. The second has been the development of legal agreements and memoranda of understanding to ensure that owners and others who are responsible for the rivers in the future are committed to the objectives of restoration and sign up to long-term safeguards of the achievements.

An 'Institutional Study' was contracted, and it concluded (RRP, 1994a,b) that there are many public sector and voluntary organisations which have a measure of responsibility or interest in the advancement of river restoration. However, all but RRP were reported to be restricted to particular aspects or areas of restoration by reason of statute or constitution. The lack of any national policy specifically aimed at river restoration and the general lack of available finance further restricts progress that may be made. Pooling of both powers and resources was concluded to be the most fruitful way ahead.

For any rehabilitation works to be successfully carried out, and the objectives subsequently recognised in future management strategies, agreements and consents need to be obtained during the project development. Agencies responsible for flood defence are not the only organisations with interests; landowners and occupiers are also critical, as are co-funders who need to be sure their resources are being utilised in a sustainable manner.

Promotion of the 'Life' project pioneered the development of 'model' agreements for promoting river restoration projects. The development of the Memoranda of Understanding and the Legal Agreements took considerable time and effort but are expected to be of benefit to future partnership river restoration projects even though no other schemes are likely to be identical. The former are single documents for each site, signed by all the participants, which identify commitments to budget contributions, outputs, project management and decision-making responsibilities. The latter are numerous, covering such aspects as individual consultants' and contractors' liabilities, to those organisations who take responsibility for land and river management in the future long after the restoration works are completed.

PHASE 3: RESTORATION NETWORK

The primary aims of Phase 3 of RRP's work is to provide a UK focal point through which the practical development of river restoration schemes can be promoted by efficient transfer of relevant information and experiences together with facilitating direct contact with those working on similar problems or schemes. This requires efficient collation and dissemination of knowledge about effective river restoration achievements and techniques. So far a system has been developed and is being used nationally by other organisations. It has been used to determine the extent of river and floodplain rehabilitation activity within the UK (Holmes, 1998).

RRP fulfilled a valuable informal networking process of keeping interested personnel informed of its own activities, but only since it became RRC in 1998 has it been able to dedicate time and resources to broadening this to a comprehensive network relating to work of others. It has also supported the development of a Europe-wide restoration network that is based in Denmark – The European Centre for River Restoration (ECRR), which produced its first newsletter in December 1996.

PHASE 4: EDUCATION, TRAINING AND PUBLICATIONS

The work by RRP has already resulted in several important reports, publications and films relating to river restoration activities, institutional frameworks and site selection procedures (RRP, 1993a,b,c, 1994a,b,c). Several papers describing the administration and project management, design and execution, and the monitoring programme of the 'Life' project were given at the 1996 Silkeborg conference on river restoration (see Hansen and Madsen, 1998). Technical reports and other outputs of the EU 'Life' programme have been provided for the EU and partners. These include the following which are of major interest to those responsible for managing rivers and protecting floodplain assets within whole river systems:

- soil conditions and topographical changes
- hydrological changes and hydraulic performance
- mass balance of nutrients in the streams and nutrient reductions on the floodplain
- sediment monitoring within the channel and mass balance on the floodplain
- changes in ochre precipitation in the river and on the floodplain (the River Breda only)
- cost/benefit assessments of works carried out
- a handbook on best practice on river restoration and a manual on restoration techniques utilised on the demonstration sites
- a video on practical aspects of river restoration

Many ecological studies have also been undertaken which are attempting to determine how biota respond to restored habitats within the rivers and floodplains. The programme has included detailed pre- and post-construction surveys as well as impact assessments during construction. Examples of studies reported on include the following:

- invertebrate and fish population changes following restoration measures
- vegetation changes in the rivers and on the floodplain
- monitoring results of changes to otter and bird populations

The studies described above, in association with public perception studies, comprise the majority of publications that aim to provide vital objective technical information required by decision-makers. This area of 'education' has previously had a paucity of information to support those promoting restoration of more naturally functioning systems.

The 'Life' sites have also provided the focal point to enable RRP to fulfil a much wider range of education and training needs. The experiences of those involved with RRP, and the 'Life' sites, has resulted in numerous articles on the project sites being featured in internal magazines and new sheets. RRP personnel have presented more than 20 lectures on river restoration in the UK and Europe in the past two years and have participated in many workshops aiming to promote wider involvement in restoration and to demonstrate the benefits. The 'Life' sites have been featured many times in articles in local newspapers, and national newspapers and television have also given good coverage. In the 12 month period May 1996 to April 1997, RRP hosted more than 50 visits to the Rivers Cole and Skerne; many of these had more than 25 personnel present. Visitors to the sites have covered a wide range of education and training interests, including those responsible for, or with an interest in, the following:

- the development of National Curriculum packs
- designing bio-engineering revetments
- undergraduate and graduate courses relating to river management, geography, ecology
- research for PhDs and other higher degrees

Many international conference delegates have had organised excursions to the restoration sites, as have many international scientists interested in a wide variety of aspects covered by the project. Recently, flood defence committees, who have responsibility for allocation of funds relating to management decisions on rivers, have visited both sites. Whilst the demonstration sites afford ideal opportunities to educate a whole range of interested parties on the practical and technical aspects of river restoration, the experiences gained through delivering the project also provide valuable education opportunities. RRP has supported many new river restoration projects, not least two others in the UK backed by EU money. These are the TAMAR 2000 project and the World Wide Fund for Nature's Wild Rivers Project in Scotland; both aim to show how land-use and catchment management can fundamentally affect the environmental and economic value of rivers.

THE DEMONSTRATION RIVER RESTORATION PROJECT

In the autumn of 1992, RRP, in conjunction with Danish colleagues from South Jutland County Council, made an application to the EU 'Life' programme for funding to establish demonstration river restoration sites in the two partner countries. The partnership was forged because the Danes lead the way in Europe for re-meandering previously straightened rivers whilst the UK (as in many other countries) needed to build upon the previous piecemeal approach of local rehabilitations which were increasingly being undertaken.

For RRP the term 'restoration' is important in that it conveys a visionary target of pristine rivers that are wholly returned to an undisturbed state requiring no management. In practice this target can be rarely achieved so restoration becomes a compromise of sustaining economies and the natural environment. The selected demonstration sites needed to respect this, but show a wide range of the restoration measures that can be applied in different situations and demonstrate their effects on the total environment. The application was successful, resulting in a large demonstration site in Denmark and two in the UK.

The work on the two UK sites aimed to demonstrate a minimum of three key things:

(i) how to work with others in selecting suitable sites and develop partnerships with organisations which bring with them their professional expertise and enable change to take place through the legal powers, statutes and responsibilities each has;
(ii) what is required to investigate opportunities and constraints at suitable restoration sites, the professional design of the programme of proposed works, and the execution and demonstration on site of these changes;
(iii) monitoring of the effects of the changes on a whole range of aspects from invertebrates of the river and floodplain to the public's perception and economic consequences of the changes.

The process of selecting the demonstration sites (one rural and one urban) began late in 1993. For practical reasons about twenty candidate sites were investigated, these being considered to offer opportunities for using a variety of the restoration techniques that would need to be demonstrated. The sites were chosen through consultation with many interested organisations, the final selection of the favoured sites involving the key statutory organisations that would be funding partners.

The procedure for selecting sites used the six broadly defined parameters (RRP, 1994c) outlined below:

- *Aims*: the site must have the potential to achieve the aims of river restoration (including in-channel and floodplain) where wildlife, landscape, recreation, water quality, fisheries, amenity and other local interests can be improved without detriment to flood defence or other needs.
- *Technical*: the site must suffer from a variety of degradations that can be reversed, measured, and promoted with confidence in the future. Reversal must be technically achievable, measurable and sustainable.
- *Funding*: adequate funding from partnerships must ensure long- and short-term economic viability.
- *Ownership*: owners and occupiers must be fully committed to ensure objectives can be achieved without lengthy negotiations.
- *Promotional*: the site-specific project must serve to support RRP's wider aims of being a catalyst, through demonstration, for increasing the rate and extent of river rehabilitation elsewhere, and advancing knowledge and understanding of restoration techniques.
- *Risks*: as far as is possible, the risk of failure must be minimal. For instance, in the context of being a demonstration site, sites should not be chosen if the risk of future serious pollution incidents is great or if the risk of land-slips or other catastrophic events are greater than at other sites which offer the same potential for reversal of damage.

Having followed this procedure, the River Cole near Swindon and the River Skerne in Darlington were the sites selected (see Figure 8.1 which shows their location, and that of the Danish site, the River Breda). The River Breda, which flows through grasslands in South Jutland, had already been selected prior to the 'Life' application. The UK sites were selected because they offered many opportunities for floodplain wetland restoration, re-meandering of straightened courses, *in situ* improvements to degraded river channels and banks, and many others. Land ownership was also secure in both cases, the former being owned by the National Trust and the latter by Darlington Borough Council.

The Cole is a tributary of the Thames, within a mixed clay, chalk and sandy limestone catchment, covering 129 km². It naturally has a flashy flow regime, but this is exacerbated by urbanisation of the headwaters around Swindon. The low gradient results in a relatively low energy river incapable of naturally reversing the effects of previous widening, deepening and straightening. The restoration site is on National Trust land on the Oxfordshire/Wiltshire border at Coleshill. The site comprises more than 2 km of river and approximately 50 ha of floodplain in mixed agricultural use. Prior to restoration the course of the river was almost entirely artificial. Straightening first occurred over 350 years ago to feed a mill, whilst less than 30 years ago the river was further deepened in places to improve agricultural drainage.

The Skerne is a slightly larger catchment of 250 km² draining a clay/alluvium catchment. The demonstration site is wholly within Darlington Borough Council ownership, being the two kilometres upstream of the historic Stephenson Bridge (depicted on the present British £5 note). As with the Cole, the channel was virtually confined to an artificial course in an urban area where public open space is an important amenity in over half the site. The whole area has been greatly modified as a result of industrialisation and

THE RIVER RESTORATION PROJECT

Figure 8.1 Location of the restoration sites

much of the floodplain has been eliminated by tipping and developments. Prior to restoration it was essentially a straight, featureless channel. The river has a long history of pollution and poor water quality which is gradually improving.

DEVELOPING VISIONARY RESTORATION PLANS

Details of the project management measures that were put in place for the development and implementation of the project are reported in papers published in the proceedings of the 1996 European River Restoration Conference held in Denmark (ERRC, 1997). A key element of the development of the project was the formation of 'working groups' for the Cole and Skerne. These groups were made up of personnel representing the full range of interests associated with the sites, spanning technical specialists on hydraulic modelling to those undertaking pre-scheme surveys of invertebrates, plants, landscape, public perception, etc. The two groups developed the survey protocol and generated ideas for restoration which were presented to 'project boards' – the members of these boards being representatives of the funders and landowners who would determine what would be done at each site.

The two working groups identified huge numbers of desirable restorations which, if implemented, would have exceeded the available budget. These were developed into 'vision plans' for the two sites to enable choices to be made regarding what could be achieved. Those that most met the objectives of the whole project, and were affordable within the available budget, were taken forward for detailed design as part of the 'core scheme'. Individual elements affecting flood risk were designed in most detail, costed and their hydraulic performance modelled. From this, 'core schemes' were refined and drawings prepared for contract tendering.

Since a key requirement of the Cole and Skerne projects was to demonstrate 'benefits' of restoration, a pre- and post-monitoring programme was put in place which is the most extensive of its kind undertaken in the UK (and in parallel with studies on the Danish site too). As the vision plan was being developed, a wide range of surveys were being undertaken. These included the following:

- full topographical surveys and river cross-sections
- geomorphology catchment audits
- a water quality survey, looking at a wide range of aspects such as nutrient and sediment reductions due to the works
- landscape assessments
- a complete range of studies on aquatic and floodplain biota (e.g. fish, birds, invertebrates, plants) and many other types of ecological survey
- hydraulic modelling to test the effects of proposals on flood conveyance at the sites and the benefits downstream from intercepting storm runoff, improving floodplain storage and reduced erosion
- community perception – the views of the community on the proposals, followed by assessment once implemented

RESTORATION WORKS: RIVER COLE

Figure 8.2 gives a summary of the works undertaken on the Cole, as well as some photographs of the river prior to works being undertaken, in progress, and on completion. The majority of the work was completed in the summer of 1995. As this work was completed within budget, the large contingency set aside for unforeseen problems was spent in 1996 on refining channel form and on additional works such as restoring the large backwater meander below the off-take to the newly restored channel. The most important restoration achievements in addition to this included the following:

- raised water levels in the mill channel enabling better control of the water regime and providing good depth within the restored meander loop under ancient willows;
- the majority of the flow being diverted through a new sinuous channel above the bridge on the old course of the river enabling periodic flooding of the old floodplain for flood storage and silt deposition as well as restoration of wetlands subject to periodic, uncontrolled, water levels;

THE RIVER RESTORATION PROJECT

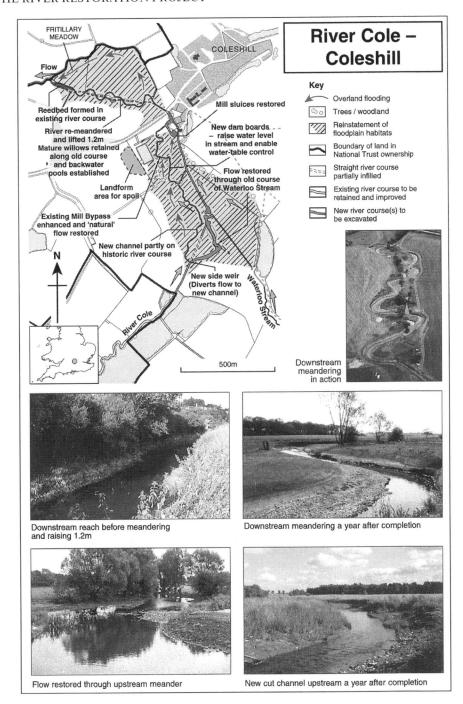

Figure 8.2 Summary of the works undertaken on the River Cole, near Swindon, UK

- a controlled water regime on the right bank upstream of the mill to enable restoration of water-table dependent habitats;
- increased flooding conditions downstream of the bridge to enable recovery of the fritillary meadow and conversion of arable fields to herb-rich floodplain grassland (achieved by reducing channel conveyance through bed raising and smaller cross-sections);
- re-meandering of the previously deepened, straightened and featureless channel downstream of the bridge, the new bed level being over 1 m above the former one and the sinuous channel developing pool–riffle–run sequences, cliffs and point bars at meanders, and reed ledges and backwater pools incorporated into the design.

The combination of works, over almost 2 km of river, covered the most extensive range of river and floodplain restoration measures ever undertaken in the UK. The re-meandering of the river across its previous straightened course, and more than 1 m above the old bed level, was unprecedented in the UK, even though it had become commonplace many years previously in Denmark (Friberg *et al.*, 1994; Madsen, 1995; Nielsen, 1996b).

RESTORATION WORKS: RIVER SKERNE

Figure 8.3 shows the main area of the Skerne site as it is now, compared with its previous condition. The core area, as shown, has undergone a dramatic transformation, with many less major activities being achieved downstream of here. Even within the area shown, more than half the river length was so constrained that only small-scale channel and bank works could be undertaken. This consisted mainly of topsoil stripping to remove the nutrient-enriched dredgings from the top of the bank and re-profiling banks to shallower slopes to enable a wetland edge to be established for emergent plants. In a straight section flow sinuosity was introduced by the use of two types of deflectors; these are drowned out in floods but should create some sustainable habitat diversity in the future. The most important feature of the physical restoration works is the creation of four large meanders, backwaters and wetland scrapes within the only area of floodplain where the river is not constrained on both sides by either tipping or services.

The aerial photograph shows the newly meandering river with its extensive backwaters formed within the old straight channel. Meandering here, in contrast to the lower part of the Cole, could be only achieved by creating bends on one side of the valley floor since sewer mains and gas pipes ran very close to the bank on the northern side. The problems facing river restoration in Darlington are very typical of many other urban locations. The ability to meander such a river on only one side of the valley gave rise to an opportunity to demonstrate the use of a variety of river revetment techniques to protect the services on the opposite bank. Each meander had a different type of soft 'bio' revetment using live vegetation in a high risk location; the degree of protection varies in different locations on each bend depending on the risk to erosion. The Environment Agency, responsible for flood protection, provided research and development money to increase the scope of work that could be undertaken (River Restoration Centre, 1998).

In addition to the huge benefits to the physical habitat restoration of the Skerne, the river has benefited from improvements in water quality, landscape and public amenity. Around 15 visually intrusive surface water outfalls with concrete headwalls were replaced

THE RIVER RESTORATION PROJECT

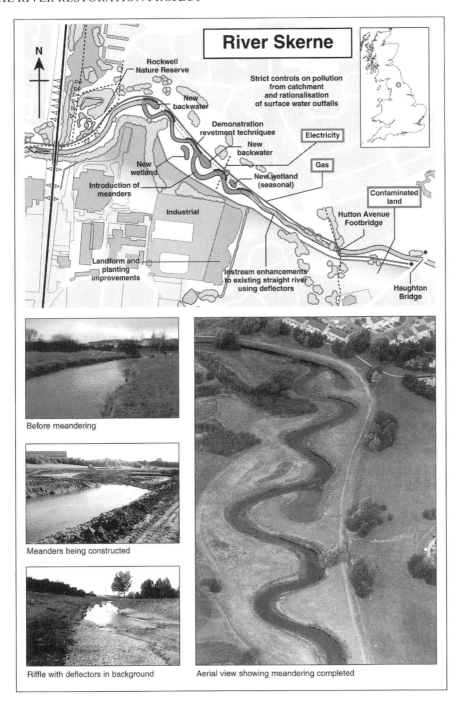

Figure 8.3 Summary of the works undertaken on the River Skerne, Darlington, UK

by invisible underwater outfalls discharging from inspection interceptor chambers buried in the bank. This was undertaken by Northumbrian Water, responsible for sewage treatment and surface runoff. Darlington Borough Council wished to improve facilities for the many local people who use the parkland, and had indicated their long-term desire to provide disabled access to, across and along the whole river. There were insufficient funds within the 'Life' funding to enable all the desired river works to take place, let alone those relating to improved access. However, a successful bid to the Heritage Lottery Fund enabled all the proposed meanders and river improvements to be constructed, and during 1997 a footpath link was built which connects the historic railway bridge to a circular walk around the core restoration reach where Darlington Borough Council have built a new footbridge.

THE FUTURE OF RRP

RRP has now developed from its project-led base to a new phase that concentrates on information dissemination, education and advice that will support further work on river restoration in the UK and link such activities with those being promoted elsewhere. To reflect this shift in emphasis, RRP has been re-named The River Restoration Centre. The broad aims of the RRC are to do the following:

- act as a national centre for enquiries for information on river restoration (information packs, media, technical and scientific enquiries, publications, manual of restoration techniques, display materials, site visits, lectures and talks, news sheets and annual reports, a restoration network of site and contact names, updates on international research on river restoration activities);
- provide specialist advice and assistance to those actively involved in river restoration, or who are wishing to embark upon such activities for the first time (training for practitioners, support for developing vision plans for new projects, advice on legal, institutional and operational aspects of river restoration, monitoring programmes);
- act as a catalyst for promoting/encouraging future national projects which advance knowledge on the benefits of, and techniques required for, river restoration.

The plans for the RRC are supported by many of the institutions that co-funded the 'Life' project. The proposals for the RRC are also linked to a number of major long-term monitoring and research activities which are under-way, or planned, linked to the 'Life' sites. These include detailed assessments of geomorphological adjustments and sediment transport, hydraulic performance, public perception and long-term responses by biota. The plans for the RRC thus aim to link the aspirations of broadening the knowledge and influence of river restoration with the 'living laboratory and classroom' that the capital investment in the demonstration sites has brought.

ACKNOWLEDGEMENTS

The RRP could not have been formed without British Coal Opencast's initial funding and likewise activities reported on above could not have been undertaken without funding and/or general support from the EU 'Life' Fund, NRA/Environment Agency, the Heritage Lottery Fund, Northumbrian Water, English Nature, the Countryside Commission, The National Trust, Darlington Borough Council, Scottish Natural Heritage, the Department of Agriculture Northern Ireland and other sponsors. The author is grateful to RRP's officers for their help in providing information for this chapter, and to Mogens Bjorn Nielsen for his project management of the 'Life' project.

REFERENCES

Boon, P.J., Calow, P. and Petts, G.E. (1992) *River Conservation and Management.* John Wiley, Chichester.

Brookes, A., Gregory, K.J. and Dawson, F.H. (1983) An assessment of rivers channelization in England and Wales. *The Science of the Total Environment*, **27**, 97–111.

Countryside Commission (1987) *Changing River Landscapes.* Countryside Commission, Cheltenham, UK.

Friberg, N., Kronvang, B., Svendsen, L.M., Hansen, H.O. and Nielsen, M.B. (1994) Restoration of a channelized reach of the River Gelsa, Denmark: effects on the macroinvertebrate community. *Aquatic Conservation*, **4**(4), 289–296.

Gerritsen, G.J. (1992) Ecological rehabilitation of the River IJssel. In *Contributions to the European Workshop – Ecological Rehabilitation of Floodplains.* Arnhem, September 1992, CHR/KHR, Lelystad, pp. 79–84.

Hansen, H.O. and Madsen, B.L. (1998) *River Restoration 196 – Session Lecture Proceedings.* NER1, Silkeborg, Denmark.

Havinga H. (1992) Floodplain restoration: a challenge for river engineering. In *Contributions to the European Workshop – Ecological Rehabilitation of Floodplains.* Arnhem, September 1992, CHR/KHR, Lelystad, pp. 95–104.

Holmes N.T.H. (1992) River restoration as an integral part of river management in England and Wales. In *Contributions to the European Workshop – Ecological Rehabilitation of Floodplains.* Arnhem, September 1992, CHR/KHR, Lelystad, pp. 165–172.

Holmes, N.T.H. (1998) *River Rehabilitation in the UK.* Addendum to Core R&D Project 477, Environment Agency, Bristol.

Jaeggi, M.N.R. (1992) Sediment regime and river restoration. In *Contributions to the European Workshop – Ecological Rehabilitation of Floodplains.* Arnhem, September 1992, CHR/KHR, Lelystad, pp. 147–156.

Kern, K. (1992) Rehabilitation of streams in south-west Germany. In Boon, P.J., Calow P. and Petts, G.E. (eds) *River Conservation and Management.* John Wiley, Chichester.

Litjens, G.J.J.M. (1992) World WildLife Fund reanimates delta of Rhine. In *Contributions to the European Workshop – Ecological Rehabilitation of Floodplains.* Arnhem, September 1992, CHR/KHR, Lelystad, pp. 105–106.

Madsen, B.L. (1995) *Danish Watercourses – Ten Years with the New Watercourse Act – Collected Examples of Maintenance and Restoration.* Danish EPA, Miljonyt 11, Copenhagen.

Nielsen, M.B. (1996a) River restoration: report of a major EU 'Life' demonstration project. *Aquatic Conservation*, **6**, 187–190.

Nielsen, M.B. (1996b) Lowland stream restoration in Denmark. In Brooks, A. and Shields, D.F. (eds) *River Channel Restoration: Guiding Principles for Sustainable Projects.* John Wiley, Chichester, pp. 269–290.

Nielsen, M.B., Ottesen O. and Petersen, B.D. (1990) Restaurering af Gelsa ved Bevtoft. *Vand & Miljo*, **4**, 123–125.

River Restoration Project (1993a) *Feasibility Study.* RRP, Huntingdon.

River Restoration Project (1993b) *Feasibility Study – Brochure.* RRP, Huntingdon.

River Restoration Project (1993c) *The River Restoration Business Plan.* RRP, Huntingdon.

River Restoration Project (1994a) *Institutional Aspects of River Restoration.* RRP, Huntingdon.

River Restoration Project (1994b) *Institutional Aspects of River Restoration – Summary.* RRP, Huntingdon.

River Restoration Project (1994c) *Site Appraisal for River Restoration*. RRP, Huntingdon.
River Restoration Centre (1998) *Revetment Techniques used on the River Skerne Restoration Project*. R&D Technical Report W83, Environment Agency, Bristol.
Rojacz, H. (1992) The River Leitha – rehabilitation from 0.0km to 15km. In *Contributions to the European Workshop – Ecological Rehabilitation of Floodplains*. Arnhem, September 1992, CHR/KHR, Lelystad, pp. 9–12.
RSPB/NRA/WT (1994) *The New Rivers & Wildlife Handbook*. Royal Society for the Protection of Birds, Sandy, Bedfordshire.
Sear, D.A. (1994) River restoration and geomorphology. *Aquatic Conservation*, 4(2), 169–177.
Williams, G. and Bowers, J.K. (1987) Land drainage and birds in England and Wales. *RSPB Conservation Review*, 1, 25–30. RSPB, Sandy, Bedfordshire.

9 Strategic Approaches to River Rehabilitation: The River Leen and the River Derwent, UK

DAVID HICKIE

Environment Agency, Midlands Region, Solihull, UK

BACKGROUND

Since 1975, the Severn-Trent Water Authority and its successors, the Severn-Trent Region of the National Rivers Authority and the Environment Agency (as of 1 April 1996) have implemented many river enhancement projects across the catchments of the rivers Severn and Trent in the UK. From 1975 to 1997, over 1500 individual projects have been undertaken, the majority of which have been opportunistic in nature and implemented in association with flood defence maintenance projects. A few projects, however, have taken a more strategic approach. The Upper Avon Enhancement Project sought to target the upper Avon catchment in a strategic way, restoring, where possible, the natural habitats, sinuosity and channel profile of the river. This was achieved in co-operation with flood defence maintenance works and backed up by funding (Purseglove, 1988).

Up to 1988, the funding for these projects came entirely from the flood defence budget, with the exception of a few individual projects. For example, the Otter Habitat Enhancement Project, designed to help restore tree and scrub cover along the upper reaches of the River Severn and its tributaries, was funded as a small-scale capital project from region-wide water authority capital sources. Since 1988, all flood defence works in the Severn-Trent region have had funding available for associated conservation enhancement works. A nominal 5% of the total regional flood defence maintenance budget has been available for expenditure on the most appropriate river enhancement projects, as selected by in-house conservation staff.

The legislation which led to the creation of the National Rivers Authority, the Water Act 1989, and subsequently the Environment Agency, included the new wider duties to promote conservation and to take account of recreation issues, together with the duties to conserve and enhance the environment. This led to the creation of two new functions within the new authority: conservation and recreation (Heaton, 1993).

In the Severn-Trent Region of the NRA, these new duties were managed by a Conservation and Recreation Section. An integrated approach to both conservation and recreation duties was developed, because of the very close links and interdependence of ecology, landscape and cultural heritage, the recreational experience, and use of waters and the associated land (NRA Severn-Trent, 1989). The new Conservation and Recre-

Rehabilitation of Rivers: Principles and Implementation. Edited by L. C. de Waal, A. R. G. Large and P. M. Wade.
© 1998 John Wiley & Sons Ltd.

ation Section, operating as part of the Fisheries, Conservation and Recreation Department, now had independent funding for conservation and recreation promotional activities.

In the summer of 1989, the new separate funding for the promotion of conservation was targeted at a number of regional initiatives. At this point in time, the majority of the work on rivers across the Severn-Trent Region had been in the rural areas, and therefore a natural progression was to now look at the opportunities for the enhancement of urban watercourses. For many years the water authority had supported piecemeal enhancements along the River Cole, Birmingham. This work, known as Project Kingfisher, was led by Birmingham City Council, the major landholder along the lower section of this urban river. Its success highlighted the advantages of developing a collaborative approach to urban river enhancement works. Furthermore, it was identified that there is the need to work to a strategic masterplan in order to ensure that individual projects fit into the larger picture, assisting in the rehabilitation of the whole river system.

The majority of our urban rivers have been quite radically altered from their natural courses and traditional channel morphology (Cosgrove and Petts, 1990). Whilst the problems of channel rehabilitation could well be greater with our urban rivers than with rural rivers, the potential influence and control on development from the planning process, and the relatively large extent of land in public authority ownership, increases the potential for a more direct influence on the development and management of the river channel and the adjacent land within the river corridor. Restoration of urban rivers to their pre-development state is an impractical aim, whereas the rehabilitation of a watercourse to a more semi-natural state, in the context of the existing and future land-use constraints is a more practical goal to strive for. A strategic approach, rather than a piecemeal individual and opportunistic approach was considered more likely to succeed and be the most effective use of the new NRA funding and staff resources (NRA Severn-Trent, 1989).

STRATEGIC APPROACH – OBJECTIVES

The overall objective within the Severn-Trent Region was to use the NRA's river-based conservation expertise and the provision of initial funding to co-ordinate the preparation of masterplans for a number of urban rivers. These would identify constraints, potential for enhancement, partnership opportunities and policies, which could be implemented by a variety of agencies, including the NRA, either as short-term or long-term goals.

Three urban river courses were initially selected as being suitable for this new strategic initiative. They were selected from a list of urban watercourses the Severn-Trent Region had already been working on by implementing minor specific enhancements. The selected rivers were the River Derwent (Derby), the River Leen (Nottingham) and the River Trent (Stoke). After further consultation with the relevant local planning authorities, the idea of the NRA helping to co-ordinate a conservation masterplan for these stretches of river was well supported in Derby and Nottingham. However, a masterplan for the River Trent flowing through the centre of Stoke was already planned to be undertaken by Nature Conservancy Council (NCC – now known as English Nature) for Stoke City Council. The NRA, therefore, started the Strategic Urban Rivers Initiative with two projects, one in Derby and one in Nottingham.

STRATEGIC APPROACHES TO RIVER REHABILITATION 151

The projects were different in terms of size and characteristics of the watercourses, but similar with regards to the enthusiasm and support of the local planning authorities, private sector organisations and voluntary groups that would develop these new initiatives.

RIVER LEEN, NOTTINGHAM

The Leen is a small mainly urban watercourse approximately 3–5 m wide. Its total length is only 20 km from its source in the grounds of Newstead Abbey, 14 km north of Nottingham, to its confluence with the River Trent near the centre of Nottingham (see Figures 9.1 and 9.2). The watercourse has been canalised and overmaintained in many places. In the past, both industry and housing schemes have turned their backs on the river. In some locations the watercourse has been culverted and completely built over.

Internal NRA reports show the biological quality of the upper reaches to be very good, with crayfish (*Astacus pallipes*) present at Newstead Abbey. This high quality river then deteriorates rapidly on reaching the outskirts of the urban area near Hucknall, and is reported to be of low quality all the way to the confluence with the Trent. However, unpublished surveywork by the Nottinghamshire Wildlife Trust had identified the importance of the river as a wildlife corridor running from the countryside to the city centre, and the Trust has developed a number of nature reserves alongside the river.

Figure 9.1 River Leen, Nottingham

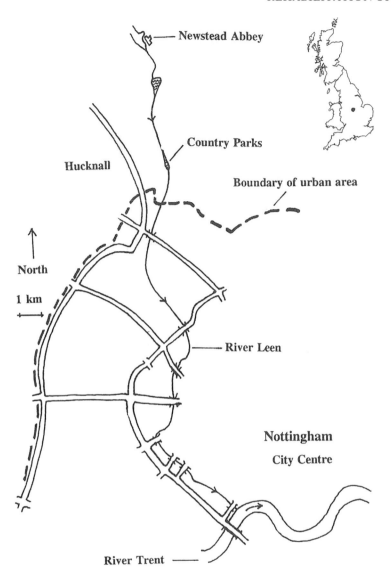

Figure 9.2 Map of River Leen Project (Nottingham, UK)

The river rises in the rural undulating, well-wooded countryside to the north of Nottingham and flows southwards to the Trent (Figure 9.2). The corridor links many old mill sites, new country parks, nature reserves, urban green spaces and parks, which have been developed in recent years by the local authorities and wildlife groups. The potential to develop this corridor effectively and to rehabilitate the urban part of the river, made it an ideal candidate for the NRA's pilot work on river masterplans.

Three separate local authorities have interests along the River Leen – Nottingham County Council, Nottingham City Council and Gedling Borough Council – all three owning large areas of open spaces along the length of the Leen. A large number of private industries are also located along the river, one of the largest being the Royal Ordnance Factory where one of the initial enhancement projects started.

MASTERPLAN – PHASE 1

The first phase of the project commenced in 1989, with the NRA commissioning Sherwood Wildlife Consultants to undertake the works (Myhill, 1989). The reach covered the section from the confluence with the Trent to Bulwell Forest on the edge of the urban area. The brief for the masterplan included a detailed survey of the river and the associated corridor, and the identification of the characteristics, constraints and opportunities at a strategic level, together with some detailed examples of how the river could be enhanced. The baseline survey included a description of underlying geology, soils, geomorphology and ecology. The ecological survey included work on habitat, and flora and fauna, including fisheries. The land use and existing related policies, landscape characteristics and historical context, including mill systems and canalisation, of the river catchment were also noted. The river was described in terms of hydrology, including discharges, flooding regimes, low flows and runoff characteristics, and water quality.

The masterplan determined and analysed the constraints and opportunities relating to the rehabilitation of the river as perceived by the NRA, other government agencies (such as the Countryside Commission and the Sports Council), local planning authorities, business and industry (including new developers) and non-governmental organisations (such as Nottinghamshire Wildlife Trust, the Civic Trust, the Council for the Protection of Rural England and the Ramblers Association).

Key constraints and opportunities included the following:

- NRA:
  - Water resources: quantity/low flows/abstraction
  - Water quality: poor quality in urban areas
  - Flood defence: need to maintain channel capacity
  - Fisheries and conservation: ecology and habitats
  - Recreation: routes and access
- Industry and new developers:
  - Superstores/factories/farms
- Local authorities:
  - Development control
  - Promotion of recreation and a green corridor

The masterplan outlined the following strategic objectives (Myhill, 1989):

As part of the NRA's policy to promote a co-ordinated approach to conservation and recreation along river systems, a masterplan has been drawn up to identify the needs and opportunities along this most important corridor.

All agencies involved in the River Leen corridor, including local authorities and the Nottingham Wildlife Trust, are co-operating to enhance the river corridor directly by planning improvement schemes and indirectly by influencing any new development to include for conservation and recreation opportunities.

The NRA intends to use this innovative scheme to promote similar initiatives along other urban watercourses across the Region.

Phase 1 of the masterplan identified the potential to create a continuous linear park along the Leen corridor. The central concept of this park was a continuous walkway from the River Trent northwards into the countryside, linking together many areas of public open space and conservation areas. The masterplan also identified opportunities for river channel rehabilitation to a more natural channel form, with variations in bank and channel profile. The limitations of existing development and services running riparian to the river did, however, restrict many such potential opportunities.

A steering group was formed to manage the development of the project, which included representatives from the NRA, the Recreation and Planning Departments of Nottingham City Council and the Nottinghamshire Wildlife Trust.

MASTERPLAN – PHASE 2

Phase 2 of the masterplan was then developed by Sherwood Wildlife Consultants, running from the outskirts of Nottingham at Bulwell to the source of the Leen, in the grounds of Newstead Abbey, near Kirby-in-Ashfield. A similar set of strategic objectives and opportunities for rehabilitation were identified, as in Phase 1.

Following the consultation with all the agencies involved, the River Leen Steering Group was set up, and on Friday 1 December 1989, Douglas Hogg, the Minister for Industry and Enterprise, launched the new River Leen Environmental Regeneration Project. By the summer of 1990, many enhancement projects were already under way. In a letter to Nottingham City Council, dated 7 June 1990, Sherwood Wildlife Consultants, highlighted that the project had 'already begun to show real results: river-widenings, weirs and other variations in the channel, wildflower–grass mix reseeding and reduced mowing regimes by the NRA; and tree and shrub planting schemes to replace mown banks.' In the years since 1990, many other enhancement projects along the Leen have been generated. At the northern end of the Leen, the Greenwood Community Forest project has included the northern section of the Leen corridor in the area of its activity.

Projects such as the Leen Cycleway, new footpaths, and both the Greater Nottingham Light Rapid Transport Project (Mott MacDonald, 1991), and City Challenge 2 (a bid for additional central government funding) (Nottingham City Council, 1993), together with numerous other development proposals, could in the worst case have resulted in the continued degradation of the River Leen. The existence of the River Leen Masterplan has enabled such projects to provide the required funding and to design the features necessary to enhance the river within the planned framework of the strategic overview developed by the masterplan.

A Project Officer was appointed in April 1994, with joint funding from the NRA, Nottingham City Council, the Countryside Commission and Gedling Borough Council. This was to ensure that the strategic initiatives could be developed and more projects implemented to enable the continued enhancement of the River Leen.

RIVER DERWENT, DERBY

A 12 km stretch of the River Derwent flowing through the City of Derby was selected as the second initial project. The River Derwent provided an example of a large urban river in contrast to the small scale of the Leen watercourse. The river has been controlled in the past through a series of weirs, and old meanders have been cut off (Figures 9.3 and 9.4). It was decided to put together a masterplan just for this stretch of the river and not for the whole length of the River Derwent, because of its relatively long length of 88 km, most of which is rural in nature. The section of the river chosen had major pressures from new developments, and existing industries such as Rolls Royce, British Gas, Severn-Trent Water plc (sewage works), British Rail Engineering and by far the largest, Courtaulds (a chemical works) (Figure 9.2). Derby City Council is a major landholder along the river, with a number of parks, and other areas for which redevelopment was planned.

The City Council had to develop a local plan for the south of Derby and obvious potential existed as part of this planning process to develop new conservation and recreation initiatives along the Derwent from the centre of Derby to the outskirts, both to the north (parks, residential areas and water meadows) and to the south (industry and parks). The NRA saw this as an opportunity to promote the masterplan approach to river enhancement works, which could be used to help resolve some existing complex conservation and potential recreation conflicts in Derby. As a result, the NRA and Derby City Council became joint lead bodies in what became known as Project Riverlife. The aim was to develop and promote a strategic conservation and recreation masterplan for the

Figure 9.3 River Derwent, Derby

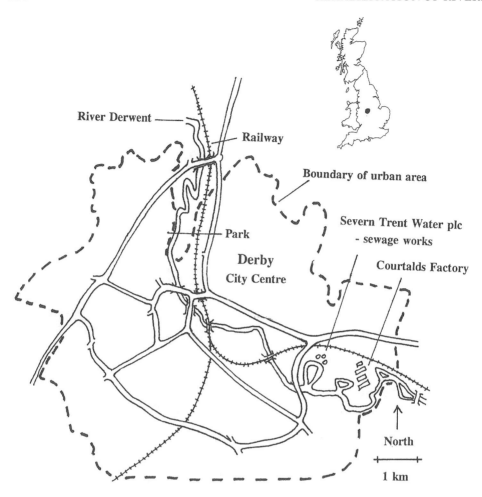

Figure 9.4 Map of River Derwent Project (Derby, UK)

river corridor. Courtaulds, a major landowner on the southern side of the city, already had a small conservation area within their own site and were keen to participate in Project Riverlife, recognising the opportunity for good public relations, in such a potentially high-profile collaborative project.

In 1989, the NRA commissioned a consultant to prepare a strategic masterplan for the river corridor to the northern and southern limits of the industrial and residential development of the city (Conlon, 1989). The remit of this consultancy work was similar to the River Leen project. The development of the masterplan was managed by a Project Riverlife Working Party, which included local authority officers, an NRA representative, Courtaulds, Derbyshire Wildlife Trust, other local conservation groups, and representatives from the local recreation users bodies such as angling, canoeing and cycling. The detailed enhancement and rehabilitation works were developed and promoted by a Project Officer jointly funded by the main agencies involved in the project. The majority of works were riverside habitat enhancements with no major opportunities to alter the main channel because of the relatively large size of the river. A route for a new cycleway and walkway was identified, giving better access along the river corridor whilst protecting some more ecologically sensitive areas from too much human disturbance.

STRATEGIC APPROACH – CONCLUSIONS

A strategic approach to river rehabilitation in the urban environment provides a framework around which the Environment Agency and other agencies such as local authorities, business, industry and wildlife groups can seek to influence other agencies to help realise the strategic vision of the future development of the river corridor. Failure to develop a strategic vision or *Leitbild* for river rehabilitation projects will limit the confidence of other agencies in the collaborative goal to rehabilitate a river system. Both the Leen and Derwent are examples that have spawned other allied environmental projects which in turn can both directly and indirectly enhance and support the work towards achieving this strategic aim of rehabilitation.

A strategic vision is required to help influence the most important urban constraint on development, that of the planning system and the development of Local Plans. In the past, rivers have helped shape the form of the landscape and the pattern of developments. The recent land-use and land-management policies have tended to constrain and shape the form of many river courses and their management today, e.g. the straightening and canalisation of long lengths of the River Leen. If we are to enhance or even rehabilitate our rivers, we must try to reverse this trend. A strategic vision, defining how the river should develop (which could be an evolutionary vision), can directly influence future land-use and management decisions. Collaborative strategic masterplans can provide this vision and help it to become a reality. In the Severn-Trent Region, the Leen and Derwent projects have provided a model for future implementation of the Environment Agency's strategic aims.

THE FUTURE

Following the use of strategic approaches to river rehabilitation in the early 1990s, the NRA produced integrated Catchment Management Plans for all river catchments. With the formation of the Environment Agency in 1996, these are now called Local Environment Agency Plans (LEAPs). These define issues, identify bodies responsible for resolving these issues, and outline options for future action. For example, the River Avon Catchment Management Plan has identified the need to produce integrated recreation plans for specific sections of river, and to address the many nature conservation and recreation conflicts along its length as well as wider environmental issues.

With limited funds and resources it is important to identify the strategic context of a river system first, before committing resources to specific river rehabilitation works. The LEAPs now provide this overview which can provide a framework within which the rehabilitation programmes can be implemented.

The Environment Agency's experience with collaborative rehabilitation projects has shown that the strategic approach, involving relevant agencies and organisations is the most sensible direction for future rehabilitation projects in the UK.

REFERENCES

Conlon, P. (1989) *River Derwent, Derby – Masterplan*. National Rivers Authority, Severn-Trent Region, Solihull.
Cosgrove, D. and Petts, G. (1990) *Water, Engineering and Landscape*. Belhaven Press, London.
Heaton, A. (1993) Conservation and the National Rivers Authority. In Goldsmith, F. and Warren, A. (eds) *Conservation in Progress*. John Wiley, Chichester.
Mott Macdonald (1991) *Greater Nottingham Light Rapid Transport System, Environmental Statement*. Nottingham City Council, Nottingham.
Myhill, N. (1989) *River Leen Masterplan*. National Rivers Authority, Severn-Trent Region, Solihull.
National Rivers Authority Severn-Trent Region (1989) *Regional Conservation and Recreation Policy*. National Rivers Authority, Severn-Trent, Solihull.
Nottingham City Council (1993) *City Challenge 2*. Nottingham City Council, Nottingham.
Purseglove, J. (1988) *Taming the Flood*. Oxford University Press, Oxford.

10 Problems Associated with the Degradation of Rivers in the North East Region, England, and Initiatives to Achieve Rehabilitation

JOHN R. PYGOTT[1] and ANDREW R. G. LARGE[2]
[1] *Environment Agency, North East Region, Leeds, UK*
[2] *Department of Geography, University of Newcastle upon Tyne, UK*

INTRODUCTION

Perhaps the biggest change for water management in England and Wales in the 1990s was the formation of the Environment Agency in April 1996 under the provisions of the Environment Act 1996, which superseded both the Water Act 1989 and the Water Resources Act 1991, as well as other consolidating legislation. This new body is formed from the amalgamation of the then National Rivers Authority (NRA), Her Majesty's Inspectorate of Pollution (HMIP) and the Waste Regulation Authorities. The Agency was created as a multifunctional environmental regulatory body with a duty under the 1996 Act to 'promote the conservation and enhancement of natural beauty and amenity of inland waters and land associated with such waters; [and] conservation of flora and fauna which are dependent on the aquatic environment'. As well as the Department of the Environment, the Agency is responsible to the Ministry of Agriculture, Fisheries and Food (MAFF), which has specific responsibilities for flood defence and fisheries management. In addition to flood defence and fisheries management, the Agency carries out the following functions in England and Wales: water resources management and integrated pollution control of land, air and water. In addition, it controls some river navigations, whilst also having statutory duties relating to the conservation and enhancement of the natural environment and the promotion of recreation contained in the Environment Act 1996. Gardiner (1992) outlined three fundamental challenges to the National Rivers Authority or NRA (one of the predecessor bodies which form the current Agency) in terms of fulfilling its role as 'guardians' of the river environments of England and Wales. It firstly had to minimise the impact of its own developments and activities; secondly it needed to exercise control over land use within the catchments it managed; and thirdly it had a duty to enhance the water environment. Gardiner (1992) highlighted three further objectives: (i) to retain land protected by Green Belt, SSSI or other designations, (ii) to conserve water-table dependent habitats, and (iii) to protect open land for flood storage, thus defining a 'beaded' river corridor complemented by connected patches of open land conserved in

Rehabilitation of Rivers: Principles and Implementation. Edited by L. C. de Waal, A. R. G. Large and P. M. Wade.
© 1998 John Wiley & Sons Ltd.

accordance with specific objectives. In R&D research commissioned and funded by the NRA, Large and Petts (1992) used a similar analogy in relation to the design of buffer zones in the riparian landscape. Between 1993 and 1996, the NRA undertook over two hundred collaborative projects across its regions which related in some way to conservation of rivers. Over this period, an increasing number of these related to rehabilitation of degraded river habitats. The Agency is active in the field of innovative management of rivers with a primary focus being on integrated catchment management (e.g. Brookes, 1988; Gardiner, 1991, 1992; Newson, 1992). In carrying out its conservation responsibilities, the Agency collaborates with other statutory conservation bodies, e.g. English Nature, the Countryside Council for Wales and the Joint Nature Conservation Committee. It has been closely involved with English Nature, the Countryside Council for Wales and Scottish Natural Heritage in the development of SERCON, a computerised system for evaluating rivers for conservation (National Rivers Authority, 1994). Liaison has also covered the Special Areas of Conservation which will form part of Europe's Natura 2000 sites to be established under the European Union Habitats and Species Directive.

In assessing the condition of the water resources under its jurisdiction, the Agency closely monitors trends in the biological and chemical status of rivers using well-established techniques. In terms of water quality alone, the Agency is responsible for some 20 European Union directives that have a direct environmental monitoring requirement and believes that pollution prevention is of key importance in achieving improvements to the water environment. The Agency advocates the use of Local Environment Agency Plans (LEAPs) as a vehicle for implementation of water quality objectives, which in turn will assist in maintaining biodiversity of water bodies under its control. It has long been recognised that rivers and a range of other water bodies play an integral role in maintaining the biodiversity of species and habitat in England and Wales (National Rivers Authority, 1994); helping to maintain and enhance this resource is a key responsibility of the Agency as part of its contribution to the UK's Biodiversity Action Plan.

THE RIDINGS AREA

The Ridings Area of the North East Region of the Environment Agency covers catchments of two distinct types. In the west of the Area, the river rises in the Pennine uplands and flows down steep-sided valleys into the lowland floodplain. The lowland areas are heavily industrialised and have the main centres of population within them, with more than four million people in total living within the Ridings Area. To the east is the catchment of the River Hull and the associated system of lowland drains. This is an area of intensive agriculture and greatly modified watercourses. Figure 10.1 shows the North East Region with the main catchments outlined. The Ridings Area is the most southerly of the three Areas contained in the North East Region.

Recent years have seen an increasing interest in the protection and improvement of the environment, and this has focused attention on those areas where there are shortfalls between the actual state of the environment and expectations held by individuals and organisations. Many of the shortfalls in the quality of the environment are caused by human influence, and this has focused attention on those areas where there are shortfalls between the actual state of the environment and expectations held by individuals and organisations. Rapport *et al.* (1985) have categorised the types of stresses imposed by humans on water ecosystems into four main groupings:

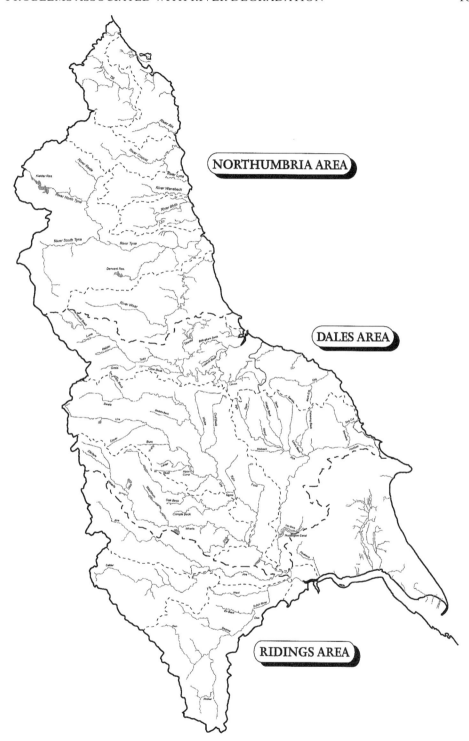

Figure 10.1 Catchment areas within the North East Region of the Environment Agency

1. discharge of pollutants
2. physical modification and re-structuring
3. introduction of exotic species
4. harvesting of renewable resources

The first three of these are pertinent to rivers in the Ridings Area and, although there may have possibly been some effects on watercourses in the Region arising from the harvesting of renewable resources, discussion here will be restricted to the first three stress types with respect to rivers in the Yorkshire area. The rivers Hull and Dearne in the Ridings Area (see Figure 10.1) will also be briefly discussed to place current restoration works occurring in the Region in context.

DISCHARGE OF POLLUTANTS

Water quality in England and Wales is monitored by the Agency and lengths of watercourse are classified according to a number of parameters such as dissolved oxygen saturation, biochemical oxygen demand and concentration of ammonia. The river quality classification runs from A (good) through to F (bad).

There are a range of ecologically damaging effects resulting from the input of sewage effluents to watercourses. In 1994, under the terms of the EC urban waste water directive, the NRA identified 33 inland areas in England and Wales where phosphorus removal/stripping will be necessary before 1998 (Turner, 1994). The impacts of sewage effluents on invertebrate communities have been well documented over many years and form the basis of a number of water quality classification systems. The impacts on aquatic vegetation and other components of the ecosystem such as fish are less well documented although some work has been undertaken in this area (Nature Conservancy Council, 1991). The impacts of sewage effluent on the overall aquatic ecosystem are also poorly understood, although there is a clear pattern of simplification of the ecosystem by the removal of those components which are most sensitive. There are also effects on the visual and aesthetic character of the rivers in the area; important considerations for the general public.

Agriculture has also brought about major problems in some parts of the North East Region. There are a number of different types of pollution risk associated with agriculture, but broadly these can be categorised into organic materials, nutrients, pesticides, oil and other contaminants and biological material such as bacteria, parasites, and the carcasses of animals (National Rivers Authority, 1992).

The 1990 survey of river water quality carried out by the National Rivers Authority showed that in the previous Yorkshire Region around 14% of the total length of rivers classified were in class 3 and 4, i.e. poor and bad quality. The vast majority of this length is contained within what is now the Southern Yorkshire Area. Most of the problems associated with water quality in Yorkshire are related to discharges of sewage effluent. Downgrading of the quality of more than 50 km of watercourse in Yorkshire following the 1990 survey was due to the effects of sewage effluents and storm sewage overflows.

THE RIVER HULL

Modifications in the form of rivers in rural areas of the Region have taken place. The River Hull valley has a long well-documented history of modification for drainage and agricultural improvement. Initial draining of the River Hull wetlands began in the 12th century and continued through until the Second World War, by which time virtually all the wetland habitats within the former floodplain area had been lost (Sheppard, 1958).

Within the catchment of the River Hull there are particular problems caused by the use of nitrogenous fertilisers on land which provided the catchment for ground and surface water systems used for potable supply. A number of Nitrate Vulnerable Zones (NVZs) are now established within the Hull catchment. This is in response to the EC Nitrate Directive (91/676) which is intended to reduce water pollution from agricultural sources. Within the NVZs, there are compulsory programmes which place restrictions on the application of organic manure, requirements for manure storage and restrictions on the application of inorganic fertiliser.

The implementation of new programmes of land drainage for agricultural improvement have by and large ceased. However, maintenance of the existing patterns of drainage and the need to protect the drained wetlands from flooding continues to present an obstacle to any initiatives to restore the functional wetlands around the River Hull. Changes in agricultural practice such as set-aside and grant-aid schemes such as Countryside Stewardship are resulting in less intensive agricultural use, but without any restoration of former habitats.

THE COAL MINING INDUSTRY

The coal mining industry in Yorkshire has also had major impacts on water quality. There have been both adverse and beneficial effects. Adverse effects have included the discharge of ochrous mine water from abandoned workings on a continuous or in some cases intermittent basis. On the beneficial side there have been instances where the dilution provided from pumped mine drainage has mitigated the damaging effects of effluents from sewage treatment works. Changes in the structure of the coal mining industry in Yorkshire, including the closure of pits and the cessation of pumping, has already begun to cause water quality problems and there is potential for large-scale ecological damage in those waters which previously benefited from dilution.

PHYSICAL MODIFICATION AND RE-STRUCTURING

In the Ridings Area, 906 km of the total length of 2080 km of watercourses are classified as 'main river' and are therefore maintained by the Agency for flood defences. Brookes *et al.* (1983) estimated the proportion of main river in Yorkshire as a whole which is channelised as being between 30 and 35%. No figures are available for other watercourses, but many of these are in urban areas or are maintained for land drainage by more then 40 Internal Drainage Boards.

Urbanisation in the Ridings Area has brought with it a series of problems for the water environment, as there has been a general trend towards the urbanisation of land in river valleys. The Countryside Commission (1987) reported a doubling of urban land in river valleys between the Second World War and 1980. Many of the urban areas of Yorkshire have watercourses flowing through their most built-up areas. As mentioned above, water quality is generally lower in these urban areas as a result of high population densities and the presence of industries discharging effluents into the watercourses. Flood alleviation schemes and urban development have resulted in many of these watercourses becoming heavily engineered, with extensive culverting, bank reinforcement, straightening and over-deepening to increase channel capacity. The illegal disposal of refuse and household waste into these watercourses has further compounded problems.

In general terms the modification of urban watercourses has had the effect of reducing or simplifying the range of aquatic habitats present in a given length. Diversity of channel morphology and vegetation type and cover is often much reduced. There is much evidence of the damaging impacts which this reduction has on the ecology of watercourses (Gorman and Karr, 1978; Swales, 1982; Hellawell, 1988; Mann, 1988; Petts, 1994a,b). In extreme cases the channel may be formed totally from artificial materials or enclosed by culverts. Loss of habitat diversity has resulted in the attendant loss of species diversity and in some cases has provided uniform conditions which are suitable for rapid colonisation by well adapted or aggressively competitive species. There are a number of alien plant species which have been able to exploit available niches in urban watercourses to the detriment of other native plant species. Culverts and weirs which are often a feature of urban watercourses can constitute impassable obstructions within the corridor provided by the watercourse. Obviously this is of particular significance for migratory salmonid fish, but also has implications for coarse fish and possibly other plant and animal species. Aesthetic considerations are also important since the extensively channelised or culverted watercourse is unlikely to have high recreation or amenity value. Indeed in a number of cases the form of the channel renders it unsuitable for recreation because access is impossible.

Changes have taken place in recent years in the design, construction and maintenance of Flood Defence schemes in terms of their increasing environmental sensitivity. Some of the impetus for these changes has come from the formal requirement for Environmental Impact Assessment which has resulted from the enactment of EC directives. There have also been many initiatives by the Agency Regions to incorporate environmental objectives into schemes as well as the publication of a series of advisory documents and publications on best practice. There is an increasing trend towards the re-engineering of channels to facilitate the return of habitats and species considered desirable. Much of this work has been undertaken for the benefit of fisheries and examples of the type of schemes undertaken are detailed in *The New Rivers and Wildlife Handbook* (RSPB/NRA/RSNC, 1994) and other publications (e.g. Harper and Ferguson, 1995; Brookes and Douglas Shields Jr, 1996). Brookes (1992) has outlined some examples of adjustment in re-sectioned reaches of high energy rivers in Northumbria, pointing out that schemes suffered little or no narrowing and consequently required little maintenance due to the long lengths of reach involved.

THE RIVER DEARNE

An example of a recent scheme aimed at enhancing and rehabilitating the river resources in the Ridings Area is the ongoing project on the River Dearne, between Barnsley and Doncaster. This project is being carried out by the Agency in association with the local authorities. The River Dearne has a long history of poor water quality associated with discharges of sewage and effluent from a variety of heavy industrial sources. Fish populations have been sparse for more than a century. The channel has also been heavily modified and diverted due to mining subsidence and flood defence works. In the past five years, improvements on sewage treatment and the decline of heavy industry have led to great changes in water quality. The river has now reached the stage where invertebrate recolonisation is taking place and improvements in fish production are occurring in the river. A range of techniques have been used to rehabilitate this river including the following:

- re-meandering of the straightened sections with training walls
- the creation of low level in-stream berms
- extensive tree and aquatic macrophyte planting
- the creation of backwaters and off-river areas
- re-stocking of fish to supplement natural recruitment

There are a number of problems to be overcome. Water quality is in need of further improvement, especially in relation to discharges of sewage effluent. There are still long lengths of straightened channel with little diversity of physical habitat. Further physical rehabilitation works are planned for the future.

INTRODUCTION OF EXOTIC SPECIES

There is a long history in the United Kingdom of introductions of exotic species either accidentally or deliberately. The species becoming of nuisance value in the North East Region first came to the attention of botanists in the last century because of their size and attractive flowers and were in the main brought from overseas for use as ornamentals in botanical gardens. Perceptions have changed over time, and now there are growing concerns about the problems of these introduced plants invading river corridors. Plant species which have had a major impact on river corridors in the North East Region and elsewhere in the UK include *Impatiens glandulifera* Royle (Himalayan balsam), *Heracleum mantegazzianum* Sommier & Levier (giant hogweed) and *Fallopia japonica* (Houtt). *Ronse Decraene* (Japanese knotweed) and, more recently, *Crassula helmsii* (T.Kirk) Cockayne (swamp stonecrop). These species were the subject of R&D commissioned by the NRA into their ecology and management along watercourses in England and Wales (de Waal *et al.*, 1994).

Impatiens glandulifera is native to the Himalayas, reproduces only by seeds and possesses an explosive seed capsule (Perrins *et al.*, 1990; Pysek & Prach, 1994). These seeds are transported by water and thus the plant is able to colonise river banks and shoals downstream. Die-back of the plant during autumn exposes banks, increasing susceptibility to erosion. Although alien, it is completely naturalised in Britain with a major habitat being the banks of many types of waterways (Beerling and Perrins, 1993). *Fallopia japonica* and *Heracleum mantegazzianum* are both listed under Schedule 9 of the Wildlife and Countryside Act 1981, Section 14. Under this legislation it is an offence 'to plant or otherwise cause to grow in the wild' these listed alien species. *F. japonica*, introduced into Britain sometime in the 1840s (Beerling *et al.*, 1994), has successfully colonised river banks and is widespread and abundant in England and Wales both in rural and urban situations. Extensive thickets can cause in-stream blockages in winter and the plant's competitiveness and ability to spread rapidly endangers native species (Roblin, 1994). *Heracleum mantegazzianum*, introduced into the UK in 1893 as an ornamental plant (Sampson, 1994), has spread along rivers replacing native flora and leaving bare ground exposed to erosion in winter (Williamson and Forbes, 1982). This spread probably reflects the dependence of the plant upon seed dispersal through the medium of flowing water (Caffery, 1994). According to Roblin (1994), the distribution of the plant has increased rapidly, perhaps by a factor of 40 over the last 50 years.

In an attempt to standardise approaches to invasive species across the Regions, and to formulate good management practices which minimise the risk to flood defence and which enhance the conservation and recreation value of river corridors, the NRA commissioned research into the biology, ecology and existing management of *I. Glandulifera*, *F. japonica*, *H. mantegazzianum* and *Crassula helmsii*. An additional species, *Azolla filiculoides* Lam. (water fern), identified as a problem species in the North East Region (Roblin, 1994) is also currently under investigation to examine the extent of the problem and to suggest possible strategies for its control. It is likely, however, that control in much of Yorkshire will not be easily attained because of the extent to which these species have already spread.

Introduced animals are also proving to be a problem for native species in the North East Region, with reports of damage to the native fauna and habitats caused by mink and non-native species of crayfish, such as American signal crayfish. Predation of native fish by the introduced zander has also been reported (RSPB/NRA/RSNC, 1994).

INITIATIVES TO ADDRESS THE PROBLEM OF DEGRADED RIVERS

There is much evidence that improvements in river water quality can result in a range of economic benefits (Fisher and Raucher, 1984). Public perception of the river corridor and water quality has also been investigated (Burrows and House, 1989; Green and Turnstall, 1992). It is clear that there is a great demand for attractive river corridors which are accessible and offer the potential for recreation. Objectives for the improvement of water quality have been set up by the Agency. Flood defence work is being tackled with an increasing awareness of both impacts of the environment and the potential to undertake environmental improvements. It is against this background that the Agency was formed and continues to address the problems of degraded river corridors.

Agricultural pollution problems have been the subject of a great deal of work by the Agency. Much of this work was summarised in a 1992 report, *The influence of agriculture on the quality of natural waters in England and Wales* (National Rivers Authority, 1992). Recommendations in this report included the promotion of a structured approach to farm waste disposal, a review of existing discharge consents for farms, additional monitoring of the occurrence of pesticides in water, increasing farm visits to provide guidance, investigations into the use of riparian buffer zones to protect watercourses, and increased emphasis on liaison and education. Practical guidance for farmers and landowners has also been promoted, for example the *Code of Good Agricultural Practice for the Protection of Water* (Ministry of Agriculture, Fisheries and Food, 1991).

Boon (1992) summarises the case for a series of different management strategies for rivers along a scale of decreasing conservation value from pristine to almost totally degraded. For those rivers retaining much of their natural or semi-natural systems a strategy based on preservation is suggested. At the other end of the spectrum, the least natural, most modified rivers are put forward as requiring strategies ranging from mitigation to restoration depending on the potential success in addressing the causes of degradation. In the most extreme cases, Boon (1992) suggests that a strategy based on dereliction is the only option. In the Ridings Area there has been a tendency in the past for strategies of dereliction to be adopted for those rivers with the most severe water quality and habitat damage problems. Restoration strategies have generally been targeted towards those rivers where ecological damage has been very much less severe. Rapidly improving water quality in some of the worst rivers has seen a move towards adopting restoration strategies much more widely. This has been brought about to a large extent by the realisation that improving water quality alone, without associated improvements in physical structure and habitat integrity, will not achieve the expected recovery of the biota and aesthetic components of the system. Perceptions are changing however, and R&D commissioned by the NRA (and subsequently by the Agency) examining techniques of potential use in aiding rehabilitation of degraded river environments will provide the impetus for rehabilitation and restoration strategies into the 21st century.

FROM CATCHMENT MANAGEMENT TO LEAPS

Between 1990 and the formation of the Agency in 1996, there was an increasing tendency to approach management of the water environment in an integrated manner via a system of catchment-based management plans. These plans integrated all functions of the then NRA, and set out strategic views on the overall management of the water environment for catchments. Wide-scale consultation and the integration of the objectives of other organisations were key features of this process. Since 1996, these principles have been taken forward in the form of integrated Local Environment Agency Plans (LEAPs).

Rehabilitation of degraded rivers is a central feature of the LEAPs being produced in the Ridings Area. These LEAPs set the agenda for rehabilitation in the next four to five years and can be categorised as having the following priorities for the water environment:

- a need for continued water quality improvement
- opportunities for habitat improvement in the most degraded rivers
- sensitivity in water management
- recognition of the amenity and recreation potential of rivers
- the need for development control to protect the water environment
- removal of physical obstructions to fish passage
- improved management of water resources in general

CONCLUSIONS

There has been a history of damage and degradation to the river environment in the North East Region extending over several hundred years, with extensive problems resulting from agricultural intensification and the development of major centres of population and industry. Recent years have seen an integrated approach to these problems adopted initially by the NRA, and taken up by the Environment Agency as they face the challenge of enhancing the river environment.

Practical works have been undertaken at a number of sites, principally to address the historical fisheries problems that have affected Yorkshire rivers over the past two centuries. Other work has concentrated on providing suitable habitats and features for other targets species such as otters. In the future, work will be carried out in a more strategic manner with greater emphasis on the fitting of individual environmental improvement schemes into a series of overall objectives. The fundamental aim of river catchment management is to conserve, enhance and where appropriate restore or rehabilitate the total river environment through effective land and resource planning across the area. In the North East Region there will be an increasing move away from the piecemeal, *ad hoc* and reactive style of work towards a more strategic and proactive approach. This will reflect the move towards integrated environmental management which provide a means for identifying, facilitating and keeping under review potential changes to river corridors in the Region. As Gardiner (1992) has similarly concluded for the Thames Region of the Agency, the drive towards conservation and restoration of river corridors in the North East Region will depend on the success of incorporating Agency objectives into statutory land-use planning and development control. A programme of LEAPs with their associated action plans will provide the main vehicle for accomplishing these strategic objectives.

ACKNOWLEDGEMENTS

The views expressed in this chapter are those of the authors and not necessarily shared by the Environment Agency.

REFERENCES

Beerling, D.J. and Perrins, J.M. (1993) Biological flora of the British Isles: *Impatiens glandulifera* Royle (*Impatiens roylei* Walp). *Journal of Ecology*, **81**, 367–382.

Beerling, D.J., Bailey, J.P. and Conolly, A.P. (1994). Biological flora of the British Isles: *Fallopia japonica* (Houtt.) Ronse Decraene (*Reynoutria japonica* Houtt.; *Polygonum cuspidatum* Seib. & Zucc.). *Journal of Ecology*, **82**, 959–980.

Boon, P.J. (1992) Essential elements in the case for river conservation. In Boon, P.J., Calow, P. and Petts, G.E. (eds) *River Conservation and Management*. John Wiley, Chichester, pp. 10–33.

Brookes, A. (1988) *Channelized Rivers: Perspectives for Environmental Management*. John Wiley, Chichester.

Brookes, A. (1992) Recovery and restoration of some engineered British river channels. In Boon, P.J., Calow, P. and Petts, G.E. (eds) *River Conservation and Management*. John Wiley, Chichester, pp. 337–352.

Brookes, A. and Douglas Shields Jr, F. (eds) (1996) *River Channel Restoration: Guiding Principles for Sustainable Projects*. John Wiley, Chichester.

Brookes, A., Gregory, K.J. and Dawson, F.H. (1983) An assessment of river channelization in England and Wales. *The Science of the Total Environment*, **27**, 97–111.

Burrows, A. and House, M.A. (1989) Public perception of water quality and the use of rivers for recreation. In Liakari, H. (ed.) *River Basin Management*, Volume 5. Pergamon Press, Oxford, pp. 371–379.

Caffery, J.M. (1994) Spread and management of *Heracleum mantegazzianum* (Giant hogweed) along Irish river corridors. In de Waal, L.C., Child, L.E., Wade, P.M. and Brock, J.H. (eds) *Ecology and Management of Invasive Riverside Plants*. John Wiley, Chichester, pp. 67–76.

Countryside Commission (1987) *Changing River Landscapes*. Countryside Commission, Cheltenham.

De Waal, L.C., Child, L.E., Wade, P.M. and Brock, J.H. (eds) (1994) *Ecology and Management of Invasive Riverside Plants.* John Wiley, Chichester.

Environment Act 1996 (1996) HMSO, London.

Fisher, A. and Raucher, R. (1984) Intrinsic benefits of improved water quality: conceptual and empirical perspectives. *Advances in Applied Micro-Economics,* **3**, 37–66.

Gardiner, J.L. (ed.) (1991) *River Projects and Conservation: A Manual for Holistic Appraisal.* John Wiley, Chichester.

Gardiner, J.L. (1992) Catchment planning: the way forward for river protection in the UK. In Boon, P.J., Calow, P. and Petts, G.E. (eds) *River Conservation and Management.* John Wiley, Chichester, pp. 397–406.

Gorman, O.T. and Karr, J.R. (1978) Habitat structure and stream fish communities. *Ecology,* **59**, 507–515.

Green, C.H. and Tunstall, S.M. (1992) The amenity and environmental value of river corridors in Britain. In Boon, P.J., Calow, P. and Petts, G.E. (eds) *River Conservation and Management.* John Wiley, Chichester, pp. 425–442.

Harper, D.M. and Ferguson, A.J.D. (eds) (1995) *The Ecological Basis for River Management.* John Wiley, Chichester.

Hellawell, J.M. (1988) River regulation and nature conservation. *Regulated Rivers: Research & Management,* **2**, 425–443.

Large, A.R.G. and Petts, G.E. (1992) *Buffer Zones for Conservation of Rivers and Bankside Habitats.* NRA R&D Project Record 340/5/Y, Department of Geography, Loughborough University of Technology.

Mann, R.H.K. (1988) Fish and fisheries of regulated rivers in the UK. *Regulated Rivers: Research and Management,* **2**, 411–424.

Ministry of Agriculture, Fisheries and Food (1991) *Code of Good Agricultural Practice for the Protection of Water.* MAFF Welsh Office, HMSO, London.

National Rivers Authority (1992) *The Influence of Agriculture on the Quality of Natural Waters in England and Wales.* National Rivers Authority.

National Rivers Authority (1994) *Annual Report and Accounts 1993/94.*

Nature Conservancy Council (1991) *Nature Conservation and Pollution from Farm Wastes.* Nature Conservancy Council, Northminster House, Peterborough.

Newson, M.D. (1992) River conservation and catchment management: a UK perspective. In Boon, P.J., Calow, P. and Petts, G.E. (eds) *River Conservation and Management.* John Wiley, Chichester, pp. 386–396.

Perrins, J., Fitter, A. and Williamson, M. (1990) What makes *Impatiens glandulifera* invasive? In Palmer, J. (ed.). *The Biology and Control of Invasive Plants.* British Ecological Society, University of Wales, Cardiff, pp. 8–33.

Petts, G.E. (1994a) Rivers: dynamic components of catchment ecosystems. In Calow, P. and Petts, G.E. (eds) *The Rivers Handbook,* Volume 2. Blackwell, Oxford, pp. 3–22.

Petts, G.E. (1994b). Large scale river regulation. In Roberts, C.N. (ed.) *The Changing Global Environment.* Blackwell, Oxford, pp. 262–284.

Pysek, P. and Prach, K. (1994) How important are rivers for supporting plant invasions? In de Waal, L.C., Child, L.E., Wade, P.M. and Brock, J.H. (eds) *Ecology and Management of Invasive Riverside Plants.* John Wiley, Chichester, pp. 19–26.

Rapport, D.J., Reiger, H.A. and Hutchinson, T.C. (1985) Ecosystem behaviour under stress. *American Naturalist,* **125**, 617–640.

Roblin, L. (1994) Alien invasive weeds – an example of the National Rivers Authority sponsored research. In de Waal, L.C., Child, L.E., Wade, P.M. and Brock, J.H. (eds) *Ecology and Management of Invasive Riverside Plants.* John Wiley, Chichester, pp. 189–193.

RSPB/NRA/RSNC (1994) *The New Rivers and Wildlife Handbook.* Royal Society for the Protection of Birds, National Rivers Authority and Royal Society for Nature Conservation.

Sampson, C. (1994) Cost and impact of control methods used against *Heracleum mantegazzianum* (Giant hogweed) and the case for instigating a biological control programme. In de Waal, L.C., Child, L.E., Wade, P.M. and Brock, J.H. (eds) *Ecology and Management of Invasive Riverside Plants.* John Wiley, Chichester, pp. 55–665.

Sheppard, J.A. (1958) *The Draining of the Hull Valley.* East Yorkshire Local History Society.

Swales, S. (1982) Environmental effects of river channel works used in land drainage improvements. *Journal of Environmental Management*, **14**, 103–126.
Turner, A. (1994) Whiter than a whitewash? *Water Bulletin*, **632**, 13–14.
Water Act 1989. HMSO, London.
Water Resources Act 1991. HMSO, London.
Wildlife and Countryside Act 1981. HMSO, London.
Williamson, J.A. and Forbes, J.C. (1982) Giant hogweed (*Heracleum mantegazzianum*): its spread and control with glyphosate in amenity areas. In *Weeds*. Proceedings of the British Crop Protection Conference, pp. 967–972.

11 Rehabilitation of the North American Great Lakes Watershed: Past and Future

FRANK BUTTERWORTH[1] and ROBBIN HOUGH[2]
[1]*Institute for River Research International, Rochester, MI, USA*
[2]*Oakland University, Rochester, MI, USA*

INTRODUCTION

The management and abuse of rivers in North America, and the attempts to rehabilitate these rivers, are not unlike practices elsewhere in the world. However, a unique rehabilitation effort, unprecedented in scale or approach, is being undertaken in the Great Lakes–St Lawrence River watershed. In this chapter the system and its impairments are briefly described, and the rehabilitation model in use is outlined and reviewed critically. The past emphasis on the chemistry of fresh water and a failure to appreciate the scale and complexity of the waters has badly directed remedial efforts which should focus on the biology and the emergent properties of this freshwater system under continued stressful economic use. A process for river basin monitoring is suggested, and three strategies for more effective and efficient rehabilitation programmes are offered.

Two other heavily managed river systems in North America that compete in scale with the Great Lakes system have significantly different goals. The Columbia River system has an area of $671\,353$ km^2, a flow rate of 5543 m^3 s^{-1} and a length of 1998 km. Management is intense where the salmon fishery and electrical power industry are the major considerations instead of pollution (Lee, 1989). The 1980 Northwest Power Act established the Northwestern Power Planning Council, comprising a wide array of stakeholders and interested parties which try to maintain a sustainable balance of the needs of the hydroelectric power users and the anadromous fish. This management activity is substantial and has escaped the large federal budget cuts because it is well funded by hydroelectric power users.

The Mississippi–Missouri–Ohio River system, the largest in North America with an average flow rate into the Gulf of Mexico of $17\,000$ m^3 s^{-1}, drains a watershed of 3.24×10^6 km^2. Nationally, basin commissions (Frisch and Chaney, 1981) composed of a broad array of stakeholders to regulate the management of the system have ceased to exist because of budgetary priorities. Since and because of the great flood of 1993, a loosely organised and unofficial confederation of agencies have converged to address issues such as wetlands, environmental restoration, and floodplain management at agricultural and social levels (Myers and White, 1993; Galloway, 1994). These organisations include the US Environmental Protection Agency and the US Army Corps of

---

Rehabilitation of Rivers: Principles and Implementation. Edited by L. C. de Waal, A. R. G. Large and P. M. Wade.
© 1998 John Wiley & Sons Ltd.

Engineers and non-governmental organisations such as American Rivers, the Nature Conservancy and the World Wildlife Fund.

However, truly broad-based river management programmes in North America, with the exception of the Great Lakes System, are presently underemphasised. Indeed the Great Lakes rehabilitation effort is reaching a critical point in its history. Both technological and organisational breakthroughs will be necessary for its survival. The lessons learned in the coming decade could serve as a model for other major rehabilitation efforts.

GREAT LAKES BASIN AS A RIVER SYSTEM

The five Great Lakes, Superior, Michigan, Huron, Erie and Ontario, and their tributaries are connected by rivers or connecting channels and all ultimately empty into the Atlantic Ocean by the St Lawrence River. The St Lawrence has an average flow rate of 7100 m^3 s^{-1} which represents the total drainage of approximately 1 295 000 km^2 from the Great Lakes watershed and an approximately equal area in the remainder of the St Lawrence Basin. Thus, in one way, the Great Lakes can be considered a 3059 km long river stretching from the St Louis River in Minnesota to the Gulf of St Lawrence where the Lakes are considered as biogeographical provinces (Dynesius and Nilsson, 1994). Of course the Great Lakes also represent a very large volume of water (over 22 000 km^3) having several important consequences. First, the Great Lakes comprise, by some estimates, 20% of the world's fresh water which is retained for significant periods of time. Lake Superior, the largest (nearly 54% of the total volume), has a retention time of 191 years. Lake Michigan, the second largest, will take 99 years to replace its contents (see Table 11.1; Quinn, 1992). Secondly, whereas spring floods may flush ordinary rivers clean, lake turnover in the Great Lakes Basin can return sediment contaminants to the surface waters, becoming more bio-available again and again, and the lake waters are changed only over time periods best measured in decades.

The shores of and tributaries to the Great Lakes and St Lawrence River are heavily industrialised, and with the development of the lock system of the St Lawrence Seaway, the waters form one of the world's major seaports. The entire watershed supports seven major urban centres and their many satellite cities. Milwaukee, Chicago, Detroit, Cleveland, Buffalo, Toronto and Montreal are the regional trade centres for a population of some 40 million. Living, as represented by participation in these manufacturing-based economies, requires the creation of products which use vast quantities of fresh water. The form and pressures of these demands have been described in detail elsewhere (Hough,

Table 11.1 Retention times and volumes of the Great Lakes (Quinn, 1992)

	Retention time (years)	Volume (km^3)
Superior	191	12 200
Michigan	99	4920
Huron	22	3540
Erie	2.6	480
Ontario	6	1640
Total		22 780

1992). Combined industrial and household consumption requires an estimated 6.8 m^3 of fresh water per person per day. About 30% of the total use is by agriculture and about 12% is accounted for by public waste water treatment facilities. Five industries with their own water treatment facilities – electrical generation, metals, petroleum, chemicals and paper – account for much of the remaining 58% of water use (Statistical Abstract, 1987). In summary, the total annual requirement of fresh water from the Great Lakes for human activity at that time was about 100 km^3.

If the total annual rainfall over the basin were 1295 km^3, and one assumes that 50% of the rainfall evaporates, transpires or is absorbed into the groundwaters, the remaining 648 km^3 will flow on through the basin. Clearly the annual requirements for human use have reached a level that is nearly 15% of the annual rainfall.

In many places in the basin, the annual rainfall in the local catchment basin is not nearly sufficient to support the concentration of these activities. Table 11.2 shows the relative usage and flow computed from rainfall (rainflow) in five of the basin cities. The difference between water use and rainflow can be made up by extending the reach of the collector system or mining the fresh waters of the lakes. In the short term, the mining approach is far less expensive. However, two historical notes emphasise the importance of this mining activity to the economic fortunes of the Great Lakes Basin. The population of the City of Chicago stood virtually still for a number of years (Hough, 1972). Only when the flow of the Chicago River was reversed to send the effluents of the city to the Mississippi Basin could growth resume by mining fresh water from Lake Michigan. At present, Chicago's freshwater consumption is more than ten times the annual rainfall. In a similar vein, Oakland County just north of Detroit in south-eastern Michigan is now one of the fastest growing and wealthiest, large-population counties in the US. Its growth was stopped short from 1957 until a large connection to the waters of Lake Huron and to the Detroit sewer system could be completed. As in the case of Chicago, the catchments of the county simply could not support added growth without the water mining activity. In this case, there is no easy release to another basin so the effluents of all of south-eastern Michigan are returned to the Detroit River from which they flow to Lake Erie. But the impact of these mining activities is not understood well enough to enable the kind of long-range forecasting which should be part of the restoration effort. Compare Postei et al. (1996) for a similar analysis on a global scale.

In summary, a volume of water which rivals the annual rainfall is being extracted and returned to the lakes. To the extent that these returned waters are polluted, the contaminants will impact the system for decades to come. As will be discussed in this chapter, there is a second impact of this extraction and return of waste water that must be

Table 11.2 Usage and rainflow in selected catchments*

City	Area (km^2)	Population	Rain (mm)	Usage	Rainflow	Deficit
Buffalo	112	328 123	920	26	3	23
Chicago	608	2 783 726	840	219	16	203
Cleveland	211	505 616	890	40	6	34
Detroit	363	1 027 974	790	81	9	72
Milwaukee	256	628 088	790	49	6	43

*Usage, rainflow and deficit are measured in cubic metres per second.

considered. If the water being returned to the lakes is made biologically barren by waste-water treatment processes, it may well have a significant impact on the ecosystem.

REMEDIAL ACTION IN THE GREAT LAKES

Recognising the economic importance of the massive amount of fresh water that lies in the Great Lakes watershed, the governments of Canada and the United States in 1909 established the Boundary Waters Treaty. In order to administer the problems associated with the protection of this critical resource the two countries created the International Joint Commission (IJC). One of their first challenges in 1912 was the problem of water-borne disease, but more recently, chemical pollution has attracted most of their attention. In 1972, after the Great Lakes Water Quality Agreement was signed, the IJC identified (Figure 11.1) a series of water quality problem areas known as Areas of Concern (AOCs). In 1987 the IJC developed the process of Remedial Action Plans (RAPs) which became an organised way to address pollution problems as well as to educate citizens in the AOCs about the biology, geology and human activities in these contaminated riverine communities. The plan was to have the RAP written and developed by a Public Advisory Council (PAC) composed of State or Provincial agencies together with local stakeholders, including a wide array of specialists and representatives of business, local government, industry, education, sportsmen, shippers and environmentalists.

The IJC instructed each PAC in an AOC to evaluate 14 impaired beneficial uses (Table 11.3). To address these impairments the RAP process involves three stages. Stage I describes the environmental problems, the beneficial uses that are impaired and the degree and geographical extent of the impairments, and the definition of causes and sources of pollution. Stage II outlines remedial and regulatory goals, current measures; alternative and additional measures; identifies who should remediate, and finally lists the impairments that cannot be restored, giving reasons why. Stage III begins when, in the eyes of the particular PAC and the IJC, restoration is complete and provides the process for evaluation of its effectiveness and the surveillance measures needed to track the effectiveness.

Members of the PAC, with the help of technical advisors, form Technical Work Groups (TWGs) to deal with such central issues as (1) identifying and reducing point-source pollution, (2) identifying and reducing non-point-source pollution, (3) eliminating combined sewer overflows (CSOs), (4) removing or remediating contaminated sediments, and (5) restoring and creating habitat. The explicit rationale is that if the sources of pollution in items (1) to (3) can be reduced or stopped, the sediments can be remediated and the habitat can be restored or created, and the impairments will be delisted. Each TWG will come up with prescribed actions to meet these goals. Finally the PAC will write and publish the RAP document which will spell out the stages, processes and milestones for successful remediation (for more information, see Hartig and Zarull, 1992).

In addition to the PAC/RAP process, grass roots activist groups are working towards local solutions to more global issues. These groups have been loosely organised and led by larger organisations such as Greenpeace, which for example has been spearheading a campaign for the ban on chlorine and to promote the concepts of 'zero discharge' into, as well as 'virtual elimination' from, the Great Lakes waters. Universities, colleges, government agencies, and specialised institutes and centres are playing an important role in river

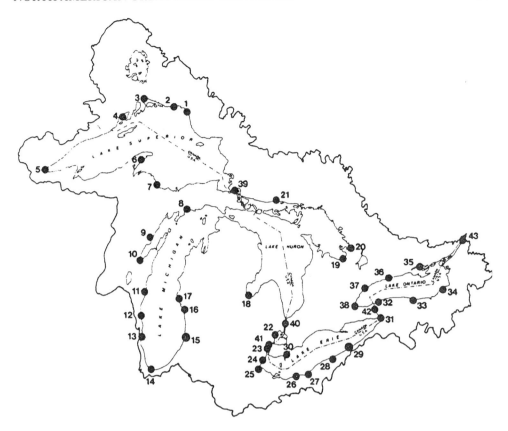

Figure 11.1 Great Lakes basin showing the Areas of Concern (AOCs). Lake Superior: 1. Peninsula Harbour, 2. Jackfish Bay, 3. Nipigon Bay, 4. Thunder Bay, 5. St Louis Bay/River, 6. Torch Lake, 7. Deer Lake-Carp Creek/River. Lake Michigan: 8. Manistique River, 9. Menominee River, 10. Fox River/Southern Green Bay, 11. Sheboygan River, 12. Milwaukee Estuary, 13. Waukegan Harbour, 14. Grand Calumet River/Indiana Harbour Canal, 15. Kalamazoo River, 16. Muskegan River, 17. White Lake. Lake Huron: 18. Saginaw River/Saginaw Bay, 19. Collingwood Harbour, 20. Severn Sound, 21. Spanish River Mouth. Lake Erie: 22. Clinton River, 23. Rouge River, 24. River Raisin, 25. Maumee River, 26. Black River, 27. Cuyahoga River, 28. Ashtabula River, 29. Presque Isle Bay, 30. Wheatley Harbour. Lake Ontario: 31. Buffalo River, 32. Eighteen Mile Creek, 33. Rochester Embayment, 34. Oswego River, 35. Bay of Quinte, 36. Port Hope, 37. Metro Toronto, 38. Hamilton Harbour. Connecting Channels: 39. St Marys River, 40. St Clair River, 41. Detroit River, 42. Niagara River, 43. St Lawrence River (Cornwall/Messena). Reproduced by permission of the International Joint Commission

management partly through specific programmes such as the US Environmental Protection Agency's Great Lakes Initiative (GLI), Assessment and Remediation of Contaminated Sediments (ARCS) and the Southeast Michigan Initiative (SEMI), or individual projects on hydrology, ecology, toxicology, biomonitor development, and remediation. These programmes and projects are often co-ordinated through national meetings such as the International Association of Great Lakes Research which meets annually at key locations in the basin.

Table 11.3 The 14 impairments to the Great Lakes Watershed as listed by the International Joint Commission

1. Restrictions on fish and wildlife consumption	When contaminant levels in fish or wildlife populations exceed current standards, objectives or guidelines or public health advisories are in effect for human consumption of fish or wildlife. Contaminants levels must be due to input from the watershed.
2. Tainting of fish and wildlife flavour	When ambient water quality standards, objectives or guidelines, for the anthropogenic substances known to cause tainting, are being exceeded or survey results have identified tainting of fish or wildlife flavour.
3. Degraded fish and wildlife populations	When management programmes have identified degraded fish or wildlife populations due to a cause within the watershed, or when bioassays confirm significant toxicity from water column or sediment contaminants.
4. Fish tumours or other deformities	When the incidence rates of fish tumours or other deformities exceed the rates at unimpacted control sites or when surveys confirm the presence of neoplastic or preneoplastic tumours in bullheads or suckers.
5. Bird or animal deformities or reproductive problems	When surveys confirm the presence of deformities or reproductive problems in sentinel wildlife.
6. Degradation of benthos	When the benthic macroinvertebrate community structure significantly diverges from unimpacted control sites or when sediment toxicity is significantly higher than controls.
7. Restrictions of dredging activities	When there are restrictions on dredging or disposal due to contaminant levels in the sediments.
8. Eutrophication or undesirable algae	When there are persistent water quality problems attributed to cultural eutrophication.
9. Restrictions on drinking water consumption or taste or odour problems	When treated drinking water: (1) exceeds standards, objectives or guidelines for disease organisms hazardous/toxic chemicals. or radioactive substances; (2) taste and odour problems are present; (3) the treatment required for raw water is beyond the standard treatment for the Great Lakes area.
10. Beach closings	When waters commonly used for full or partial body contact recreation exceed the standards, objectives or guidelines for such use.
11. Degradation of aesthetic	When any substance in water produces persistent objectionable deposit, colour, turbidity or odour.
12. Added cost to agriculture or industry	When additional treatment is required prior to use.
13. Degradation of plankton populations	When populations significantly differ from unimpacted control sites.
14. Loss of fish and wildlife habitat	When fish and wildlife management goals have not been met as a result of habitat loss due to perturbation of the physical, chemical or biological integrity.

CRITIQUE OF THE RAP PROCESS

Every two years the IJC holds a public forum to evaluate the progress of the 43 RAPs. At the biennial meeting in December 1993, at Windsor, Ontario (International Joint Commission, 1994), the IJC held a Remedial Action Plan forum, 'Sustaining the Momentum of the RAP'. Here the RAP progress and also the process were widely criticised. After eight years, RAP progress has been woefully lacking. Of the 43 AOCs, seven still have no Stage I; only two AOCs have completed Stage II, and these received poor evaluations from the IJC. Generally this lack of progress was thought to be due to the present PAC/RAP process. It was suggested that a clear timetable for clean up is needed: the identity of those who are responsible for clean-up action and their role must be announced; a clear money commitment from identified sources is required; 'enforcement of promises' language must be present in the RAP documents; a prevention plan with a money agreement is required; and an ecosystem approach including land, air and water media is needed (Jackson, 1993). Thus the goals set out in the three stages will never be adequately met unless the above needs are met. Recognising that attaining these goals is more difficult than was originally anticipated, various state agencies that help run the RAPs in the Great Lakes watershed are developing plans to make the RAPs achieve more realistic goals. For example, the Department of Natural Resources in the State of Michigan, recognising the lack of progress, has shortened the process to two years where all three stages are addressed simultaneously. Made to be a 'snapshot' in time, progress in the new system will be assessed biennially when the next PAC takes over.

There is an even more basic problem with the RAP system: the issue of complexity. The Great Lakes–St Lawrence Basin is a complex system composed of a large number of agents interacting in a large number of ways. There are many variables, and it is not often clear which are the controlling ones and which are passively occurring ones; or what are the causes, and what are the effects. In order to begin to grasp this and other complex phenomena, a new scientific discipline has formed: the science of complexity, which Waldrop (1992) has defined metaphorically in the subtitle of his book as 'the emerging science at the edge of chaos and order'. Complex systems are spontaneous, self-organising, and adaptive, bringing 'order and chaos into a special kind of balance' (Waldrop, 1992). Abused river systems such as the Great Lakes–St Lawrence Basin will spontaneously self organise and adapt, seeking a balance. But will it be the balance we want?

The threads of interdisciplinary learning that are accumulating in a variety of fields place emphasis on the importance of the emergent properties of such complex systems (Langton, 1990). Waldrop (1992), in discussing P. W. Anderson's contribution to the idea of emergent properties, writes: 'At each level of complexity, entirely new properties appear. And at each stage, entirely new laws, concepts, and generalizations are necessary, requiring inspiration and creativity to just as great a degree as the previous one'. It follows then that in order to begin to understand the complexity of river rehabilitation, there has to be a consideration and understanding of the emergent properties of the stressed river system.

What are the emergent properties of a lake system that has been continuously 'mined' for 100 years? The 'tailings' of these mining operations, which rival rainfall in their scale, are waters that may be chemically 'clean' but must be biologically barren because of the present treatment processes which they have undergone. What do we know about the ability of these tailings to support life? For example, lake trout have disappeared from all but Lake Superior (which sustains the lowest levels of human activity). Commercial fishing is a shadow of its former levels. As is well known, aquatic life is based on a simple food web, but where in the web have the failures occurred? There are a number of possible starting points: algal reproduction, zooplankton, whitefish larvae, or the habitat and spawning grounds for the large fish, to name a few (Cota and Sullivan, 1990; Gerasimova and Lebedeva, 1990; Vanderploeg et al., 1992). Another better understood disturbance is the skewing of the normal ecological quality of the Laguna Madre in Corpus Christi, Texas, where in 1990 a brown tide enhanced by polluting nutrients shut down the fishery for nearly five years by disrupting normal planktonic processes and curtailing growth of aquatic vegetation (Buskey, 1995). A very recent example of a complex system balancing between order and chaos is the historic and explosive growth of aquatic plants in 1994 in Lake St Clair in the Huron–Erie Corridor, causing major losses in recreation revenues. The emergent properties of this nearly overwhelming plant growth may involve a combination of increased water clarity because of the recent zebra mussel infestation and massive loadings of unregulated salts (see: http://www.edict.com/monitor/) (Hough, 1998). It is believed that, because some dominant fish species are no longer reproducing and the zebra mussel population has stabilised, a new ecosystem will emerge in Lake St Clair.

These examples demonstrate the spontaneous adaptation to a balance between order and chaos and also demonstrate the need to understand the emergent properties of these systems so that water managers can strive toward the appropriate balance. This balance and the understanding of emergent properties of these complex systems requires a new approach to monitoring; one that transcends the current, chemically based approach to what is really a biological system.

For instance, the total pollution discharge in items (1) to (4) listed above (point- and non-point-source pollution, combined sewer overflows (CSOs), and contaminated sediments) carries forward the assumption that the problem is largely a chemistry problem. From this it follows that a zero discharge policy will result in the restoration of the Great Lakes. To date the most important visible impact in the history of the RAP process has been to reduce point-source pollution, simply by issuing state-regulated, discharge permits to municipal and industrial waste-water treatment facilities. As these pollution permits have become more and more stringent, the toxic output has understandably been reduced. However, the success of the process in dealing with the other three causes is far less clear. Furthermore, as noted earlier, the Great Lakes Basin is a system without the ability to clean or flush itself as other rivers can. More importantly, it is a massive, slowly changing biological system ultimately based on planktonic processes for which current water-mining activities make little allowance. Under the current RAP process, progress will be problematic.

CRITIQUE OF CURRENTLY EMPLOYED MONITORING

The current system of surveillance of the health of the Great Lakes waters is totally inadequate.

CRITIQUE OF CHEMICAL MONITORING SYSTEMS

The EPA follows 28 toxic bio-accumulative chemicals in the Great Lakes. However, organisms in nature do not confront single compounds, but complex mixtures instead. Thus the chemical and biological monitoring of water and sediments must test for complex mixtures of contaminants, which raises two more problems.

Firstly, it is extremely difficult to predict the toxicity of mixtures. Using the method of structure activity relationships one can predict the toxicity of untested but structurally related compounds. But models that predict the toxicity of complex mixtures of pollutants do not exist. For example, Fahrig (1987, 1992, 1994) has shown that various mixtures of several known toxic chemicals including dioxin will yield a variety of genetic effects (mutagenesis, recombinogenesis, carcinogenicity, antimutagenesis, etc.) depending on the types and concentrations of the compounds used, with no obvious causal relationship. Our own studies (Pandey et al., 1995; McGowan, 1996) indicate that mixtures of varying concentrations of dichlorobiphenyl, benzo[a]pyrene, and arsenic give results that cannot be predicted from the effects of the compounds tested singly. Indeed, recombinogenesis, the newly realised form of pollution-induced genotoxicity, may have far more repercussions on human and ecological health than mutation alone (Butterworth et al., 1999). Each mixture will have its own toxic profile, and the only effective way to evaluate these profiles is to use biomonitors that test for specific endpoints. Thus in addition to chemical discharge permits, point-source discharge should also be rated for toxicity by a suite of biomonitors.

Second, chemical analysis for environmental contaminants is limited by cost to a relatively low number of compounds. Because of the permitting process only those chemicals get attention. However, it is generally understood that there are a much larger number of anthropogenic compounds in the environment. The discovery of the toxicity of polybrominated biphenyls (PBBs) in the State of Michigan illustrates the point. At one time environmental PBBs were not tested for routinely, indeed were unknown, in environmental surveys until PBBs were accidentally mixed with dairy cattle feed. Thus with a varied and complex mixture of compounds, many of which are unknown, the only hope for an adequate warning system is the application of biomonitors.

EVALUATION OF BIOMONITORING SYSTEMS

Biomonitors have long been used to detect threats to the health of the environment and humans (NRC, 1991). The general principle of biomonitoring is that animal and plant test systems share the same environment as humans, and will, as a function of their smaller size and higher rate of growth, recognise an environmental threat more rapidly than a human, yet will have the same sort of endpoint and the same mechanism of action. Biomonitoring systems used in the Great Lakes Basin are in an early stage of development, are not part of the permitting process, and are poorly co-ordinated with other remedial activities in the watershed. Biomonitoring systems now in use in the Great Lakes (Table 11.4) fall into four general categories: bio-availability/bio-accumulation, human health advisory, toxicity, and field survey. Bio-availability/bio-accumulation, which seeks to determine whether pollutants are stored by organisms, has become less necessary, since the phenomenon can be predicted on the basis of chemical structure of

Table 11.4 Some currently used biomonitors in the Great Lakes broken down into four classes: BB (bioaccumulation/bioavailability), HHA (human health advisory), FS (field survey) and TOX (toxicity) (largely from the Upper Great Lakes Connecting Channel Study by the IJC (1988))

Animal/plant system tested	Biomonitoring system	References
Phytoplankton, bacteria and plants	FS, TOX	(Munawar et al., 1985; Schloesser et al., 1986; White et al., 1987; Edsall et al., 1988; Edwards et al., 1989; Giesy et al., 1988b; Manny et al. 1988)
Zooplankton	FS, TOX	(White et al., 1987; Giesy et al., 1988a; Manny et al., 1988)
Benthic macroinvertebrates	BB, FS, TOX	(Vaughn and Harlow, 1965; Kauss and Hamdy, 1985; Pugsley et al., 1985; Freitag, 1987; Giesy et al., 1987; White et al., 1987); Edsall et al., 1988; Kreis, 1988; Thornley and Hamdy, 1984; Rosiu et al., 1989)
Water fowl, raptors, mammals	FS, TOX	(Scharf et al., 1978; Blokpoel and Tessier, 1986; Gilbertson, 1988; McNicholl, 1989b)
Fish: carnivorous, benthic, planktivorous, caged, and tumours	BB, HHA, TOX, FS	(MacCubbin, 1987; Black et al., 1988; Manny et al., 1988; Kreis, 1988)

the pollutant and the medium it occupies. On the other hand, human health advisories are essential for anglers and consumers of fish, and give a current status of biomagnification in the environment. The primary purpose of toxicity monitoring is to evaluate complex mixtures of contaminants, but this use is rarely emphasised. Field surveys will continue to be valuable, giving the general ecological status of the communities sampled, but with several caveats. They are extremely expensive to perform and can take years before results are available.

Three different RAPs were selected to evaluate their Stage I use of biomonitors. The Green Bay and the Fox River RAP (WDNR, 1988) represents the bay north of Chicago on the western shore of Lake Michigan. Biomonitors were recommended for a significant aspect of this RAP in seven impairments. The Cuyahoga River (the river that caught fire) is the major river flowing through the Cleveland, Ohio metro area into the southern shore of Lake Erie. The Stage I RAP report (OEPA, 1992) has the benefit of being the most recent, yet there is little improvement over Green Bay's report. Indeed the use of biomonitors is not as extensive as in the Green Bay RAP and tends to use *in situ*, sentinel organisms exclusively. The Detroit River is a highly polluted connecting channel between Lake Huron and Lake St Clair on the north and Lake Erie on the south. This RAP (MDNR/OME, 1991) listed nine impaired beneficial uses and two environmental concerns, and although many biomonitoring systems were cited, taken largely from work (Table 11.4) sponsored by the International Joint Commission (1988), in the Stage I document, no biomonitors were recommended.

In conclusion these three RAPs were aware of biomonitoring, and suggested using known systems; but none was convinced that biomonitors were essential. Also concerns about chemical accuracy were not presented nor were questions raised about the efficacy of the current group of biomonitors. Not suprisingly, there is no evidence that any of the RAPs were aware of the water-mining problem. Clearly if the RAP process is to succeed, RAPs must be rewritten with the idea in mind that the systems are too complex for the original tools and thinking.

Many of these currently used monitors should now be reconsidered for replacement with newer generations of tests. For instance, carnivorous fish are used for bio-accumulation and bio-availability data as well as consumption advisories for humans, but serious problems abound. The general method is to extract the fish tissues and measure the toxic pollutants extracted using gas chromatography/mass spectroscopy or high performance liquid chromatography. But this approach is expensive with the strong potential for unreliability due to severe limitations in extraction methodology and inadequate sample size. Also, because the fish have relatively large ranges, geographical resolution will be poor. Thus in terms of industrial and agricultural spills and non-point-source pollution, not much can be learned. Benthic omnivorous fish are also popular biomonitors used to study bio-accumulation and bio-availability in sediments for human health advisories and the condition of the sediment. Here a limited range of the animals gives better geographical resolution for locating the pollutant source. On the plus side for both fish groups there is year-round availability giving opportunity to look at seasonal uptake, and the information is becoming part of a growing database. However, the chemical analysis problems are still there. Thus using fish to monitor pollution is not the correct system because of the huge variability encountered probably due to vagaries in sampling, extraction and metabolism.

The field survey for species diversity in various habitats gives a comprehensive, ecosystem view of the sampling area. Although valuable information is obtained, the test is costly, requiring highly specialised personnel and long time periods. Such a study has been done for the Detroit River sediments with three separate sampling periods over the past 22 years (Farara and Burt, 1993). Certain toxic chemicals (13 heavy metals, 16 PAHs, PCBs, and 5 DDT-type pesticides) and 176 animal taxa were investigated in 77 sites along the 30 km length of the river. The study showed that the quality of the river has deteriorated over time despite the efforts to reduce the point-source pollution rate. This type of study gives a fine historical view, but more rapid answers are needed to detect species diversity changes. An additional flaw is that populations of benthic macroinvertebrates do become pollution-tolerant (Klerks and Levinton, 1989; Luoma and Ho, 1993) suggesting the biological impairment may be much worse. Thus a faster and more reliable indicator is needed.

Certain biomonitoring forms such as plants and air monitors are completely absent from basin-wide studies. These omissions are particularly critical. Knowledge about the toxicity to organisms that form the fundamental trophic level is essential. Also as a significant amount of non-point-source pollution is airborne, developing and using biological surveillance systems to measure the source and evaluate the effect of polluted air is essential. The RAP process must recruit such systems that are already available (Butterworth *et al.*, 1995, 1999). The following state-of-the-art examples could be employed immediately as a multi-endpoint biomonitoring suite, capable of sampling all three media and involving both animal, plant and microbial test systems.

Chironomids, the benthic, aquatic insect larvae, are traditionally used in field surveys or to measure bio-accumulation of pollutants in sediments. Aside from the accuracy problem of chemical analysis, they are potentially excellent biomonitors because their small range gives good geographic resolution and a sizable database exists. In a recently developed toxicity assay, Dickman et al. (1992) and Hudson and Ciborowski (1995) use a morphological endpoint within the organism itself, i.e. structural abnormalities in the animals' mouth parts, which occur in response to the toxic chemicals in the sediment. These are easily identified, and can be validated chemically in the laboratory. An attractive feature is that pristine organisms can be reared in the laboratory and exposed to samples of toxic sediments or experimental matrices spiked with toxic materials in the laboratory independently of the weather.

The *Elodea/Ceratophyllum* test, an aquatic plant biomonitor, has been developed by Tracy et al. (1995). Both plants manifest developmental–growth abnormalities, called developmental instability, when stressed by pollutants. Rootless *Ceratophyllum* is capable of detecting low concentrations of pollutants in the water column, whereas *Elodea* is a rooted plant and can also monitor the sediments.

Both the above test systems can have multiple endpoints by using them also to detect pollution effects on DNA structure and on stress-protein production. The comet test developed by Singh et al. (1991) and Tice (1995) detects DNA breaks in cell nuclei in organisms exposed to pollutants. The test is in the process of being validated with various pollutants and pollutant mixtures (R. R. Tice, pers. comm.). It is inexpensive, easy to use, takes only a day to complete, and is quantifiable. Goering (1995) describes the use of stress proteins as a biomarker for pollution effects. Here extracts of the tissues were simply run on gels and stained for specific proteins that had been shown to be secreted very soon after exposure to pollutants. All higher organisms have stress genes that secrete proteins in response to environmental stress on the organism. The author indicates this test has been designed currently to test for heavy metal exposure, but the potential is tremendous to test other stressors and for molecular biology applications. For example, the stress genes can be linked to reporter genes in genetically engineered test organisms that would give a visible response obviating the above laboratory procedures.

To give the above systems more meaning and validity the suite could test the water and sediment in which the above systems are living. One example of such a biomonitor are specific bacteria that lose their ability to fluoresce when challenged with toxic substances. This system, called the Microtox® assay, is easy to use, has a simple, quantifiable endpoint, and has reasonably good validation (Jacobs et al., 1993; Galli et al., 1994), although as yet the mechanism of action is unclear. The company that markets Microtox® (Microbics Corp., Carlsbad, CA, USA) also sells a bacterial genotoxicity kit Mutatox® which is a quicker but more expensive alternative to the Ames assay (Maron and Ames, 1983). Either genotoxicity assay used in combination with Microtox® will give a rapid, comprehensive picture of the biological health of water and sediments.

Unfortunately the genetic information derived from microbes is relatively narrow, not telling us much about sensitivities of higher organisms to pollutants. One of the best monitors of environmentally induced genetic damage in higher organisms is the TRAD-MN test or *Tradescantia* micronucleus assay (Ma et al., 1992). The unique aspect of this whole-organism test is that it is one of the few assays that involve higher plants and is a good monitor of water, extracted sediment (or soil) and air pollution (Ruiz et al., 1992). Chromosomal breaks appear in the inflorescence cutting in response to pollutants. It is easy to use and requires only several days to complete. It is very inexpensive, requiring only a modest microscopy laboratory and a lightbox for winter flowering (but a greenhouse is better). It quantitatively measures chromosome breaks which are easily observable even by minimally trained personnel. Thus in a week or two a large amount of biological information can be obtained from a simple suite of tests.

STRATEGIES FOR THE FUTURE

THE EMERGENT PROPERTIES OF AQUATIC SYSTEMS

There are some 734 river basin segments on the planet which can be described as economically bounded and partially self-sufficient entities. Fresh, unpolluted water is critical to their existence. But many of these systems are improperly managed (Dynesius and Nilsson, 1994). For a long while, the challenge to the scientific community has been set forth as finding ways to purify the available water for human and industrial consumption. That challenge is critically flawed as it relates to the global system. The challenge fails to understand the emergent properties of the systems on which the structure of aquatic life is based. Science must turn to the development of a new understanding of those living systems as seen in the rivers and lakes.

There are three essential strategies which give promise of an effective and efficient rehabilitation model.

A comprehensive biomonitoring strategy

It is concluded that a comprehensive biomonitoring strategy is essential to attain the goals set out in the RAP process. Successful biomonitoring will give prompt feedback regarding the biological condition of the waters, the watershed and the air. All of the 14 impairment categories (Table 11.3) depend upon healthy water. Chemical analysis is important but it cannot identify the effect of complex mixtures of chemicals. Previously toxicological studies focus on the toxicity of single compounds, but organisms in the environmental setting are exposed to complex mixtures of substances that may or may not be harmful singly. However, the biomonitoring already in use must be questioned as to its feasibility, since present systems are costly, time-consuming and too often inaccurate.

Clearly a standard for biomonitoring must be established. Although all areas of concern (AOCs) are unique, the need for standardisation of organism, methodology and rationale will be essential because all RAPs will have basis for comparison, resource of problem solving and the possibility of constructing a large and useful database that will carry us through Stages II and III. Only with standardisation can the methods and systems be refined so that emerging technologies will reduce costs and increase efficiency of the process.

For these reasons, the long-term approach to biomonitoring is to select and test suites of biomonitoring systems that will measure the various endpoints and toxicity levels of any habitat quickly and cheaply in all AOCs. Here each AOC will be characterised or profiled both chemically and biologically in a readily accessible database. New profiles will be superimposed so that emerging trends can be ascertained, and new spills can be recorded and evaluated as the chemical and toxicological profiles change.

An emerging technology strategy

There are technologies emerging that will support such an ambitious approach to monitoring. Present chemical methods employed in the Great Lakes waters are not accurate enough because of sampling error: the levels of contaminants are often too low to measure directly. However, recently developed instrumentation is becoming available. For example, utilising specific absorptive surfaces and columns (smart surfaces) promote chemo-accumulation from very large volumes of water, thus solving the 'too dilute sample' problem (Butterworth et al., 1999). First using pentafilters to obtain the suspended PCBs and next XAD resins in columns to accumulate the dissolved PCBs, Swackhamer and Skogland (1991), avoiding the fish intermediate. 'Smart surface' chemistry also needs to be developed for lower cost, higher reliability chemical analysis of organisms and tissues.

Networks of automated, chemical monitoring stations supported by automatic data logging technology and coupled by cellular telephone technology are capable of continuous surveillance covering each AOC (Butterworth et al., 1999). Thus co-ordinated, basin-wide, chemical and biological baselines could be established. Pulse discharges and other 'events' could be detected by above-normal-triggering devices. A pilot station is now in place on the St Clair River delta (see: http://www.edict.com/monitor/).

Suites of various, newly developed biomonitors that exploit the biology of the systems must be employed to cover critical toxicity endpoints: immunological, physiological, developmental, biochemical, behavioural and genetic. The turn-around times would be significantly shorter, and a variety of life forms have to be used, including mammals, fish, invertebrates, insects and plants to test all media, land, water and air. Eventually real-time biomonitors must be developed so that chemical and biological data can be collected synchronously.

An organisational communications strategy

Parallel to the development of more effective and efficient monitoring strategies should be an effort to make the 43 AOCs real partners in the development effort.

As many of the AOC organisations have recognised, educational efforts must accompany technical efforts. People must have stories, rules-of-thumb, metaphors, analogies and roles to play. Only these can guide their thinking about complex systems. Although some progress has been made, the critical guides have just not been developed. Pretty posters do not advance understanding – riveting concepts do!

If coupled by the Internet or the developing 'Information Superhighway', cheap, accurate and timely measurements can be undertaken in all 43 AOCs simultaneously and the results communicated immediately to and from all AOCs. The Great Lakes Information Network is already on line and can be accessed at glin-talk@great-lakes.net. Such collaborative efforts can support a growth in understanding of riverine problems that will be unparalleled.

CONCLUSION

This chapter has provided a brief overview of river rehabilitation in the United States. In the context thus placed, the Great Lakes are considered as an important river system in which a large international rehabilitation activity is being carried out. The processes that have been developed, the technical working groups and the monitoring systems in use are described and evaluated. Overall, it is concluded that there are significant shortcomings in the manner in which the rehabilitation problem has been conceptualised. These shortcomings are largely due to an emphasis on the chemical properties of the system rather than the biological properties. A threefold strategy has been advanced as potentially contributing to the improvement of the rehabilitation effort.

The following areas of further research activities could make important contributions to river and lake rehabilitation. In the broadest view, a conception of biologically 'healthy' fresh water should be sought. Just as there are profiles of the probable dissolved chemicals in water, profiles of probable living systems must be developed. Secondly, efforts to exploit the biology of biomonitoring systems up to and including real-time monitoring of life forms should be pursued vigorously.

REFERENCES

Black, J.J., Dymerski, P.P. and Zapisek, W.F. (1988) Fish tumor pathology and aromatic hydrocarbon pollution in a Great Lakes estuary. In Afgan, B.K. and Mackay, D. (eds) *Hydrocarbons and Halogenated Hydrocarbons in the Environment*. pp. 559–565.

Blokpoel, H. and Tessier, G.D. (1986) The ring-billed gull in Ontario Canada: a review of a new problem species. Canadian Wildlife Service, Occasional Paper No. 57, Cat. No. CW69-1/57-1986E.

Buskey, E. (1995) Texas brown tide: a classic case of nutrient pulse added to a disturbed environment. *Estuarine Research Federation Newsletter*, **21**, 1–2.

Butterworth, F.M., Corkum, L.D. and Guzmán-Rincón, J. (ed.) (1995) *Biomonitors and Biomarkers as Indicators of Environmental Change*. Plenum, New York.

Butterworth, F.M., Pandey, P., McGowan, R., Ali-Sadat, S. and Walia, S.K. (1995) Genotoxicity of polychlorinated biphenyls (PCBs): recombinogenesis by transformation products. *Mut. Res.*, **342**, 61–69.

Butterworth, F.M., Gonsebatt, M.E. and Gunatilaka, A. (1999) *Biomonitors and Biomarkers as Indicators of Environmental Change*. **Vol. 2.** Plenum Press, New York.

Cota, G.F. and Sullivan, C.W. (1990) Photoadaptation, growth and production of Bottom Ice Algae in the Antarctic. *Journal of Phycol.*, **26**, 399–410.

Dickman, M., Brindle, I. and Benson, M. (1992) Evidence of teratogens in sediments of the Niagara River watershed as reflected by chironomid (Diptera: Chironomidae) deformities. *Journal of Great Lakes Research*, **18**, 467–480.

Dynesius, M. and Nilsson, C. (1994) Fragmentation and flow regulation of river systems in the northern third of the world. *Science*, **266**, 753–762.

Edsall, T.A., Manny, B.A. and Raphael, C.N. (1988) The St Clair River and Lake St Clair: an ecological profile. *US Fish and Wildlife Service Biological Report*, **85**, 1–187.

Edwards, C.J., Hudson, P.L., Duffy, W.G., Nepszy, S.J., McNabb, C.D., Haas, R.C., Liston, C.R. Manny, B.A. and Busch, W.N.J. (1989) Hydrological, morphometrical and biological characteristics of the connecting rivers of the international Great Lakes: a review. In Dodge, D.P. (ed.) *Proceedings of the International Large River Symposium*. Canadian Spec. Publ. Fish. Aquat. Sci. **106**, pp. 240–264.

Fahrig, R. (1987) Effects of bile acids on the mutagenicity and recombinogenicity of triethylene melamine in yeast strains MP1 and D61. *M. Arch. Toxicol.*, **60**, 192–197.

Fahrig, R. (1992) Tests for recombinogens in mammals in vivo. *Mut. Res.*, **184**, 177–183.

Fahrig, R. (1994) Genetic effects of dioxins in the spot test with mice. *Environmental Health Perspectives*, **101**, 257–261.

Farara, D.G. and Burt, A.J. (1993) Environmental Assessment of Detroit River sediments and benthic macroinvertebrate communities – 1991. *Report to the Ontario Ministry of Environment and Energy* by Beak Consultants Limited, Brampton, Ontario, Volumes I and II.

Freitag, T. (1987) Unionids of the Detroit River, unpublished document, US Army Corps of Engineers, Detroit District.

Frisch, J. and Chaney, T. (1981) Current status of the replacements for Title II River Basin Planning Commissions. In Allee, D.J., Dworsky, L.B. and North, R.M. (eds) *Unified River Basin Management – Stage II*. American Water Resources Association, Minneapolis, pp. 327–337.

Galli, R., Munz, C.D. and Scholtz, R. (1994) Evaluation and application of aquatic toxicity tests: use of Microtox® test for the prediction of toxicity based upon concentrations of contaminants in soil. *Hydrolobiologia*, **273**, 179–189.

Galloway, G. (ed.) (1994) Sharing the Challenge: Flood plain Management into the 21st Century. Government Printing Office Document, Washington, DC.

Gerasimova, T.N. and Lebedeva, L.I. (1990) The use of production indices of rotifers for evaluating water quality, Vodnye Resursy, January–February, pp. 155–162.

Giesy, J.P., Graney, R.L., Newsted, J.L. and Rosiu, C.J. (1987) Toxicity of in-place pollutants to benthic invertebrates. Interim Report to US EPA, LLRS, Grosse Isle.

Giesy, J.P., Graney, R.L., Newsted, J.L., Rosiu, C.J., Benda, A., Kreis, R.G., Horvath, F.J. (1988a) Comparison of three sediment bioassay methods using Detroit River sediments. *Environ. Toxicol. Chem.*, **7**, 483–498.

Giesy, J.P., Rosiu, R.J., Graney, R.L., Newsted, J.L., Benda, A., Kreis Jr., R.G. and Horvath, F.J. (1988b) Toxicity of Detroit River sediment interstitial water to the bacterium *Photobacterium phosphoreum*. Journal of Great Lakes Research, **14**, 502–513.

Gilbertson, M. (1988) Restoring the Great Lakes – Means and ends. *Canadian Field Naturalist*, **102**, 555–557.

Goering, P.L. (1995) Stress proteins: molecular biomarkers of chemical exposure and toxicity. In Butterworth, F.M. (ed.) *Biomonitors and Biomarkers as Indicators of Environmental Change*. Plenum, New York.

Hartig, J.H. and Zarull, M.A. (1992) *Under RAPs: Toward a Grass Roots Ecological Democracy in the Great Lakes Basin*. University of Michigan Press, Ann Arbor.

Hough, R.R. (1972) Toward a general systems model of growth and development. In Hanika, F.de P. (ed.) *Advances in Cybernetics and Systems Research*. Transcripta Books, London.

Hough, R.R. (1992) Time and space: an economic model. In Frank, A.U. *et al.* (eds) *Theories and Methods of Spatio-Temporal Reasoning in Geographic Space*. Springer-Verlag, New York, pp. 59–67.

Hough, R.R. (1998) An endangered living system: the reformation of the Lower Great Lakes. *Journal of Biological Systems*, **6**, 35–48.

Hudson, L.A. and Ciborowski, J.J.H. (1995) Larvae of Chironomid (Diptera) as indicators of sediment toxicity and genotoxicity. In Butterworth, F.M. (ed.) *Biomonitors and Biomarkers as Indicators of Environmental Change*. Plenum, New York, pp. 45–57.

International Joint Commission (IJC) (1988) *Upper Great Lakes Connecting Channels Study*, Volumes I and II, Environment Canada (Ottawa) and US Environmental Protection Agency, Washington, and IJC, Windsor, Ontario.

International Joint Commission (IJC) (1994) Seventh biennial report on Great Lakes water quality, IJC, Windsor, Ontario.

Jackson, J. (1993) A summary of remarks made at the Biennial Meeting of the IJC a remedial action plan forum, 'Sustaining the Momentum of the RAP', December 1993, Windsor, Ontario.

Jacobs, M.W., Coates, J.A., Delfino, J.J., Bitton, G., Davis, W.M. and Garcia, K.L. (1993) Comparison of sediment extract Microtox® toxicity with semi-volatile organic priority pollutant concentrations. *Arch. Environ. Contam. Toxicol.*, **24**, 461–468.

Kauss, P.B. and Hamdy, U.S. (1985) Biological monitoring of organochlorine contaminants in the

St Clair and Detroit Rivers using introduced clams, *Elliptio complanatus. Journal of Great Lakes Research*, **11**, 247–263.
Klerks, P.L. and Levinton, J.S. (1989) Effects of heavy metals in a polluted aquatic ecosystem. In Levin, S.A. *et al.* (eds) *Ecotoxicology: Problems and Approaches.* Springer-Verlag, New York.
Kreis, R.G. Jr. (ed.) (1988) Integrated study of exposure and biological effects of in-place sediment pollutants in the Detroit River, Michigan: an Upper Great Lakes Connecting Channel. US EPA, Office of Research and Development Report, ERL-Duluth, MN and LLRS-Grosse Isle, MI.
Langton, C.G. (1990) Computation at the edge of chaos: phase transitions and emergent computation, Physica D, nonlinear computation, 42/1, p. 12.
Lee, K.N. (1989) The Columbia River Basin: experimenting with sustainability. *Environment*, **31**, 6–33.
Luoma, S.N. and Ho, K.T. (1993) Appropriate uses of marine and estuarine sediment bioassays. In Calow, P. (ed.) *Handbook of Ecotoxicology*, Volume 1. Blackwell Scientific, Oxford, pp. 193–226.
Ma, Te-Hsiu, Jianhua Xu, Wenjie Xia, Zudon Jong, Weichi Sun and Guanheng Lin (1992) Proficiency of the Tradescantia-micronucleus image analysis system for scoring micronucleus frequencies and data analysis. *Mut. Res.*, **270**, 39–44.
MacCubbin, A.E. (1987) Biological effects of in-place pollutants: neoplasia in fish and related causal factors in the Detroit River system, interim report to the US EPA.
Manny, B.A., Edsall, T.A. and Jawarski, E. (1988) The Detroit River, Michigan: an ecological profile biological report. US Fish and Wildlife Service, US Department of Interior, Contrib. No. 683, Nat. Fisheries Res. Ctr.-Great Lakes, Ann Arbor, MI.
Maron, D.M. and Ames, B.N. (1983) Revised methods for the *Salmonella* mutagenicity test. *Mut. Res.*, **113**, 173–215.
McGowan, R.M. (1996) Genotoxicity of polychlorinated biphenyls, polycyclic aromatic hydrocarbons, and heavy metals in a eukaryotic system. Thesis. Oakland University.
McNicholl, M.K. (1989a) Aspects of herring gull breeding biology on Fighting Island, Ontario in 1988. Unpublished report to Canadian Wildlife Service.
McNicholl, M.K. (1989b) Bird observations on Fighting Island, Detroit River, Spring 1988. Unpublished manuscript.
MDNR/OME (1991) *Detroit River Remedial Action Plan – Stage 1.* Michigan Department of Natural Resources, Lansing and Ontario Ministry of the Environment, Sarnia.
Munawar, M., Thomas, R.L., Norwood, W. and Mudroch, A. (1985) Toxicity of Detroit River sediment-bound contaminants to ultraplankton. *Journal of Great Lakes Research*, **11**, 264–274.
Myers, M.F. and White, G.F. (1993) The Challenge of the Mississippi Flood. *Environment*, **35**, 6–35.
National Research Council (NRC) (1991) *Animals as Sentinels of Environmental Health Hazards.* Committee on Animals as Monitors of Environmental Hazards, National Academy Press, Washington, DC.
OEPA (1992) *Cuyahoga River Remedial Action Plan Stage One Report.* Ohio Environmental Protection Agency, Cleveland.
Pandey, P., McGowen, R.M., Vogel, E. and Butterworth, F.M. (1995) The SMART eye-spot assay: an indicator of genotoxicity of complex mixtures of PCBs and PAHs. In Butterworth, F. M. (ed.) *Biomonitors and Biomarkers as Indicators of Environmental Change.* Plenum, New York.
Postel, S.L., Daily, G.C. and Ehrlich, P.R. (1996) Human appropriation of renewable fresh water. *Science*, **271**, 75–788.
Pugsley, C.W., Hebert, P.D., Wood, G.W., Brotea, G. and Obal, T.W. (1985) Distribution of contaminants in clams and sediment from the Huron Erie corridor. I-PCBs and octachlorostyrene. *Journal of Great Lakes Research*, **11**, 275–289.
Quinn, F.H. (1992) Hydraulic residence times for the Laurentian Great Lakes. *Journal of Great Lakes Research*, **18**, 22–28.
Rosiu, D.J., Giesy, J.P. and Kreis, Jr., R.G. (1989) Toxicity of vertical sediments in the Trenton Channel, Detroit River, Michigan, to *Chironomus tentans* (Insecta: Chironomidae). *Journal of Great Lakes Research*, **15**(4), 570–580.
Ruiz, E.F., Rabago, V.M.E., Lecona, S.U., Perez, A.B. and Ma, T.-H. (1992) *Tradescantia*-micronucleus (TRAD-MN) bioassay on clastogenicity of wastewater and in situ monitoring. *Mut. Res.*, **270**, 45–51.

Scharf, W.C., Shugart, G.W. and Chamberlain, M.L. (1978) Colonial birds nesting on man-made and natural sites in the US Great Lakes. US Army Engineer Waterways Experiment Station, Vicksburg, Mississippi.

Schloesser, D.W., Hudson, P.L. and Nichols, S.J. (1986) Distribution and habitat of *Nitellopsis obtusa* (Characeae) in the Laurentian Great Lakes. *Hydrobiologia*, **133**, 91–96.

Singh, N.K., Tice, R.R. and Stephens, R.E. (1991) A microgel electrophoresis technique for the direct quantitation of DNA damage and repair in individual fibroblasts cultured on microscope slides. *Mut. Res.*, **252**, 289–293.

Statistical Abstract of the United States (1987) US Department of Commerce, Government Printing Office, Washington, DC.

Swackhamer, D.L. and Skogland, R.S. (1991) Bioaccumulation of PCBs by phytoplankton: kinetics versus equilibrium. *Environ. Toxicol. Chem.*, **12**, 831–838.

Thornley, S. and Hamdy, Y. (1984) *An Assessment of the Bottom Fauna and Sediments of the Detroit River*. Ontario Ministry of the Environment, Toronto.

Tice, R.A. (1995) Applications of the single cell gel assay to environmental biomonitoring for genotoxic pollutants. In Butterworth, F.M. (ed.) *Biomonitors and Biomarkers as Indicators of Environmental Change*. Plenum, New York.

Tracy, M., Freeman, D.C., Emlen, J.M., Graham, J.H. and Hough, R.A. (1995) Developmental instability as a biomonitor of environmental stress: an illustration using aquatic plants and macroalgae. In Butterworth, F.M. (ed.) *Biomonitors and Biomarkers as Indicators of Environmental Change*. Plenum, New York.

Vanderploeg, H.A., Bolsenga, S.J., Fahenstiel, G.L., Liebig, J.R. and Gardner, W.S. (1992) Plankton ecology in an ice-covered bay of Lake Michigan: utilization of a winter phytoplankton bloom by reproducing copepods. *Hydrobiologia*, **243**, 175–183.

Vaughn, R.D. and Harlow, G.L. (1965) Report on pollution of Detroit River, Michigan waters of Lake Erie and their tributaries. US Department of Health, Education and Welfare, Public Health Service, Division of Water Supply and Pollution Control.

Waldrop, M.M. (1992) *Complexity: The Emerging Science at the Edge of Order and Chaos*. Simon and Shuster, New York.

WDNR (1988) *Lower Green Bay Remedial Action Plan*. Publ-WR-175-87-REV88, Wisconsin Department of Natural Resources, Madison.

White, D.S., Bowers, J., Jude, D., Moll, R., Hendricks, S., Mansfield, P. and Flexner, M. (1987) Exposure and biological effects of in-place pollutants. Interim report to US Environmental Protection Agency, Large Lakes Research Station, Grosse Isle, MI.

12 Four Decades of Sustained Use, of Degradation and of Rehabilitation in Various Streams of Toronto, Canada

GORDON A. WICHERT and HENRY A. REGIER

Institute for Environmental Studies, University of Toronto, Ontario, Canada

DESCRIPTION OF THE TORONTO AREA WATERSHED

The Toronto Area Watershed (TAW), as defined here, is on the north shore of Lake Ontario and contains five relatively major river basins – the Credit, Don, Humber and Rouge Rivers and Duffins Creek – and four smaller watersheds – Mimico, Etobicoke, Highland and Petticoat Creeks (Figure 12.1). The TAW is bounded to the west and north approximately by the Niagara Escarpment and Oak Ridges Moraine respectively. The eastern edge of the Duffins Creek watershed forms the eastern boundary of the TAW. Soils of the escarpment and moraine are composed mostly of relatively coarse sands and gravels, and the rock of the Niagara Escarpment is highly fractured, thus much infiltration and upwelling of groundwater occurs in these two physiographic regions (Chapman and Putnam, 1984). The five major rivers of the TAW arise from the numerous groundwater seeps and springs which occur along the escarpment and moraine. Groundwater in the TAW flows at about 9–10 °C year round (Meisner, 1990), thus fish of the locales in which groundwater comprises a high proportion of summer baseflow are adapted to relatively cold and clear water during the summer months and relatively warm water during the winter months. The tempering influence of groundwater is reduced in middle and lower stream reaches; fishes of these zones are adapted to warmer, more turbid conditions during the summer.

The first European settlement in the TAW began in 1749 around Humber Bay (Jackes, 1948). The TAW was only lightly used by indigenous people at that time and had relatively pristine natural conditions, including heavily forested areas, marshes, cold and clear-flowing streams with numerous beaver ponds (Coventry, 1948) and rivers with brook trout (*Salvelinus fontinalis*) and Atlantic salmon (*Salmo salar*) (Huntsman, 1944; Parson, 1973). By the 1840s most of the Toronto area had been cleared by forest cutting for urban development and agriculture, with only a few undisturbed areas remaining, usually in the hilly headwaters and some wetland areas (ODPD, 1947, 1948, 1950, 1956a,b). At present, several stream reaches of the Upper Credit River and Duffins Creek still resemble their pristine states and some coastal marshes are not strongly modified (Stephenson, 1990a,b). Since the 1940s some agricultural and open land has been developed for human settlement. Most urbanisation has radiated northward from Lake Ontario

Rehabilitation of Rivers: Principles and Implementation. Edited by L. C. de Waal, A. R. G. Large and P. M. Wade.
© 1998 John Wiley & Sons Ltd.

but some urbanisation also occurred in several centres north of Metropolitan Toronto, notably in Georgetown, Orangeville, Richmond Hill and Markham (Figure 12.1).

DATA SOURCES OF PREVIOUS STUDIES

'Sampling stations' are particular stream locations where fish collections or temperature measurements were made. Several sampling stations over a stream reach of several kilometres long and with similar environmental conditions throughout, comprise a locale.

Most of the fish data came from two relatively comprehensive studies of the TAW. The Ontario Department of Planning and Development (ODPD) surveyed over 500 sampling stations in the TAW between 1946 and 1954. (H. A. Regier was on survey teams that collected about half of these data.) In both 1984 and 1985, Steedman (1987) sampled over 200 stations in the TAW. About 180 stations sampled by ODPD were in close proximity to about 190 stations sampled by Steedman and were thus suitable for comparative analyses. For time series analyses, additional data sets from the Upper Credit River and Duffins Creek were also used. These sets were collected by Reed (1968), Falls (unpublished data for 1966–1991), Johnson and Owen (1966), Ontario Department of Lands and Forest (1971), Ontario Ministry of Natural Resources (formerly called the Ontario Department of Lands and Forests) (unpublished data for 1973 and 1978), Wainio *et al.* (1975), Martin (1984), MTRCA (unpublished data for 1982–1989) and Wichert (unpublished data for 1989–1992).

During seasonally warm and dry periods of the summers of 1991–1993, water temperature data, here termed 'observed maximum stream habitat temperatures' (OMHT), were obtained at numerous stations in the Upper Credit, the east branch of the Humber and Duffins Creek. In order to collect stream temperatures when they were at their highest and when baseflow was at its lowest, streams were sampled after periods when no rain had occurred and air temperatures had exceeded 25°C for five or more previous days. In the summer of 1992 the air temperature did not go above 25°C on five consecutive days so streams were sampled in late summer during what was expected to be the warmest time of the summer. By subsequently examining 1992 weather reports it became apparent, after the fact, that OMHT data had been collected during the warmest time of that summer. Means of OMHT estimates were calculated for the three years and used as OMHT in subsequent analyses.

DESCRIPTION OF PREVIOUS STUDIES

Various studies have used fish as indicators (ODPD, 1947, 1948, 1950, 1956a,b; Hallam, 1959; Steedman, 1987, 1988; Wichert, 1994, 1995a,b) to relate changes in stream ecosystems in the Toronto area to urbanisation and some other stresses. Following the stress-response approach (see Loftus and Regier, 1972), changes in the fish associations of the TAW may be associated with some large-scale, ecosystem-level changes in water quality and water temperature and some other stresses acting separately or together. The dynamics of stream ecosystems represent part of the land–stream interface of the land–stream–bay–lake–river continuum elucidated by Regier and Kay (1996).

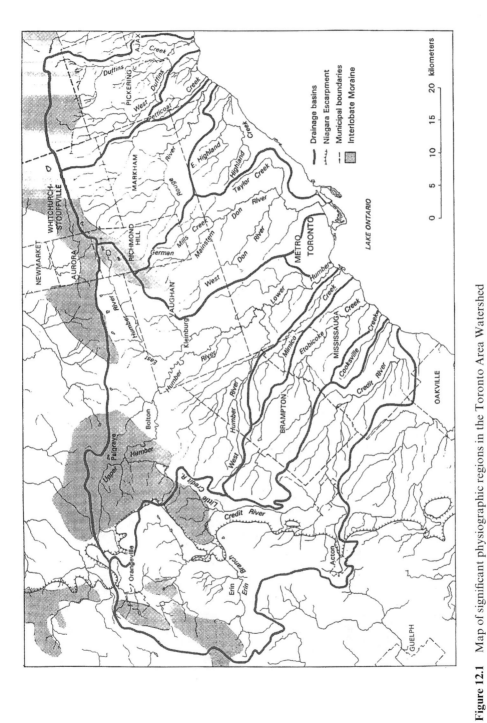

Figure 12.1 Map of significant physiographic regions in the Toronto Area Watershed

A series of interrelated studies was undertaken to use data collected in the past to infer changes in the ecological state of streams in the Toronto area and to test hypotheses about causes of those changes. Steedman (1987, 1988) collected comparable data in each of the summers of 1984 and 1985 for stations in all the streams of the TAW and analysed differences between locales within that time period. In this study we have added a temporal dimension in that we have used earlier and some subsequent data to focus also on differences over several decades within locales and within sets of locales as they were subjected to changing stress regimes. The motivation and methods of all the stream surveys were generally similar, but some differences exist. These differences have necessitated adjusting some qualitative data to make them more comparable for statistical purposes. The rather simple indicator species approach used in the 1940s and 1950s (see Hallam, 1959) did not appear to make full use of the available information. The Index of Biotic Integrity (IBI) (Karr et al., 1986; Steedman, 1987, 1988) – unless it is simplified strongly (Wichert, 1991) – required more information than was available from early surveys. Thus two new indices (Wichert, 1994, 1995b; Wichert and Lin, 1996) were used in this study. These indices make use of more of the information collected during the early surveys than did the early reports, and make it comparable with information collected during more recent surveys.

A Species Association Tolerance Index with respect to Water Quality (SATI-WQ) was created to assess the effects of remediation of sewage outfalls on the fish association below these point-sources of stress (Wichert, 1995b). (The various SATI and WSATI indices mentioned here are defined in the methods section below.) SATI-WQ scores were compared upstream and downstream and before and after major alterations to Water Pollution Control Plants (WPCPs). At all eight locales where comparisons of the fish association were made, land use below the outfall remained relatively constant, either settled or not settled, through time. The SATI-WQ scores increased at seven of eight locales which were downstream of altered WPCP outfalls implying that waste-water remediation allowed sensitive species to colonise and persist in downstream reaches. Urbanisation increased above altered outfalls at six of eight locales; SATI-WQ scores decreased at each of these six locales implying that urbanisation had a negative influence on the fish association.

Wichert (1994) used the Weighted Species Association Tolerance Index with respect to Water Quality (WSATI-WQ) to compare changes through time in ecological conditions at 11 selected smaller sub-watersheds in Toronto area streams. WSATI-WQ scores were generally largest at relatively undisturbed locales and became progressively smaller for locales which deviated from an undisturbed state. In some urban locales tolerant fish species have been collected during recent surveys, i.e. since the mid-1980s, where no fish was found in the past, thus modern urbanisation stresses appear to be less harmful to aquatic ecosystems than were urbanisation stresses of the past.

A WSATI-WT was developed to measure the 'behaviour' of an interactive set of species with respect to temperature (Wichert, 1995a; Wichert and Lin, 1996). The WSATI-WT was tested in the field at two locales, one on the Credit River and the other on the Humber River, at the confluence of a cold tributary with the warmer mainstream. Three relatively distinct temperature zones occur at each of these locales; a positive relationship was found between the WSATI-WT score for the fish association and the observed stream temperature in each temperature zone at both locales in each of two years.

The ODPD (1947, 1948, 1950, 1956a,b) classified streams into three categories referring to maximum summer water temperatures – cold, cool and warm. These categories were

Table 12.1 Description and measurement scales for environmental factors used for stratification purposes or as continuous variables in this study

Factor	Score	Definition
Physiography	1	Not near (> 1 km) the Oak Ridges Moraine or Niagara Escarpment
	2	On or near (< 1 km) the Oak Ridges Moraine or near the edge of the Niagara Escarpment
Riparian vegetation (%)		Proportion of stream length with vegetation along both banks as indicated by 1:50 000 scale topographical maps published around 1950* and in 1984
Land use	1	Urban: the stream reach below the next upstream sampling station was 'built up'
	2	Suburban: the stream reach below the next upstream sampling station was partly 'built up' and no urban reach occurred within 5 km upstream
	3	Rural: the stream reach below the next upstream sampling station consisted of one or more low intensity uses such as extensive agricultural, open areas, small woodlots, scattered homes, small settlement, areas of aggregate extraction, and no urban or suburban reach occurred within 5 km upstream
	4	Undisturbed: $\geq 85\%$ of the stream length above the sampling station was wooded and thought to be unchanged in that respect since pre-European contact
ODPD stream temperatures	1	Cold: suitable temperature for trout species throughout summer months (13–18 °C)
	2	Cool: suitable temperature for trout species in spring and fall but not summer (18–24 °C)
	3	Warm: suitable temperature for centrarchid species (> 24 °C)
	4	Intermittent to completely dry during summer months
Observed maximum habitat temperatures (°C)		Direct stream temperatures collected during seasonally hot and dry periods of the summers of 1991–1993

*Topographical maps for the TAW were published over several years, i.e. 1948–1952; topographical maps for the entire TAW were published in 1984.

inferred from data on the presence and absence of indicator fish and some invertebrate species which were collected from TAW streams from 1946–1954. The relationship between the SATI-WT scores based on the ODPD fish data and ST (the ODPD stream temperatures defined in Table 12.1) from the whole TAW are summarised by the following regression relationship:

SATI-WT = 17.1 + 1.55 ST, $\quad r = 0.48, \quad p < 0.01, \quad n = 188$

WSATI-WT scores calculated using relative abundance type of data collected in 1984–1985 were positively correlated with observed maximum habitat temperature, OMHT, data collected during warm periods in the summers of 1991–1993 (Wichert and Lin, 1996). Thus the WSATI-WT appears to be useful in measuring integrated temperature effects on a fish association.

In the present study larger ecosystems or basins within the entire TAW were used as the units of analysis; thus the indices developed in Wichert (1994, 1995a,b) were applied at larger spatial scales.

- Relationships were investigated between fish associations and some environmental factors affected by human activities for three sub-watersheds in each of two time periods some three to four decades apart.
- Changes through time at three relatively site-specific locations – headwater (data collected from 1970 to 1979), a middle reach (data collected from 1966 to 1991) and near the mouth (data collected from 1966 to 1984) – on Duffins Creek were assessed.
- Changes through time in the fish association with respect to water quality and water temperature in two sub-watersheds were assessed, using data collected periodically in addition to the relatively complete data sets for the periods 1946–1954 and 1984–1985.
- Comparable sampling stations across the whole TAW were stratified by stream temperature and land use. Relationships between the fish association and these environmental and human cultural factors were assessed in two time periods: 1946–1954 and 1984–1985.

METHODS OF ASSESSING CHARACTERISTICS OF A FISH ASSOCIATION

Various approaches have been taken to assess fish associations with respect to water quality and water temperature. The Ontario Department of Planning and Development (ODPD, 1947–1956; Hallam, 1959) used sensitive fish and invertebrates as indicators of habitat with a high water quality and to classify stream reaches as 'cold', 'cool' or 'warm' fish habitat in summer. 'Cold' meant able to support trout species during the summer months; 'cool' meant suitable for trout species during spring and fall but not the summer; and 'warm' meant suitable for centrarchid species (ODPD, 1947–1956). These and other environmental factors are explained in Table 12.1. (See Regier *et al.* (1996) for a more complete review of the interrelationships between fish ecology and their habitat temperatures.)

The Index of Biotic Integrity (IBI) was developed by Karr *et al.* (1986) initially for use in streams of the US Mid-West. Steedman (1987, 1988) modified the IBI for use in the streams of southern Ontario. Wichert (1991) simplified Steedman's version of the IBI to compare the fish associations of the TAW in 1946–1954 with those in 1984–1985, as a feasibility study for the present work. The IBI was designed to assess the fish association primarily with respect to water quality, not water temperature and not habitat structure.

For reasons stated earlier, a new set of indices was created, as described below.

$$\text{SATI-WQ} = \frac{1}{N}\sum_{i=1}^{N} \text{STS}_i$$

where STS_i is the species tolerance score as a measure of the relative intolerance of species i to water quality degradation typical of the relatively non-industrialised Toronto Area Waters; and N is the number of species found. Note that all of the survey data available for purposes of this index specified the presence or absence of different species but not all survey data specified the relative abundance of species.

The SATI-WQ measures the degree of intolerance of the fish association to reductions in water quality with respect to the sensitivity of each species to four stresses: low dissolved oxygen, high chlorine, high turbidity and destruction of physical habitat. The STS for each species is the arithmetic mean of the four numbers. Fish species of the TAW are ranked with respect to these criteria in Table 12.2.

Table 12.2 Tolerance of different species to habitat criteria and the species tolerance score (STS) and final temperature preferendum (FTP)

Species	Chlorine	Dissolved oxygen	Turbidity	Habitat destruction	STS	FTP
Brook trout	3	3	3	3	12	15.3
Brown trout	3	3	3	3	12	17.5
Rainbow trout	3	3	3	2	11	19.3
Rainbow darter	3	3	3	2	11	19.8
Northern pike	2	3	3	3	11	19.5
Rosyface shiner	3	3	3	2	11	26.8
Smallmouth bass	2	3	3	3	11	28
Slimy sculpin	2	3	2	3	10	10.7
Mottled sculpin	2	3	2	3	10	16.5
Iowa darter	3	2	3	2	10	nda
Redside dace	2	3	3	2	10	nda
River chub	2	3	2	2	10	21.8[a]
Rock bass	3	2	3	2	10	26.6
American brook lamprey	2	3	2	2	9	nda
Largemouth bass	1	3	2	3	9	30
Pumpkinseed	1	2	3	3	9	30.3
Fantail darter	2	2	2	3	9	20.3
Blackside darter	2	2	2	3	9	nda
Logperch	2	2	2	3	9	nda
Yellow perch	2	2	2	3	9	21.6
Golden shiner	2	2	3	2	9	23.7
Stonecat	2	3	2	2	9	25.1
Common shiner	2	3	2	2	9	22.1[a]
Emerald shiner	2	3	2	2	9	23.0
Trout-perch	2	2	2	2	8	16
Sand shiner	2	1	3	2	8	nda
Northern hogsucker	1	2	2	3	8	26.6
Horneyhead chub	2	2	2	2	8	nda
Northern redbelly dace	2	2	2	2	8	25.3
Pearl dace	2	2	2	2	8	nda
Finescale dace	2	2	2	2	8	23.9[a]
Blacknose shiner	2	2	2	2	8	nda
Spottail shiner	2	2	2	2	8	28.5
Alewife	2	2	2	1	7	16.0
Spotfin shiner	2	2	1	2	7	29.5
Central mudminnow	1	1	2	3	7	nda
Brook stickleback	2	1	2	2	7	21.4[a]
Johnny darter	2	2	1	2	7	22.9
Longnose dace	1	3	2	1	7	22.7[a]
White sucker	1	2	1	2	6	24.1
American eel	1	2	1	2	6	24.9[a]
Creek chub	2	2	2	1	6	21.0[a]
Fathead minnow	2	2	1	1	6	27.9
Bluntnose minnow	2	2	1	1	6	28.1
Blacknose dace	1	2	1	1	5	19.4[a]
Brown bullhead	2	1	1	1	5	26.6
Goldfish	1	1	1	1	4	28.1
Carp	1	1	1	1	4	31.3
No fish	0	0	0	0	0	

The numbers 1 to 3 reflect informed judgement based on available literature and were not inferred from distributional data used in the present study: 1 = highly tolerant and 3 = highly intolerant of the relevant condition; nda means no data available. Maximum possible STS score for each species is 12. Common names for fish species are consistent with those listed in Robins *et al.* (1991). FTP for each species is from Wichert (1995a).
[a] FTP estimated using regression relationships in Wichert (1995a).

The SATI-WQ was modified for data on the relative abundance of species at a locale to yield a Weighted Species Association Tolerance Index with respect to Water Quality (WSATI-WQ):

$$\text{WSATI-WQ} = \sum_{i=1}^{N} A_i \text{STS}_i$$

where STS_i is the STS for species i, as above;

$$A_i = n_i / \sum_{i=1}^{N} n_i$$

which represents relative abundance of each species as a percentage of total catch as estimated from data obtained by using relatively non-selective gear or a combination of gear to provide a relatively unbiased sample; n_i is the number of individuals of species i; and N is the number of species found.

Similar to the SATI-WQ, the SATI-WT was defined as:

$$\text{SATI-WT} = \frac{1}{N} \sum_{i=1}^{N} \text{FTP}_i$$

where FTP_i is final temperature preferendum of each species i; and N is the number of species found. A Weighted Species Association Tolerance Index with respect to Water Temperature (WSATI-WT) was also developed (Wichert, 1995a,b):

$$\text{WSATI-WT} = \sum_{i=1}^{N} A_i \, \text{FTP}_i$$

where FTP_i is the FTP for species i, as above; and A_i, n_i and N are as with WSATI-WQ.

The final temperature preferendum, FTP, for each of the fish species is shown in Table 12.2; other critical temperatures for fish found in streams of the TAW and nearshore waters of Lake Ontario are reported in Wichert (1995a).

Comparisons using relative abundance data for both water quality and water temperature indices were expected to provide greater spatial–ecological resolution than presence/absence data alone. Most sets of fish data collected in the TAW before 1965 were of the presence/absence type so data were transformed to unweighted scores to use them with the weighted indices. Wichert (1995a) reported the following theoretical constraints in the water quality model: WSATI-WQ and SATI-WQ scores are undefined when $0 < \text{index score} < 4$; also when SATI-WQ scores are 0, 4 or 12, WSATI-WQ scores must be 0, 4 or 12 respectively because of the way the index is defined. Given these constraints and the biases introduced with various curve-fitting techniques applied in Wichert (1995a) a directly proportional relationship provides an appropriate comparison between WSATI-WQ and SATI-WQ scores, i.e.

WSATI-WQ = SATI-WQ

The water temperature model was subject to the following constraints: when SATI-WT scores were undefined (i.e. no fish were found at a sampling site), WSATI-WT scores must be undefined and when only one fish species was found at a sampling site then weighted and unweighted scores must be equal. For example, if only brook trout were found at a site both SATI-WT and WSATI-WT scores must be 15.3°C (the FTP for brook trout).

Given the theoretical constraints of the water temperature model, a directly proportional relationship appears the most defensible, i.e.

WSATI-WT = SATI-WT

Throughout this study the above relationships were used to obtain corrected WSATI-WQ and WSATI-WT scores from presence/absence data.

Wichert (1994) inferred that given similar environmental conditions, WSATI-WQ scores for a sampling station based on independent surveys as conducted here would be within 10% of each other two times out of three, and within 22% 19 times out of 20. The variance associated with the WSATI-WT using relative abundance data was estimated by comparing results using fish data collected by Steedman from the same sample stations in two consecutive years. From a comparison of the size of means and variances of stations in relatively homogeneous locales we found that variance was approximately proportional to the mean and that a logarithmic transformation would reduce this proportionality. The mean difference of Ln transformed WSATI-WT scores was 0.00473 with a standard deviation of that distribution of differences of 0.0764 (Figure 12.2). The observed distribution is approximately symmetrical, like the normal distribution, but the chi-square value is 67.6; a chi-square value ≥ 67.6 is expected less than 5 times in 1000. According to the chi-square test the distribution of the 1984 scores is slightly different

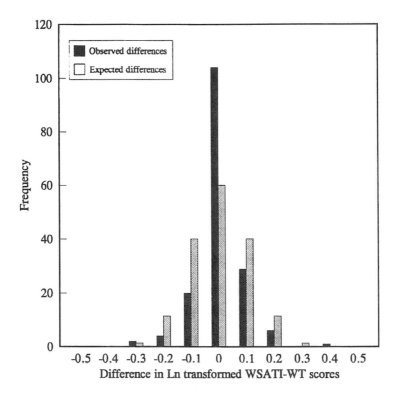

Figure 12.2 Comparison of the expected normal distribution and the observed difference in Ln transformed WSATI-WT scores from the same sampling stations in two consecutive years using data collected by Steedman (1984–1985); 1985 scores were subtracted from 1984 scores

from the 1985 distribution; by inspection it appears to have a sharper peak. Given similar environmental conditions in mid-summer, from the distribution of the actual differences in Figure 12.2 it is found that WSATI-WQ scores for a sampling station would be within 7% of each other two times out of three or within 14% 19 times out of 20.

The WSATI-WQ and WSATI-WT were based on different data on tolerances and preferences of fish species and were expected to be roughly orthogonal measures of water quality and water temperature characteristics of a fish association at relatively undisturbed sampling stations. But there appears to be a weak relationship between the temperature preferences and water quality tolerances for the set of fish species of the TAW (Figure 12.3). It is not clear how such an 'internal relationship' might contribute to a relationship between the indices themselves, but it was judged that the latter relationship would be of a similar type and with a small regression coefficient.

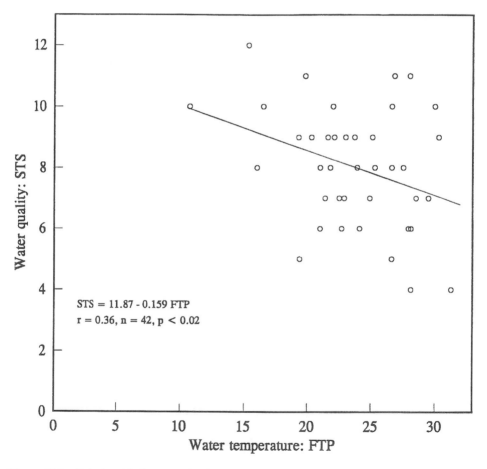

Figure 12.3 Relationship between final temperature preferendum (FTP) and species tolerance score (STS) for fish species of the TAW

ASSOCIATING CHANGES IN THE FISH ASSOCIATION WITH SOME CAUSES

SITES FROM THREE PARTS OF DUFFINS CREEK

Three sites from Duffins Creek – one in the headwaters, one in a middle reach and one near the mouth – were sampled nearly annually for periods of variable length, i.e. 8 to 20 years. Site-specific data were used to test year-to-year changes in the fish association with respect to water quality and water temperature and to test whether WSATI-WT, as a measure of the preferred temperature of a fish association, is influenced annually by maximum temperature in each summer or cumulatively by the maximum temperature over several preceding summers.

The data for this section of the study were collected by undergraduate students in a field component of a course in ecology taught by Professor J. Bruce Falls. Sampling dates were in autumn months while the dates in other surveys reported here were in summer.

In general, human effects on the stream in these three areas have been smaller than in other parts of the TAW. However, some estate development may have occurred in the vicinity of the headwater site. The middle reach site was in a multiple-use Conservation Area park. The site near the mouth was likely influenced by periodic intrusions of lake water due to surface seiches; upwelling occurs occasionally in summer along this coast. The headwater site was sampled between 1970 and 1979; the site at the mouth from 1966 to 1984; and the one in the middle reach from 1966 to 1991.

Maximum summer stream temperatures for the relevant sites were not available for the time during which the fish data were collected so air temperatures collected by Environment Canada from Toronto International Airport, about 50 km west of the sampling sites, were used as a surrogate. The average maximum air temperature was calculated for the six consecutive days with the highest daily maximum air temperature and compared with the WSATI-WT scores for the three sampling sites. It was expected that the long-term trends in air temperature and WSATI-WT would be consistent but it was not expected that the WSATI-WT scores would fluctuate annually as much as the year-to-year maximum air temperatures because of the relationships observed in Wichert and Lin (1996).

SUB-WATERSHEDS WITH A RELATIVELY LOW INTENSITY OF SUSTAINED USE

Changes through time in the fish association for the Upper Credit River and Duffins Creek basin with respect to water quality and water temperature were assessed for each of two TAW sub-watersheds that were sampled more frequently over the decades than were other sub-watersheds. On the whole, these two basins have been less influenced by cultural degradation over the past four or five decades than have other parts of the TAW. In each year that sampling occurred, all of the sampling stations in both sub-watersheds were classified by temperature ranges – 'cold' (< 18°C), 'cool' (18–24°C) and 'warm' (< 24°C) – using OMHT stream temperature data collected from 1991–1993. Mean index scores (both WSATI-WQ and WSATI-WT) were calculated for the set of stations in a temperature class in a given year. For example, in 1954, 10 stations in the upper Credit basin were assigned to a 'cold' (< 18 °C) class based on 1991–1993 stream temperature data; the mean WSATI-WQ score for these stations was 9.6 (see Figure 12.6).

By plotting mean WSATI-WQ and WSATI-WT scores for each temperature zone, trends were 'back-casted' over the preceding decades. This provides a way of using the incomplete data available to infer whether major changes in water quality and water temperature had occurred in those parts of the TAW.

WSATI-WT scores from data collected in 1954 and 1984–1985 were compared with physiography and the proportion of stream reach with riparian vegetation along both banks (PRV) in the basins of the upper Credit, east Humber and Duffins streams in order to understand the influence of these environmental factors on the observed preferred temperature of the fish association. Environmental factors and expected hypotheses are described in Tables 12.1 and 12.3. WSATI-WT scores were not compared to land use in these particular analyses since the land use at nearly all sample stations was 'rural' in 1946–1954 and has remained relatively unchanged in recent decades. A relationship between WSATI-WQ scores and physiography was not addressed with respect to the three sub-watersheds listed above because no ecological basis for such a relationship was apparent, and considering how the WSATI-WQ was defined; any apparent trend would be likely to be due to other unknown factors.

THE WHOLE TAW

For individual sampling stations across the whole TAW, the relationship between index scores and particular environmental factors was explored. The environmental factors were PRV (as defined above) and general land-use categories. About 180 stations sampled by ODPD between 1946 and 1954 were in close proximity to about 190 of the stations sampled by Steedman in 1984–1985. These sampling stations were classified using the criteria and scoring procedures described in Table 12.1. Relationships between physiography and index scores were not explored for the whole TAW because only 9.8% of the comparable sites were near the Escarpment or Oak Ridges Moraine and these have been addressed already in the analyses of the relatively undisturbed sub-watersheds.

Table 12.3 Expected relationships between scores for environmental factors (as defined in Table 12.1) and index scores

Index	Factor (as defined in Table 12.1)	Hypothesis
WSATI-WQ	Physiography	No test (see text)
	Land use	Positive relationship
	Riparian vegetation	Positive relationship
	Stream temperature	Not a strong relationship
WSATI-WT	Physiography	Low scores for sites on and near the moraine and edge of the escarpment
	Land use	Not a strong relationship
	Riparian vegetation	Negative relationship
	Stream temperature	Positive relationship

FINDINGS

SITE-SPECIFIC ANALYSES IN DUFFINS CREEK

Trends of the changes in the fish association through time with respect to water quality and water temperature at the three sites are shown in Figures 12.4 and 12.5; regression data are shown in Table 12.4. The slope of the summer six-day mean maximum air temperature was not significantly different from zero over the study period (slope coefficient was 0.08, $p = 0.13$).

The WSATI-WQ scores for the site in the middle reach and the site near the mouth showed no significant increase or decrease through time (Table 12.4) but the WSATI-WQ scores for the headwater site showed a significant decrease through time (Table 12.4). This decrease may be associated with estate development in the vicinity of the sampling site.

The WSATI-WT scores showed no significant increase or decrease at any of the three sites on Duffins Creek. The WSATI-WT scores for the headwater and middle reach site

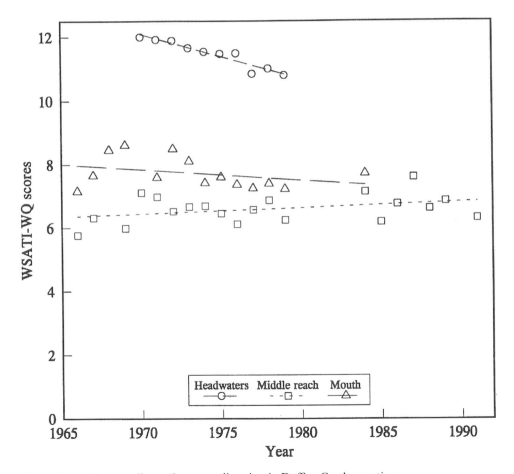

Figure 12.4 Water quality at three sampling sites in Duffins Creek over time

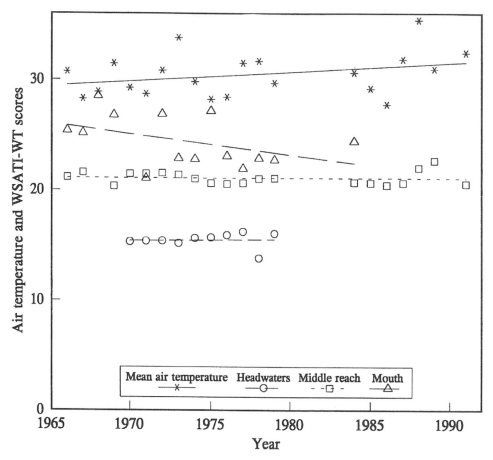

Figure 12.5 Water temperature index scores at three sampling sites in the Duffins Creek watershed over time as possibly related to six-day mean maximum temperatures

appear tightly clustered around the regression line rather than closely following the year-to-year changes in maximum summer air temperatures. This suggests that the preferred temperature of the fish association is influenced more strongly by long-term trends in maximum temperatures than by year-to-year maximum temperatures. The WSATI-WT scores for the headwater and middle reach site appear to fluctuate somewhat less than the scores for the site near the mouth. Deeper groundwater that enters the stream in headwaters and in some locales of the mid-reaches flows at temperatures of 9–10°C year round. The site at the mouth may be influenced by seiches in Lake Ontario (Kauffman *et al.*, 1992) which cause cold water from the lake to inundate Duffins Creek as far as 1 km upstream from the mouth.

SUB-WATERSHED ANALYSES

Changes in the fish association through time with respect to water quality and water temperature are shown in Figures 12.6–12.9. The mean WSATI-WQ scores within the

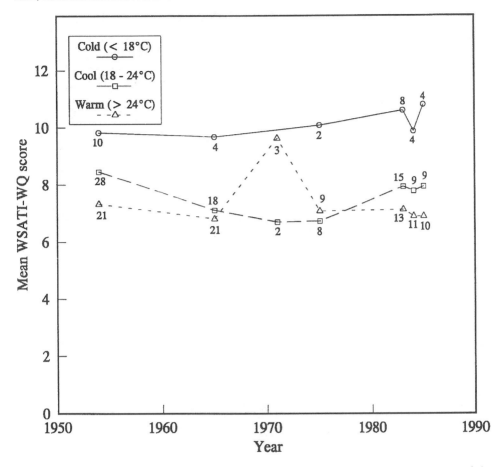

Figure 12.6 Mean WSATI-WQ scores for sampling stations in three temperature zones of the upper Credit River with temperature zones determined according to 1991–1993 observed maximum habitat temperature data. Except for data collected by Johnson and Owen in 1965 and ODLF in 1971, all data before 1982 were transformed from SATI-WT scores. Numbers in the figure refer to sample size

temperature zones in both sub-watersheds (Figures 12.6 and 12.7) showed no distinct trend or pattern through time, which is consistent with what was expected initially (Table 12.3). The mean WSATI-WQ scores for cold stream reaches in both sub-watersheds are greater than the mean scores for cool and warm reaches. This result was expected because most cold reaches are headwaters or near the Moraine and Niagara Escarpment which have not deviated from an undisturbed land-use state as much as have cool and warm sites. But the weak 'internal relationship' between these indices shown in Figure 12.3 may imply that differences between the temperature zones were overestimated in Figures 12.6 and 12.7.

The general pattern of mean WSATI-WT scores is consistent with what was expected: the smallest mean scores are found in cold zones and the largest mean scores are found in warm zones (Figures 12.8 and 12.9). Outlying points are generally based on small

Table 12.4 Results of regression of WSATI-WQ and WSATI-WT scores through time for three sampling sites – one in the headwaters, one in a middle reach and one near the mouth – on Duffins Creek

Stream reach	WSATI-WQ			WSATI-WT		
	Constant	Coefficient	p value	Constant	Coefficient	p value
Headwaters	287.6	− 0.140	< 0.001	− 9.67	0.013	0.87
Middle reach	− 31.03	0.019	0.14	23.92	− 0.001	0.94
Mouth	75.69	− 0.034	0.22	401.7	− 0.191	0.12

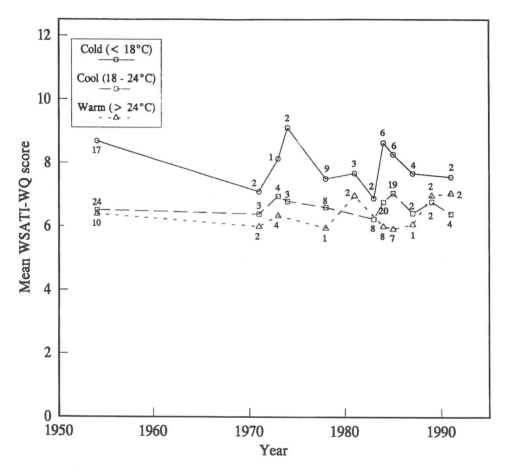

Figure 12.7 Mean WSATI-WQ scores for sampling stations in three temperature zones in Duffins Creek with temperature zones determined according to 1991–1993 observed maximum habitat temperature data. Data collected before 1974 were transformed from SATI-WQ scores. Numbers in the figure refer to sample size

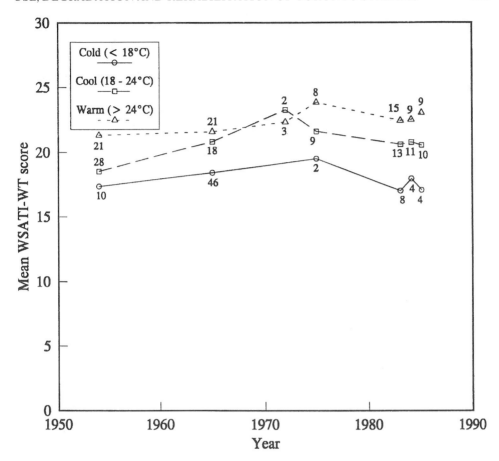

Figure 12.8 Mean WSATI-WT scores for sampling stations in three temperature zones of the upper Credit River with temperature zones determined according to 1991–1993 observed maximum habitat temperature data. Except for data collected by Johnson and Owen in 1965 and ODLF in 1971, all data before 1982 were transformed from SATI-WT scores. Numbers in the figure refer to sample size

numbers of stations. If a 'trend line' is drawn in each of the three temperature zones to connect data points associated with large sample sizes, the remaining data points associated with small sample sizes lie above and below the 'trend lines' in approximately equal numbers.

The regression data for relationships between environmental factors and WSATI-WT scores for combined data from three sub-watersheds which have not been strongly modified from an undisturbed condition – the upper Credit, east Humber and Duffins – are shown in Table 12.5. These relationships were calculated using data collected during two time periods about 30 years apart. For each comparison in both time periods, the sign of the slope coefficient was consistent with the relevant hypothesis shown in Table 12.3.

The effect of physiography on the WSATI-WT is to diminish this index by two to three units, presumably due to groundwater seepage in the Moraine and Escarpment locales.

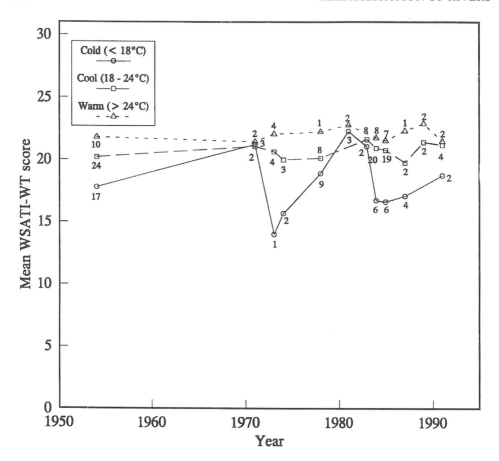

Figure 12.9 Mean WSATI-WT scores for sampling stations in three temperature zones in Duffins Creek with temperature zones determined according to 1991–1993 observed maximum habitat temperature data. Data collected before 1974 were transformed from SATI-WT scores. Numbers in the figure refer to sample size

The maximum effect of riparian vegetation, at 100% compared to 0%, is to diminish the index by one to three units, presumably due to shading of the water.

WHOLE TAW COMPARISONS

The relationships between index scores using fish data collected by ODPD (during 1946–1954) and Steedman (during 1984–1985) and two environmental factors defined in Table 12.1, land use and riparian vegetation, are shown in Table 12.6. The only relationship which differed from the initial expectations (listed in Table 12.3) was between WSATI-WT scores and land use, for which no strong relationship was expected.

The empirical relationship between WSATI-WQ and WSATI-WT scores (Figure 12.10) was steeper, according to our judgement, than might be expected based on the 'internal relationship' shown in Figure 12.3. Note that the correlation between WSATI-

Table 12.5 Results of multiple regressions between WSATI-WT scores and environmental variables for three sub-watersheds of the TAW: upper Credit, east branch of the Humber and Duffins Creek. Factors are as defined in Table 12.1

Factor (see Table 12.1)	Coefficient		p value	
	1946–1954	1984–1985	1946–1954	1984–1985
Constant	23.74	25.41	< 0.01	< 0.01
Physiography	− 2.85	− 2.24	< 0.01	< 0.01
Riparian vegetation	− 1.49	− 2.85	0.25	0.03
Overall r	0.52	0.53	< 0.01	< 0.01
Number	65	62		

Table 12.6 Results of multiple regressions between two kinds of index scores and environmental factors for the whole TAW: early data were collected in 1946–1954, recent data in 1984–1985

Factor (see Table 12.1)	Coefficient		p value	
	1946–1954	1984–1985	1946–1954	1984–1985
(a) WSATI-WQ				
Constant	2.09	4.70	0.01	< 0.01
Land use	1.52	0.705	< 0.01	< 0.01
Riparian vegetation	1.39	0.920	0.01	0.01
Overall r	0.52	0.50	< 0.01	< 0.01
Number	180	193		
(b) WSATI-WT				
Constant	27.61	25.46	< 0.01	< 0.01
Land use	− 2.16	− 0.659	< 0.01	< 0.01
Riparian vegetation	− 2.99	− 2.66	< 0.01	< 0.01
Overall r	0.57	0.48	< 0.01	< 0.01
Number	180	193		

WQ and WSATI-WT scores is similar using 1946–1954 and 1984–1985 data (Figure 12.10). The two index scores are here inferred to be negatively related to a significant degree within TAW in these two periods, and this correlation was not due solely to an internal bias.

SYNTHESIS

All comparable sampling stations in the TAW were stratified by land use around 1950 and 1984 and stream temperature zones as described in ODPD (1949–1956) studies. Mean WSATI-WQ and WSATI-WT scores for sampling stations in each of the categories are shown in Tables 12.7 and 12.8. As expected (Table 12.3), mean WSATI-WQ scores appear

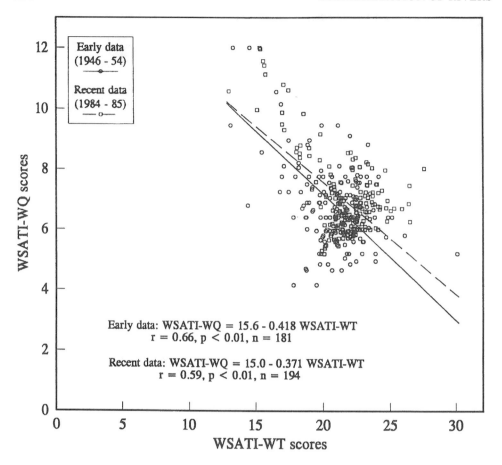

Figure 12.10 Comparison of WSATI-WQ and WSATI-WT scores for all comparable sampling stations in the TAW. Early fish data were collected by ODPD between 1946 and 1954; recent fish data were collected by Steedman in 1984–1985. Symbols that appear to be black are superimposed open symbols

to decrease as land use deviates from an undisturbed state (the horizontal dimension of Table 12.7) but no distinct pattern emerges moving from cold to warm or intermittent stream reaches within a given land-use type (the vertical dimension of Table 12.7).

Mean WSATI-WT scores for all comparable sites of the TAW appear to increase from cold to warm stream reaches, as expected (the vertical dimension of Table 12.8). Mean WSATI-WT scores appear to increase as land use deviates from an undisturbed state but this trend is apparent mainly in cold and cool stream reaches (the horizontal dimension of Table 12.8). As areas become increasingly urbanised, baseflow decreases and greater volumes of surface flow enter streams in shorter periods following rain events (Leopold, 1968; Klein, 1979). Surface flow is generally warmer than baseflow during summer months, thus stream temperatures may rise in cold and cool streams due to the flush of relatively warm surface water entering them after a rain event. Since WSATI-WT scores in cold and cool rural and suburban locales are higher than in cold and cool undisturbed

Table 12.7 Mean and standard deviation of WSATI-WQ scores for sample sites in the TAW. Sites were grouped by land use in 1946–1954 when early fish collections were made and land use in 1984–1985 when recent fish collections were made. Temperature zones were inferred by ODPD studies in 1949–1956

Temperature category	Value	Undisturbed		Rural		Suburban		Urban	
		1946–1954	1984–1985	1946–1954	1984–1985	1946–1954	1984–1985	1946–1954	1984–1985
Cold	Mean	10.47	9.19	7.55	7.63	6.33	5.95		
	SD	1.78	1.77	1.66	1.82	0.72	0.47		
	N	14	8	46	53	6	4		
Cool	Mean	9.70	9.34	7.11	6.46	7.06	6.27	7.38	5.98
	SD	3.25	3.71	0.86	0.67	0.50	0.48		
	N	2	2	51	37	5	11	1	1
Warm	Mean			7.38	7.21		6.85	1.43	6.09
	SD			0.81	0.72		0.82	3.19	0.65
	N			43	31		15	5	21
Intermittent	Mean			7.07	6.61		6.26		6.19
	SD			0.59	0.75		0.33		
	N			6	11		2		1

Table 12.8 Mean and standard deviation of WSATI-WT scores for sample sites in the TAW. Sites were grouped by land use in 1946–1954 when early fish collections were made and land use in 1984–1985 when recent fish collections were made. Temperature zones were inferred by ODPD studies in 1949–1956

Temperature category	Value	Undisturbed		Rural		Suburban		Urban	
		1946–1954	1984–1985	1946–1954	1984–1985	1946–1954	1984–1985	1946–1954	1984–1985
Cold	Mean	16.15	17.38	19.71	20.36	21.04	20.90		
	SD	2.00	2.82	2.22	2.43	1.13	0.97		
	N	14	8	46	52	6	4		
Cool	Mean	17.12	19.78	20.96	21.43	21.45	22.46	21.35	21.45
	SD	3.37	6.33	1.59	1.07	0.58	1.35		
	N	2	2	51	37	5	11	1	1
Warm	Mean			21.69	22.38	22.61	22.98	20.68	22.25
	SD			1.17	1.25	1.80	1.71	1.75	0.89
	N			45	31	10	15	2	21
Intermittent	Mean			21.46	23.91	22.41	22.09		22.89
	SD			1.50	2.75		0.27		
	N			6	20	1	4		1

Table 12.9 Mean, sample standard deviation, sample size and range of WSATI-WQ scores for four land use types in the TAW. All mean scores within a given sampling period are significantly different with respect to land use type ($p < 0.05$). For a given land use type between the two sampling periods only the difference between early and recent mean scores for urban land use is significant ($p < 0.05$). Data were collected by ODPD in 1946–1954 and Steedman in 1984–1985. The early data, unweighted SATI-WQ scores, were transformed into weighted WSATI-WQ scores

Land use	1946–1954				1984–1985 combined			
	Mean	SD	N	Range	Mean	SD	N	Range
Undisturbed	10.85	1.72	16	6.60–12.0	9.59	2.00	10	7.39–12.0
Rural	7.31	1.10	149	5.59–11.48	7.12	1.36	132	5.39–12.0
Suburban	6.69	0.73	22	5.00–8.67	6.50	0.72	32	5.31–8.12
Urban	2.79	3.39	9	0.0–7.38	5.83	1.38	24	0.0–8.23

Table 12.10 Mean, sample standard deviation, sample size and range of WSATI-WT scores for four land use types in the TAW. Values within a sampling period, with similar superscripts, are significantly different ($p < 0.05$). The difference in mean scores for a given land use type between the two sampling periods is not significant ($p < 0.05$). Data were collected by ODPD in 1946–1954 and Steedman in 1984–1985. The early data, unweighted SATI-WQ scores, were transformed into weighted WSATI-WQ scores

Land use	1946–1954				1984–1985 combined			
	Mean	SD	N	Range	Mean	SD	N	Range
Undisturbed	16.1[a,b,c]	1.84	16	14.6–20.3	17.9[c,d,e]	3.42	10	13.0–24.3
Rural	20.8[a]	1.79	149	15.3–24.7	21.4[c,d,e]	2.17	131	15.3–27.6
Suburban	21.9[b]	1.51	22	19.5–26.9	22.5[e]	1.57	32	19.9–25.9
Urban	20.9[c]	1.47	4	18.7–21.9	22.2[c,d]	0.88	24	20.5–24.1

locales (Table 12.7) we infer that the fish association has responded to the phenomenon described above. This trend is not apparent in warm locales since rivers in those locales are dominated by surface and storm sewer flows during dry and wet periods.

Mean WSATI-WQ and WSATI-WT scores for all comparable sampling stations of the TAW were classified into four major land uses and appear in Tables 12.9 and 12.10. These are 'marginal totals' from Tables 12.7 and 12.8. Mean WSATI-WQ scores calculated from data collected during 1946–1954 and 1984–1985 appear to decrease progressively as land use deviates from a pristine state (Table 12.9).

Mean WSATI-WT scores for sampling stations in undisturbed areas are significantly smaller than mean WSATI-WT scores for any other land-use category (Table 12.10). There is no clear pattern with respect to mean WSATI-WT scores for rural, suburban and urban land uses in the early or recent survey, but it appears that the highest mean WSATI-WT scores in both surveys were from suburban land-use type (Table 12.10). When results from 1946–1954 were compared with those from 1984–1985, the mean WSATI-WT scores within each land-use type were not significantly different ($p > 0.05$).

It appears that broad land-use categories as defined and used here are useful in characterising the level of stress from various land-use types. The above results are generally consistent with the findings of Weaver and Garman (1994) in a progressively urbanising watershed in Virginia. Over the past four or five decades the effects of the suite of stresses from undisturbed, rural and suburban land uses with respect to water quality have not changed appreciably (Table 12.9). The significant difference between undisturbed and rural sites, both during 1946–1954 and during the more recent collection period, suggests that stress from agriculture and other practices is well above 'background levels' and has not changed much over the past three or four decades. Effects of the suite of stresses from urban land uses have changed significantly. Management of sewage has improved in the TAW over the past four or five decades (Davey, 1985; Wichert, 1995b) and it is likely that this improvement has been responsible in part for the increase in WSATI-WQ scores observed for urban sampling stations.

From this study it appears that the WSATI-WQ and WSATI-WT can be used to assess changes in fish associations found in relatively large drainage basins. Changes in the characteristics of a fish association, with respect to water quality or water temperature, reflect changes, positive or negative, in the stress regime in a watershed.

Almost all of the data for this study were collected by surveys not designed explicitly for the use to which the data have been put here. The complexity of the analysis is due in part to the need to make data comparable. In spite of the difficulties in using such data, some interesting and useful inferences have emerged.

With respect to future monitoring and assessment of ecological conditions in Toronto streams, the present study may be taken as an empirical basis for designing better sampling and estimation methods.

ACKNOWLEDGEMENTS

Karen Ing, Nathalie LaViolette, Ping Lin, Karen Smokorowski, Crystal Thomas, Alison Thompson and Deborah Walks assisted with the fieldwork. Data collected by undergraduate students in the field component of a course in ecology were contributed by Professor J. B. Falls. Robert Steedman and John Homes provided advice. The work was funded by University of Toronto and Ontario Ministry of Colleges and Universities Fellowships to Gordon Wichert and grants to Henry Regier from the Ontario Renewable Resources Research Fund and the Donner Canadian Foundation.

REFERENCES

Chapman, L.J. and Putnam, D.F. (1984) *Physiography of Southern Ontario*, 3rd edition. Ontario Geological Survey, Special Volume 2. Accompanied by map P.2715 (coloured), scale 1:600 000.

Coventry, A.F. (1948) *The Need of River Valley Development*. Ontario Department of Planning and Development, Conservation Branch.

Davey, T. (1985) *Recollections: Water Pollution Control in Ontario*. Pollution Control Association of Ontario, Aurora, Ontario.

Hallam, J.C. (1959) Habitat and associated fauna of four species of fish in Ontario streams. *Journal of the Fisheries Research Board of Canada*, **16**, 147–173.

Huntsman, A.G. (1944) Why did Ontario salmon disappear? *Transactions of the Royal Society of*

Canada Series 3, Section V, **38**, 83–102.

Jackes, L.B. (1948) *Tales of North Toronto*. North Toronto Businessmen's Association, Toronto, Ontario.

Johnson, M.G. and Owen, G.E. (1966) *Biological Survey of the Upper Credit River*. Water Quality Surveys Branch, Ontario Ministry of the Environment, Toronto, Ontario.

Karr, J.R., Fausch, K.D., Angermeier, P.L., Yant, P.R. and Schlosser, I.J. (1986) *Assessing Biological Integrity in Running Waters: A Method and Its Rationale*. Illinois Natural History Survey, Special Publication 5, Champaign, Illinois.

Kauffman, J., Rennick, P., Regier, H.A., Holmes, J.A. and Wichert, G.A. (1992) *Metro Waterfront Study*. Metropolitan Toronto Planning Department, Toronto, Ontario.

Klein, R. D. (1979) Urbanization and stream quality impairment. *Water Resources Bulletin*, **15**, 948–963.

Leopold, L.B. (1968) *Hydrology for Urban Land Planning – A Guidebook on the Hydrologic Effects of Urban Land Use*. US Geological Survey Circular 554, Washington, DC.

Loftus, K.H. and Regier, H.A. (1972) Introduction to the proceedings of the 1971 symposium on salmonid communities in oligotrophic lakes. *Journal of the Fisheries Research Board of Canada*, **29**, 613–616.

Martin, D.K. (1984) The fishes of the Credit River: cultural effects in recent years. MSc thesis, Institute for Environmental Studies and Department of Zoology, University of Toronto.

Meisner, J.D. (1990) Potential loss of thermal habitat for brook trout, due to climatic warming, in two southern Ontario streams. *Transactions of the American Fisheries Society*, **119**, 282–291.

ODLF (1971) Credit River (north of the Forks). Ontario Department of Lands and Forests, Fisheries Inventory Unit, Stream Survey Report, Maple, Ontario.

ODPD (1947) Etobicoke Valley Report, 1947: recommendations and summary. Ontario Department of Planning and Development, Toronto, Ontario.

ODPD (1948) The Humber Valley conservation report. Ontario Department of Planning and Development, Toronto, Ontario.

ODPD (1950) The Don Valley conservation report. Ontario Department of Planning and Development, Toronto, Ontario.

ODPD (1956a) The Credit Valley conservation report. Ontario Department of Planning and Development, Toronto, Ontario.

ODPD (1956b) The Rouge, Duffins, Highland, Petticoat conservation report. Ontario Department of Planning and Development, Toronto, Ontario.

Parson, J.W. (1973) *History of Salmon in the Great Lakes, 1850–1970*. Technical Papers of the Bureau of Sport Fisheries and Wildlife, No. 68. United States Department of the Interior. Fish and Wildlife Service.

Reed, D.J. (1968) A resurvey of the fishes of the Credit River. MSc thesis, Department of Zoology, University of Toronto, Toronto, Ontario.

Regier, H.A. and Kay, J.J. (1996) An heuristic model of transformations of the aquatic ecosystems of the Great Lakes – St Lawrence River basin. *Journal of Aquatic Ecosystem Health*, **5**, 3–21.

Regier, H.A., Lin, P., Ing, K.K. and Wichert, G.A. (1996) Likely responses to climate change of fish associations in the Laurentian Great Lakes Basin: concepts, methods and findings. *Boreal Environment Research*, **1**, 1–15.

Steedman, R.J. (1987) Comparative analysis of stream degradation and rehabilitation in the Toronto area. PhD thesis, Department of Zoology, University of Toronto, Toronto, Ontario.

Steedman, R.J. (1988) Modification and assessment of an index of biotic integrity to quantify stream quality in southern Ontario. *Canadian Journal of Fisheries and Aquatic Sciences*, **45**, 492–501.

Stephenson, T.D. (1990a) Fish reproductive utilization of coastal marshes of Lake Ontario near Toronto. *Journal of Great Lakes Research*, **16**, 71–81.

Stephenson, T.D. (1990b) Significance of Toronto area wetlands for fish. In Bardecki, M.J. and Patterson, N. (eds) *Proceedings of Conference on Ontario Wetlands: Inertia or Momentum*, 20–21 October 1988. Federation of Ontario Naturalists, Don Mills, Ontario, pp. 415–423.

Wainio, A.A., Haarmeyer, G.J., Inglis, A. and Stewart, C. (1975) *Credit River Stream Survey Report*. Ontario Ministry of Natural Resources, Maple, Ontario.

Weaver, L.A. and Garman, G.C. (1994) Urbanization of a watershed and historical changes in a stream fish assemblage. *Transactions of the American Fisheries Society*, **123**, 162–172.

Wichert, G.A. (1991) The fish associations of Toronto area waters, 1948–85: major changes and some causes. MSc thesis, Department of Zoology, University of Toronto, Toronto, Ontario.

Wichert, G.A. (1994) Fish as indicators of ecological sustainability: historical sequences in Toronto area streams. *Water Pollution Research Journal of Canada*, **29**, 599–617.

Wichert, G.A. (1995a) Effects of Toronto regional development processes on the dynamics of stream ecosystems as reflected by fish associations. PhD thesis, Department of Zoology, University of Toronto, Toronto, Ontario, Canada.

Wichert, G.A. (1995b) Effects of sewage effluent remediation and urbanization on fish associations of Toronto streams. *North American Journal of Fisheries Management*, **15**, 440–456.

Wichert, G.A. and Lin, P. (1996) A species tolerance index for maximum water temperature. *Water Quality Research Journal of Canada*, **31**, 875–893.

# 13 Rehabilitation of the River Murray, Australia: Identifying Causes of Degradation and Options for Bringing the Environment into the Management Equation

ANNE JENSEN
Wetlands Management Program, Resource Management Branch, Department of Environment and Natural Resources, Adelaide, Australia

THE CONTEXT OF THE MURRAY-DARLING BASIN

The rivers of the Murray-Darling Basin vary from cool fast mountain streams to the ephemeral streams of the western arid lands. The catchment covers more than one-seventh of Australia, but 88% of the runoff is supplied from less than 10% of the area, from the mountains of the south-eastern alpine region (Figure 13.1). The major rivers provide essential water supplies to the inhabitants of the basin and also to most of the population of South Australia at its downstream end. Production from the basin exceeds A$10 billion per annum, making this one of Australia's prime agricultural resources (Jacobs, 1990).

Over the past 100 years, management located in the basin has concentrated on the need to ensure consistent water levels for navigation and water supply for agriculture across the riverine plains. The uncertainties of summer droughts and low flows have been eliminated with the construction of large headwater storages, mid-stream control structures and temporary storages, a series of weirs to control flows and five barrages to exclude seawater (Figure 13.2). Additional water resources have been diverted across the watershed divide in the Snowy Mountains, taking water from easterly-flowing streams and directing it instead into the River Murray. This additional water flows more than 2000 km westwards and provides greater security of water supply to irrigation settlements, pastures and towns of the inland regions (Jacobs, 1990).

Traditionally, management has concentrated on a secure water supply, with the River Murray Commission co-ordinating water management between the states along the River Murray from 1917 to 1988. Only since 1982 has consideration for the environment been part of the management agenda for the River Murray. Initially, environmental issues were considered in a piecemeal fashion, starting with the issues hindering continued assurance of supply and water quality. The first problems considered were rising salinity and turbidity, with concerns about nutrient levels coming later in the 1980s as the

---

Rehabilitation of Rivers: Principles and Implementation. Edited by L. C. de Waal, A. R. G. Large and P. M. Wade.
© 1998 John Wiley & Sons Ltd.

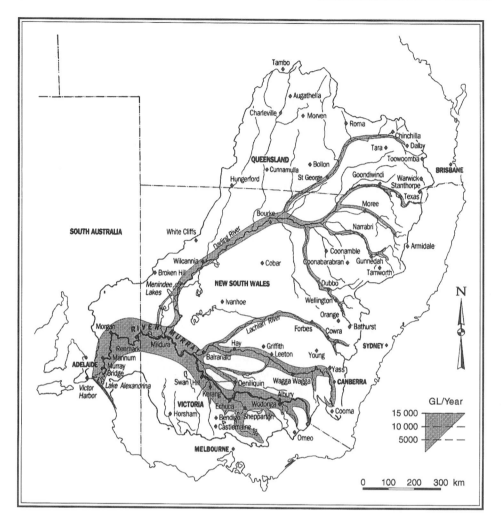

Figure 13.1 Average annual river discharges in Murray-Darling Basin subcatchments. (Reproduced by permission of the Murray-Darling Basin Commission)

frequency of blue-green algal blooms increased compared to previous occurrences (Murray-Darling Basin Ministerial Council, 1987).

The catchment approach to management was first introduced in 1985, leading to the formation in 1988 of the Murray-Darling Basin Commission, which has a brief to co-ordinate management of land, water and environmental resources in the basin. This is a major change from the original brief of the River Murray Commission (since 1917) to manage water supply in the River Murray. In 1992, Queensland also joined the Commission, bringing in the headwaters of the Darling River to complete basin representation.

The charter of the Murray-Darling Basin Ministerial Council and Commission is to 'promote and coordinate effective planning and management for the equitable, efficient and sustainable use of the land, water and environmental resources of the Murray-Darling

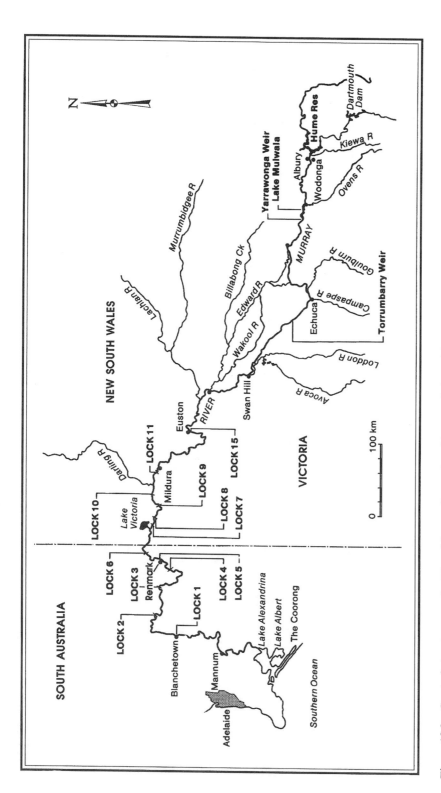

Figure 13.2 Control structures on the River Murray. Reproduced by permission of the Murray-Darling Basin Commission

Basin' (Murray-Darling Basin Ministerial Council, 1989). As the strategies and policies of the Murray-Darling Basin initiative are being developed, recognition of the environment as a legitimate user of water is finally emerging. An environmental water policy is being developed for inclusion in the package of integrated policies and strategies addressing the key management issues in the basin. In a landmark decision in 1995, the Ministerial Council introduced a 'cap' on water diversions, limiting the volume diverted to 1993–1994 levels of development. This action potentially halts the degradation caused by reduced flows and sets a baseline for environmental water provisions.

THE NEED FOR REHABILITATION

The need for rehabilitation in the Murray-Darling Basin became increasingly apparent from the early 1980s onwards, as the impacts of river regulation, agriculture and urban development became more visible and better understood. Poor recruitment in native fish (Murray-Darling Basin Commission, 1989; Pierce, 1992) and reduced habitat for waterbirds (Thompson, 1986; Pressey, 1986) are some of the signs of degradation in the ecosystem. Margules *et al.* (1990) reported less vigorous and less frequent regeneration in key floodplain plant species in the lower regulated Murray Valley compared to upstream reaches, and die-back in floodplain trees has been observed in areas with reduced overbank flows (Sharley, 1992). Walker *et al.* (1992) reported instances of severe bank slumping following accelerated flood recessions. Current work by Sheldon indicates that extremely stable weir pool levels may be limiting aquatic food chain diversity (Sheldon, pers. comm.). The major causes of degradation are the changes in the hydrological regime due to regulation and diversion of water, the effects of grazing and development on the floodplain, and the introduction of exotic plant and animal species (Jensen *et al.*, 1994a).

In the 1990s, decision-makers and managers are finally realising the need to maintain a healthy river environment in order to maintain a healthy resource for all users. Researchers and scientific managers are gathering all available ecological information to determine the needs of the environment for inclusion in management decisions. The top current environmental priority is the development of an environmental water policy and flow management strategies for the rivers of the Murray-Darling Basin. Attitudes are slowly changing but there is still resistance from agricultural and engineering interests to the concept of reserving water for environmental use, even though a healthy environment is of benefit to all users. The question now is whether the changes can be realised quickly enough to sustain the river ecosystems before irreversible damage is done to the catchment and water resources.

REHABILITATION OR RESTORATION

The theoretical debate on management of the rivers and wetlands of the Murray-Darling Basin has included discussion on the appropriate terminology to use – 'restoration' or 'rehabilitation' (Jensen *et al*, 1994a). It has been acknowledged that there are two main objections to the use of the term 'restoration'. The first objection is the difficulty of defining a precise state of undisturbed, pre-regulation conditions to which the river should be restored. There is no significant objective scientific information from before

European settlement and the beginning of river regulation. There is also the additional complication of the ever-changing dynamic and evolving nature of river ecosystems. At best, a river could only be restored to a point within an evolving and varied continuum, if that point could be defined. It would not be appropriate to try to restore the river to one fixed state, as this would be contrary to natural evolution. The second difficulty with the concept of restoration is the impossibility of restoring the pre-regulation hydrological regime, which would require the removal of the dams and weirs on the Murray-Darling system. This would be unacceptable politically, due to the social and economic investment in their continued function to ensure water supplies to over 2.4 million people.

The argument in favour of the term 'rehabilitation' is that it is appropriate to describe attempts to partially restore or simulate natural processes or key features of the natural hydrological regime. The management objectives of rehabilitation are to promote natural regeneration and breeding processes, to remove or reduce degenerative impacts, and to find the best available balance between maintaining the water supply function and sustaining a healthy river ecosystem.

The term 'rehabilitation' has therefore been adopted, in recognition that current management programmes can only restore or simulate parts of the natural hydrological regime, within strong economic, social and political constraints.

DEFINING THE NATURAL HYDROLOGICAL REGIME

The natural hydrological regime of the major rivers of the Murray-Darling Basin is highly variable, with unpredictable cycles of flood and drought (Jacobs, 1990). The Murray and Murrumbidgee Rivers tend to peak in spring to early summer, while the Darling River is more likely to flood in summer. However, peak flows vary from short sharp peaks of a few weeks to extended high flows of many months. Recessions can be rapid or attenuated. These variations are directly dependent upon the pattern of rainfall within one season and the combined effect of rainfall and flows of previous years.

Flow velocities are very slow due to the low gradients and long distances, with water taking over two months to travel from Hume Dam at Albury to the Murray Mouth, and one month from the Darling junction to the mouth (Jacobs, 1990). The pattern of rises and falls in level of the mainstream and the frequency of overbank flows in a particular river reach are thus complicated and unpredictable, although statistical frequencies have been calculated for flows of a particular volume in various reaches. However, two flows of the same volume will behave differently unless pre-flood conditions are identical.

An important feature of the hydrological regime in the floodplains of the Murray-Darling system is the cycle of wetting and drying which occurs in the wetlands. Areas with clay substrates require a period of drought to release nutrients which become bound to the waterlogged clay during extended periods of inundation (Briggs and Maher, 1985). Re-wetting releases the nutrients, which are rapidly taken up as the food chain develops through successional stages and new breeding cycles commence.

Thus definition of the natural hydrological regime for the rivers of the Murray-Darling system can only be described within broad boundaries, such as the range of levels, flow velocities, rates of rise and fall of peak flows, frequency of overbank flows, duration of floodplain inundation, water temperatures, seasonality of peak flows, depth of inundation and source of water. Reinstatement of a natural flow regime or key features of that

IMPACTS OF DEVELOPMENT AND REGULATION ON RIVERS AND WETLANDS

The high productivity, diversity and value of rivers and wetlands is now being recognised internationally, with increasing concern over the need to protect and maintain these generally undervalued areas (Boon *et al.*, 1992). Important and useful functions of rivers and floodplains include the following (Jensen *et al.*, 1994a):

- breeding, feeding and roosting habitat for aquatic biota
- flood storage and retention
- nutrient filtration and uptake into biological processes
- sediment and toxicant retention
- interaction with riparian zone
- local water supply.

These functions have been significantly disrupted in the Murray-Darling Basin by a number of factors (Jensen *et al.*, 1994a):

- regulation of mainstream and backwaters and associated changing of the flow regime
- changed seasonality of flow peaks
- blockage of floodplain flows by causeways, levee banks and structures
- grazing by stock and feral animals, particularly rabbits
- cropping on the floodplain and lake beds
- introduced fish species
- introduced weed species
- stormwater, sewage and irrigation drainage effluent disposal
- drainage runoff into wetlands, causing salinity and waterlogging problems
- rising saline groundwater or increased lateral hydraulic flows

Rehabilitation projects must identify the causes of degradation at a particular location and propose measures to control ongoing degradation as well as rehabilitation of past damage. They should specifically prescribe improved future management such as reinstatement of the natural hydrological regime. Unless these changes are controlled or counteracted, widespread degradation of river ecosystems is predicted, with serious consequences for water quality and maintenance of wildlife and integrity of the landscape.

While techniques for assigning economic values to environmental resources are still not well defined (Jensen, 1993), it is clear that there would be major social and economic costs if the river ecosystems of the rivers in the Murray-Darling Basin were seriously degraded. Greater effort is required to establish the economic values of healthy wetland and river systems in the Murray-Darling Basin, and the costs of degradation in these systems, in order to ensure balanced accounting in management decisions which affect the integrity of river ecosystems throughout the basin.

IMPACTS ON THE HYDROLOGICAL REGIME

Construction of major headwater storages and regulation of river flows through dam structures and weirs has had significant effects on the pattern of flows in the Murray-Darling Basin, particularly in the River Murray system (Jacobs, 1990). High flows from winter and spring rains are stored, and controlled releases ensure minimum flows through the summer season for irrigation use. Low flow periods now occur in winter instead of summer, with summer droughts eliminated.

The impact of regulation on the lower reaches of the River Murray below the Darling junction has been to change the variable river flows to a series of permanent relatively stable pools, with significantly reduced frequency of minor floods (Caldwell Connell Engineers, 1981; Walker, 1986) (Figures 13.2 and 13.3). Natural variations in levels in the pools have been dampened and summer flows are maintained at an artificially high level (Nicolson, 1993). The river estuary surface area has been greatly reduced, with the large terminal lakes converted to a freshwater regime through the exclusion of seawater by a series of barrages across several barrier islands. The former estuarine lakes Alexandrina and Albert have become fresh, permanent and relatively stable (Geddes and Hall, 1990).

Regulated flows are contained within the mainstream and former frequent minor floods no longer occur (Caldwell Connell Engineers, 1981). However, the weirs are unable to control flood flows of greater than 1 in 7 years frequency and these flows follow natural paths across the floodplain, subject to human-made obstructions. The frequency of these medium-sized floods, up to about 1 in 50 years return periods, has been reduced by large upstream storages and diversion of significant volumes for irrigation in the upstream states (Close, 1990).

Changed flow regimes in the River Murray have increased the proportion of permanent wetlands, while decreasing floodplain inundation. Previously, the majority of wetlands were temporary, with inundation by high spring flows once every 2–3 years. However, the impact of these changes is still being investigated, with some contradictory results. Permanent wetlands were thought to be less productive biologically than temporary wetlands because nutrients become locked into the clay substrate and are not available to the food chain (Briggs and Maher, 1985; Thompson, 1986). However, recent results have indicated greater diversity and biomass in invertebrates in now permanent wetlands compared to temporary and ephemeral wetlands on the lower Murray floodplain (Suter et al., 1994).

The lower frequency of inundation of the floodplain and the temporary wetlands on the lower Murray reduces the number of successful breeding and regeneration events in floodplain biota which respond to the flood pulse (Jensen, 1983; Walker, 1986; Sharley, 1992). Recent Victorian investigations show that permanent wetlands change in community composition, with a reduction in aquatic plants and therefore less invertebrate habitat and lower levels of productivity (Lloyd, pers. comm.; Boulton and Lloyd, 1991).

Peak flows are reduced and delayed, and peak recessions are very abrupt (Walker et al., 1991). Overbank flows are much less frequent and volumes are reduced (Caldwell Connell Engineers, 1981; Sharley, 1992). For example, peak flows in 1993, potentially in the order of a 1 in 20 years event, were reduced by approximately 43% through manipulation of two major upstream storages, to maximise water storage and to reduce material damage on the floodplains. Flows at the Darling junction peaked at 126 000 Ml day^{-1}, but an additional 95 000 Ml day^{-1} was retained in Hume and Dartmouth storages (Erdmann,

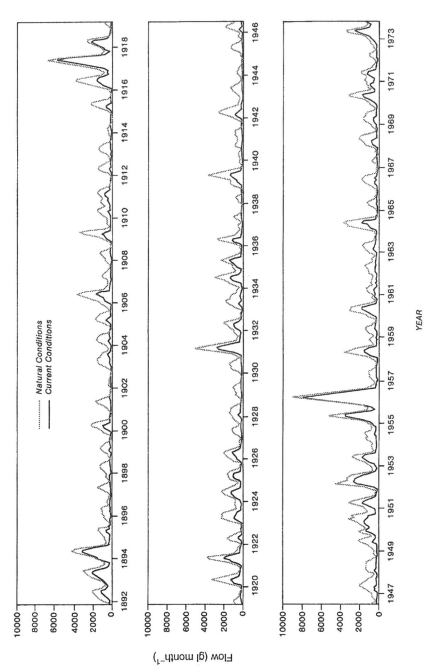

Figure 13.3 Impact of river regulation on the natural hydrograph in South Australia. Source: Mackay and Eastburn (1990). Reproduced by permission of the Murray-Darling Basin Commission

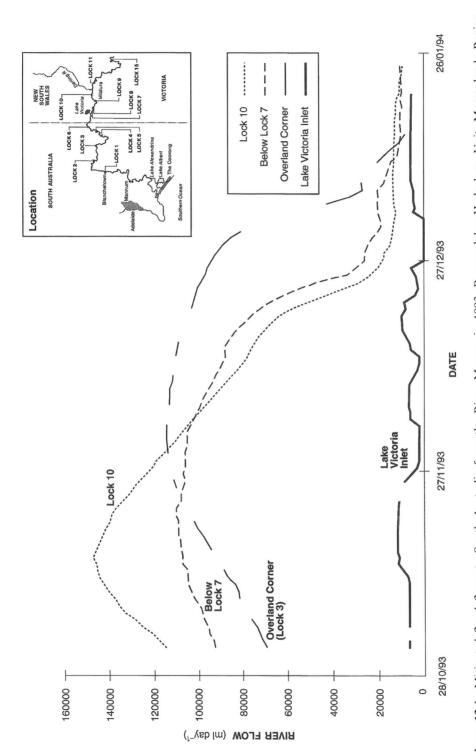

Figure 13.4 Mitigated flood flows to South Australia from the River Murray in 1993. Prepared by: Hydrology Unit, Murraylands Region, Department of Environment and Natural Resources

pers. comm.). The resulting flow to South Australia was approximately a 1 in 10 year event. In 1993, peak flows into South Australia were further mitigated by 12 000 Ml day^{-1} by diversion into Lake Victoria and later released, in order to prevent closure of a key regional road (Figure 13.4). However, this action also prevented water from reaching outlying sections of floodplain and watering populations of stressed river box trees (*Eucalyptus largiflorens*) on the Chowilla floodplain. It is estimated that overbank flows to the high value Chowilla floodplain have been reduced by river regulation from 1 in 4 to 1 in 13 (Sharley, 1992). The effect of reduced overbank flows has been compounded by rising saline groundwater reaching the root zone of the trees, resulting in widespread die-back. The rise in groundwater has been caused by a combination of hydraulic pressure from the local weir pool and increased regional groundwater inflows towards a natural sink due to vegetation clearance and increased rainfall recharge. Inundation during flood peaks is thus essential for rehabilitation of these areas.

Control of river flows has led to changes in the pattern of source water delivery to the lower River Murray. Five recent flood events have indicated the importance of source water in the hydrological regime. High flows each spring/summer since 1989 have all originated from the Murray–Murrumbidgee catchment, delivering water with higher tannin content but lower turbidity. This water has been stored in Lake Victoria and brought to South Australia in controlled releases once river flows returned to normal (Figure 13.2). This situation is in strong contrast to the low flow years of the mid-1980s, when highly turbid Darling River water was stored in Lake Victoria for release to South Australia.

The impact of the difference in source water is clear in the prolific growth of macrophytes and macroinvertebrates observed in wetlands and the mainstream itself since 1989. It is estimated that the photic zone has been increased by 60–70% due to the change in source water (Suter *et al.*, 1994). Wetlands assessed as having low value due to lack of invertebrates and macrophytes in the mid-1980s (Thompson, 1986) have been reassessed as having high value due to dense macrophyte beds and high waterbird populations (Suter *et al.*, 1994). It is clear that flow management strategies which favour Murray flows over the more turbid Darling flows would benefit the ecosystem.

IMPACTS OF INTRODUCED SPECIES

The impact of grazing pressure from domestic stock and feral animals on the floodplains has been severe, with little successful recruitment of floodplain plants and lost habitat for small mammals and reptiles (Thompson, 1986; Carter and Nicolson, 1992). Several exotic species have had significant impacts on the river ecosystems. The most notable are the European carp (*Cyprinus carpio*), the European rabbit (*Oryctolagus cuniculus*) and the European willow (*Salix* spp). The European carp, with an annual breeding cycle and few predators, has successfully spread throughout the catchment since its introduction in the 1960s. It has major impacts on wetland and stream habitats, through its suction method of feeding which stirs bottom sediments and uproots aquatic plants (Murray-Darling Basin Commission, 1989). Native fish, on the other hand, rely on floods as breeding cues, and numbers have fallen significantly due to reduced overbank flows and loss of suitable habitat, as well as competition from carp.

The rabbit has bred prolifically throughout the basin and is a major threat to any regenerating native plant species, particularly eucalypt seedlings. Rabbit control is an essential management measure for any projects attempting to promote regeneration of floodplain plants (Carter and Nicolson, 1992). European willows (*Salix alba, S. rubens*) were planted from the 1830s to mark the river mainstreams for navigation and to stabilise levee banks for irrigation on the floodplain. They have since proliferated by vegetative reproduction, particularly upstream of each weir structure where water levels are very stable. Coverage varies from 100% of mainstream banks for over 100 km of the lower Murray, to isolated clumps further upstream. They exclude native vegetation along river and stream banks and do not provide suitable habitat or woody debris for insects, birds or fish as part of the riparian ecosystem (Mitchell and Frankenberg, 1993).

REHABILITATION APPROACHES

The first essential step of successful rehabilitation techniques for sustainable river and wetland rehabilitation is to identify all factors causing degradation and to remove or control these impacts as far as possible. As already indicated, the primary causes of degradation include greatly reduced flows and flood peaks, grazing by stock and feral animals, and the competitive and disruptive effects of a range of exotic plant and animal species (Jensen *et al.*, 1994a).

The primary objective in river rehabilitation should be to maintain, restore or enhance natural floodplain processes in a sustainable way. Broad-scale rehabilitation measures, such as restoration of flow patterns, increased overbank flows and exclusion of stock from riparian zones can result in successful recruitment and regeneration in a number of riparian and aquatic species through the restoration of the conditions required for natural processes to operate (Jensen *et al.*, 1994b). However, in special cases, conditions may be managed to meet the requirements of one particular species or class of species (e.g. fish or waterbirds).

The most critical natural process is the hydrograph, and natural conditions prior to regulation should be mimicked, if possible, to provide a sequence of naturally timed peaks, inundation and recession which can provide breeding and regeneration cues for floodplain species (Murray-Darling Basin Commission, 1994).

Effective rehabilitation techniques vary throughout the basin, especially in the lower Murray, where techniques must consider increasing floodplain salinity and modification due to river regulation, changed flow regimes and irrigation impacts. In the past, rehabilitation has failed or been only mildly successful because not all factors causing degradation were identified (Jensen *et al.*, 1994a).

A set of principles has been defined for the rehabilitation of floodplain wetlands and incorporated into the Murray-Darling Basin Strategy for Managing Floodplain Wetlands (Murray-Darling Basin Commission, 1994):

- Active rehabilitation measures should be taken to maintain, restore or enhance natural floodplain wetland processes in a sustainable way.
- Successful rehabilitation of floodplain wetlands must consider all the factors causing degradation and remove or control these factors as far as possible.
- Projects must establish clear objectives for management actions.
- Projects must be monitored for effectiveness of rehabilitation techniques and the results recorded and publicised to both managers and the community.
- Rehabilitation techniques must be environmentally sound and community-owned.

ACTIVE REHABILITATION MEASURES

The options available for active rehabilitation in the Murray-Darling Basin include variations in river operating procedures to simulate portions of the natural hydrograph, controls on grazing, removal of disruptions to floodplain flow paths, and the control of introduced species. Possible variations in river operating procedures for the lower regulated River Murray include the following:

- timing of releases of water from upstream storages for environmental benefits as well as water quality objectives;
- timing of local weir manipulations to maximise flushing of poor quality water from backwaters;
- timing of local weir manipulations to affect the occurrence, extent or depth of flooding, to enhance flood peaks or to target specific wetlands;
- local weir manipulations to vary mainstream river levels in a more natural pattern to enhance littoral habitat for native species;
- manipulation of barrage openings/closings to benefit the Coorong or to keep the Murray Mouth open.

At local sites, particularly on the floodplain, a combination of procedures can be used to improve habitat value. As a general rule, maximum habitat diversity is the goal, often through the following actions (Jensen *et al.*, 1994a):

- redirection of saline inflows;
- reduced ponding of saline water;
- increased overbank flows to remove concentrated stagnant water and accumulated salts, and to trigger breeding cycles;
- control of riparian grazing;
- protection of natural regeneration of riparian species;
- planting of riparian species where appropriate;
- manipulation of water levels to benefit waterbird breeding;
- creation of additional temporary wetland habitat.

It is therefore important to clearly define management objectives before a project commences at a local site. Flow management strategies must be co-ordinated on the reach and catchment scale to avoid conflicting objectives. The most valuable management tool at present is the opportunity to learn from current and future field trials, through monitoring and feedback of results. The knowledge derived on effective rehabilitation techniques must be shared widely with other managers and the community for application across the basin. Research is also becoming more problem-oriented, with strong links to management initiatives.

DEFINING ENVIRONMENTAL MANAGEMENT REQUIREMENTS

The next essential step is to define the needs of the river and wetland ecosystems in terms of water delivery which relate to the water management system. From an engineering viewpoint, it would be preferable to deliver a set amount of water at a precise time in predictable amounts which can be included in management predictions for the next planning cycle. Unfortunately for water managers, the natural hydrograph is based on irregular, unpredictable cycles, and the environment requires floods and droughts rather than even, easily managed delivery of water. The approach being explored for the lower regulated River Murray is to identify key wetlands or reaches requiring changes in the hydrological pattern, and to attempt to identify how much of the natural hydrology could be restored through altered management of flows.

From the environmental point of view, the ecosystem required all of the available water prior to regulation. Therefore, it now requires the restoration of as much water as possible. However, it is clear that only a small proportion of the original flows can be restored. One important point which reduces potential conflict between users is that ecosystems do not require water during a drought, when water supplies are at a premium to irrigators. Conversely, floods are highly desirable to wetlands and floodplains, providing flood mitigation for settlements threatened by floodwaters. Floods can be directed into wetland areas as part of flood management strategies for mutual environmental and social benefit.

The beneficial effect of available flows can be increased by management through simulating higher flows, e.g. by raising levels at the weirs. This gives the effect of a higher flow and may allow overbank flows at key points by lifting water over the threshold of a floodplain creek. Calculations and field trials have shown that flows of 25 000–60 000 Ml day^{-1} can be managed to simulate up to 10 000 Ml day^{-1} higher flows by raising levels at a weir by 300 mm (Erdmann, pers. comm.).

The Lower Murray Flow Management Working Group develops technical options for flow management strategies for the regulated Lower River Murray. The primary objective is to increase overbank flows. It is proposed that an environmental allocation for the lower reaches of the river should take the form of reserving all flood peaks above 25 000 Ml day^{-1} flows for environmental benefit through inundation of the floodplain (Jensen and Nicholls, 1996).

An expert panel is currently assessing the environmental flow requirements for the full length of the River Murray, for inclusion in flow management strategies. The expert panel will evaluate the proposals of the Flow Management Working Group.

The definition of environmental management requirements is a top priority in wetland management and information is being compiled from a range of wetland projects across the basin. This will be combined with information on opportunities for altered flow management and co-ordinated through flow management strategies for various river reaches.

BALANCING ENVIRONMENTAL, ECONOMIC AND SOCIAL INTERESTS

The decision to change flow management will be political rather than scientific, balancing the competing views of irrigators, local government, recreational groups, conservation interests and interstate competition for water. It is essential that the best possible objective information is provided as a basis for the decision-making process, including scientific knowledge of natural processes and an acknowledgment of the economic value of natural resources.

ECONOMIC VALUE OF WETLANDS AND NATURAL RESOURCES

With all development proposals currently evaluated in economic terms, problems continue to exist with assigning a relative value to wetlands to ensure that a true evaluation is made of all the impacts or benefits of proposed projects. It is essential that the base resource is not undervalued, as can easily happen in economically based cost–benefit analyses. However, the wetland resource, in the form of the floodplains of the basin, is the basis for recreational use, landscape values, maintenance of water quality and many other uses, and its value would be hard to exaggerate.

A pilot study in Australia has attempted to assign an economic value to the 29 500 ha of the Barmah Forest, a major river red gum wetland forest on the Murray floodplain in the state of Victoria. A sample of Victorians was surveyed to indicate how much they would pay to preserve this wetland of international importance, if the choice was between preservation and total loss. Following rigorous analysis of the validity of the responses, it was calculated that the value of the Barmah Forest, with all of its conservation, recreation and landscape attributes, lies between A$76.5 million and A$97.5 million (Stone, 1991). This contingency valuation method of assigning a dollar value to cover all attributes of the wetland as one unit overcomes the problem of valuing separate attributes by different methods and attempting to add these values together. It is not a totally satisfactory solution economically or environmentally, but it is regarded as the best available political and social tool which allows consideration of the value of the resource within current economically based decision-making methods.

Other examples from the United States have assigned dollar values for one specific attribute of wetlands. For example, the value of a wetland for nitrogen retention is suggested as US$200 000 ha^{-1}, based on the cost of mechanical treatment to remove a similar amount of nitrogen from the river to allow use downstream as a water supply (Maltby, 1991). The value for flood retention is given as US$13 500 ha^{-1}, based on the cost of constructing levees for equivalent floodwater storage and protection (Maltby, 1991). The high economic values emphasise the very significant value of the natural system for more than just conservation purposes.

The example above indicates that the Barmah Forest is worth approximately A$3000 ha^{-1}. If this value is applied to the 258 140 ha of all eight Ramsar-listed wetlands of international importance in the Murray-Darling Basin, these key sites alone could have a value of around A$770 000 000 (Jensen, 1993). And there are many other wetlands of the basin to be valued as well. The figure of A$3000 ha^{-1} may still be a gross undervaluation, if compared to a marina development downstream of the Barmah Forest which has been assigned a similar value in development assessments.

The value and extent of recreational use can also be easily undervalued, because of the difficulty of demonstrating dollar returns, particularly from passive activities. These benefits must also be included in the decision-making process, alongside conservation benefits where natural areas are managed and conserved.

The most important principle is to conserve and sustain the base resource for all users, including wildlife and recreation users, as well as higher profile consumptive users. In the short term, this means supporting proposals for allocating water for environmental purposes and supporting wetland management projects. The successful implementation of a co-ordinated flow management strategy for the rivers of the basin will also require the support and co-operation of communities throughout the catchment. Every effort must

DEVELOPING FLOW MANAGEMENT STRATEGIES

Opportunities exist to enhance long-term health and maintenance of rivers and wetlands within the constraints of the present system of water management. Field trials and changes in government policy over the past ten years are already demonstrating the potential for changed operational strategies and management of individual wetlands (Wetlands Working Party, 1989; Jacobs, 1990; Jensen et al., 1994b; Ohlmeyer, 1991; Nicolson, 1993).

The process of developing more environmentally beneficial flow management procedures requires four stages:

1. definition of environmental flow requirements within regions and reaches;
2. identification of opportunities to alter flow management within the constraints of water delivery requirements and the physical constraints of the storages and control structures;
3. consultation with all user groups on the benefits and costs of various flow management options;
4. co-ordination of local and regional flow management needs within a catchment context.

Once the first three steps of this process are complete at local and regional scales, it will be possible to define flow management strategies for major rivers and regions. The potential for greatest benefit to wetland and river ecosystems lies in a co-ordinated approach to flow management strategies, which can optimise all opportunities to use available flows to meet the requirements of as many wetlands or regions as possible.

A complex analysis and decision support system will be required to balance the limited opportunities for altered flow management against environmental, irrigation and recreation requirements. Current work is directed at articulating environmental requirements and operational flexibility as the first stage in developing this management ability. It is proposed to co-ordinate flow management requirements through the proposed Sustainable Rivers Program, being developed in the Murray-Darling Basin Commission to integrate various recommendations involving flow management in relation to algal, fish and wetlands management, and irrigation and water quality management (Banens et al., 1994).

Technical information is being gathered by expert panels, with assessment of the Barwon-Darling completed and assessment of the River Murray under way. A separate panel will be convened for the Coorong estuarine lagoons. Hydraulic data are also being sought for sections of the Murray floodplain to establish flow/inundation relationships. Once the technical information is available, a major consultation exercise will be required to involve the many different user groups who would be affected by changes in river levels and flows.

FIELD TRIALS AND INVESTIGATIONS

Several projects under way in the Murray-Darling Basin are providing useful practical information about rehabilitation techniques for wetland sites, including guidelines for establishing environmental needs and monitoring of the benefits of hydrological changes. For example, guidelines established by Pressey (1987) indicate that Murray Valley wetlands require drying in late summer/autumn for a minimum of 2–4 months, at least once every 3 years, with re-flooding preferably coinciding with the spring high flows.

In South Australia an important field trial over three years included hydrological works, riparian rehabilitation and biological monitoring on six wetland sites in the Riverland region of the Murray floodplain (Jensen et al., 1994b, c). Overbank flows have been reinstated or created, stock and rabbit grazing controlled, and strategic replanting undertaken (Figure 13.5). The aim of the trial was to enhance wetland habitat value and to achieve maximum habitat biodiversity through hydrological manipulation and the implementation of broad-scale rehabilitation techniques.

At one site, the water levels are controlled to provide better quality for downstream irrigators, at the same time as making sure that ibis chicks have matured and left the nests in a major rookery on the lake. At a neighbouring lake, the regulator originally installed to prevent saline outflows from the lake is now also used to flood the dry lake bed in spring if there are extra flows in the river. This creates a rich temporary wetland which can support many thousands of ducks, swans, coots and other waterbirds through the breeding season.

All sites have been extensively inundated in five consecutive years. Results show a positive response to flooding in aquatic invertebrate populations (Suter et al., 1994) and strong natural regeneration of key salt-tolerant riparian species in rabbit-free areas (Muller and Nicolson, 1994). Four strong age classes have been identified in Murray cod (*Maccullochella peelii*) in association with overbank flows in the Chowilla floodplain region (Pierce, 1992).

However, the dominant floodplain tree species have not yet shown a positive response to the increased overbank flows at the rehabilitation sites. Some isolated areas of river red gum (*Eucalyptus camaldulensis*) germination were observed on the Murray floodplain at the Bulyong Island site following the 1990 inundation. Still, very few of these individuals survived more than 18 months, apparently due to a combination of local factors, particularly kangaroo grazing, moisture stress and soil sodicity (Jensen, in prep. 1998).

At the Disher Creek floodplain site, after five consecutive inundations, no evidence can be found of either seedling germination in the field or the presence of viable seed in the soil (Muller and Nicolson, 1994). Possible factors include residual soil salinity from the former evaporation basin on the site and lack of seed sources. It is also possible that seed is dispersed aerially rather than by water and the seed drop could occur several months after the flood recession. River red gum is also reported to drop massive amounts of seed every second year, a mechanism that would enhance the chance of coinciding with the preferred flooding frequency of 1 in 2–3 years for this species (Pressey, 1987). Further investigations will be conducted to establish potential seed sources at Disher Creek and management options to assist in the re-establishment of the key tree species.

Upstream in Victoria, the major wetland rehabilitation project is the development of the Integrated Watering Strategy for mid-Murray Wetlands (IWS Project). The IWS Project has the multiple aims of describing the hydrology and ecology of mid-Murray wetlands, developing methods for assessing the water requirements of wetlands and developing local and regional water management plans and strategies. Its key objective is to link the management of specific wetlands with river operations. The project seeks to improve understanding of hydro-ecological processes occurring within these environments for application to the management of Murray wetlands.

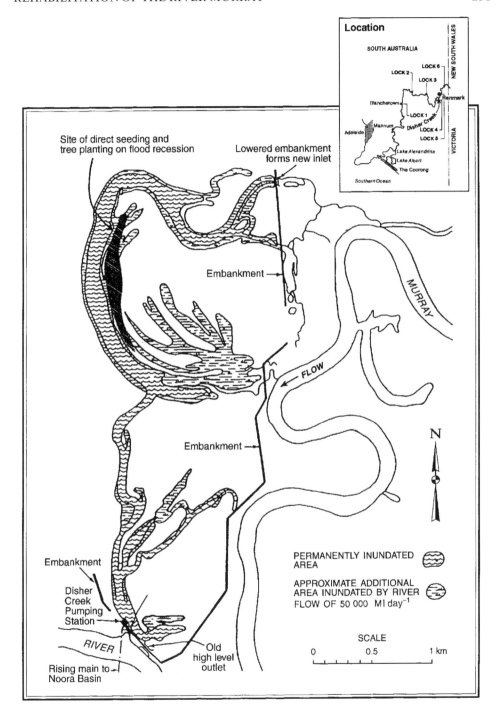

Figure 13.5 Rehabilitation works at Disher Creek Disposal Basin. Source: Jensen *et al.* (1994b)

The wetland systems included in the IWS Project, such as Barmah Forest, Gunbower Forest, Hattah Lakes, Wallpolla Island and Lindsay Island, were selected using the following criteria:

- The widest possible geographic area is included to achieve representativeness in the variety of wetland types and different effects of regulation.
- Wetlands should be of high conservation significance (based on naturalness and representativeness), need hydrological management to maintain their biological diversity and conservation status, and be amenable to hydrological manipulation using water held in storages, available from surplus flows or drawn from the irrigation supply system.

The emphasis in the New South Wales portion of the Murray-Darling Basin is currently on preventing further wetland degradation rather than on rehabilitation of degraded sites (Jensen et al., 1994a). The following case study demonstrates how important it is to first remove the cause of degradation before proceeding with rehabilitation.

The Gwydir River Valley extends from near Armidale on the northern tablelands of New South Wales to the Barwon River. The flat topography and self-mulching soils have allowed the development of about 220 000 ha of wetland. While most of this area is covered by coolabah (*Eucalyptus coolabah*) woodland, there is a core area of about 20 000 ha which is dominated by water couch (*Paspalum disticum*) and rushes (*Bolboschoenous fluviatilis*). Residents report substantial changes in the vegetation and the fauna of the watercourse country since the development of irrigated agriculture during the 1970s and 1980s. The abundance of waterbirds has declined and many areas once dominated by lush water couch are now being overtaken by exotic species like lippia (*Phyla nodiflora*).

The local community is working with the Department of Water Resources and the University of New England to develop local management plans for these wetlands (Jensen et al., 1994a). These plans will incorporate rehabilitation techniques such as revegetation, where necessary. Rehabilitation measures will not be likely to succeed unless an appropriate flooding regime can be restored.

The Department of Water Resources is preparing a water management plan that will ensure that wetland water needs are given full consideration in water management decisions. To reinstate a suitable flooding regime it will probably be necessary to restrict irrigators' access to surplus flows. An allowance from the upstream Copeton Dam may also be an option, but this will be particularly costly to the irrigation industry. On-the-ground rehabilitation measures have been postponed until these issues are resolved and more water is provided to the watercourse and the wetlands.

CONCLUSIONS

The rivers, floodplains and wetlands of the Murray-Darling Basin represent a highly valuable resource which is under threat from several factors, particularly changed water regimes, lost flood peaks, grazing and the introduction of exotic species. Co-ordinated, active management measures are required to maintain and rehabilitate these ecosystems in the interest of maintaining the general health of the natural resources which are the base for significant agricultural production, rural development and urban water supplies.

The factors causing degradation must be controlled and flood regimes must be reinstated as the first step in any rehabilitation or maintenance programmes. Larger-scale actions are required to achieve changes in flow regimes over a reach, region or whole river system. Local action at a wetland site can then maximise the benefits of increased flushing or flooding with local measures to direct flows, encourage regeneration or control grazing.

The emphasis should be on protecting and enhancing natural processes such as regeneration and breeding, allowing native species to reproduce successfully in the environment to which they are best adapted, thus utilising the power of natural systems to overcome degradation and work towards restoring ecosystem health.

The major negative impacts on wetlands in the lower Murray Valley are severe reduction of minor flood frequencies and overbank flows, grazing of riparian vegetation by stock and rabbits, and introduced fish, particularly carp. The first priorities are to reinstate the natural hydrological regime as far as possible and to encourage and protect regeneration and breeding processes.

Management of floodplains and wetlands should be based on mimicking the natural hydrograph wherever possible (within system constraints), particularly on key flood peaks which act as breeding cues for birds and fish or trigger plant regeneration. The height of the peak, duration of inundation and timing are all key factors in determining whether breeding will be successful.

The Murray-Darling Basin floodplain ecosystem is adapted to a highly variable hydrological regime, featuring extremes of conditions. It must be remembered that these extremes and short-term variations in conditions can be just as critical to the continued long-term survival of the ecosystem and individual species, as the average or predominant conditions. However, there are physical, economic, social and political constraints on opportunities to reinstate the natural hydrograph, due to current water supply commitments, and these will restrict the range of options for environmental management of water. In the lower regulated Murray, the most effective and practical opportunities for water management for environmental purposes occur during surplus and flood flows. Maximum use should be made of peaks in regulated and flood flows to increase extent, duration and height of floods to individual wetlands and across broad floodplain areas.

Operation of existing storages, weirs and floodplain structures could be adjusted to increase environmental benefits within the constraints of water delivery requirements. Procedures which should be reviewed include the process of weir reinstatement on a falling river, to mitigate the severe environmental effect of abrupt cessation of flows and dramatic falls in river level over short periods.

Natural regeneration of native species is the most effective method of broad-scale vegetation rehabilitation, and is best achieved by grazing controls combined with overbank flows or drying/wetting treatments.

Local native species are preferred for revegetation unless the site is too salinised for their survival. These species should have the best survival rates and least susceptibility to insect attack in their local habitat. Non-local native species should only be used for specific purposes, such as rehabilitation of changed habitats. Non-native species should not be used because of potential problems with uncontrolled invasions.

Introduced plant species should be discouraged and controlled. For example, European willows have a serious negative impact on riparian zones and their spread should be controlled and new plantings actively discouraged. Where possible, willows should be replaced with native riparian species. The recently formed National Carp Task Force is co-ordinating research on practical control options for this serious pest. The release in 1996 of rabbit calicivirus disease offers an unprecedented opportunity for control of this major pest, with local control work required to supplement the effect of the virus.

Extensive field trials and monitoring should be conducted, with collaboration between the different agencies throughout the catchment, combined with interaction and feedback of findings to modify management practices which contribute to the degradation of wetlands. This knowledge is also essential to understand the most effective rehabilitation techniques.

Most importantly, extensive community participation will be necessary in the process of determining what trade-offs must be made between environmental needs to sustain the health of the resource and the human demands placed on it. The outcomes will affect the entire community, both in the short term and the long term, and community support will be essential to achieve the necessary changes.

REFERENCES

Banens, R.J., Blackmore, D.J., Lawrence, B.W., Shafron, M.C. and Sharley, A.J. (1994) A program for the sustainable management of rivers of the Murray-Darling Basin, Australia. Paper presented at a joint South African/Australian conference on Environmental Health Indicators and Classification of Rivers, Capetown, February 1994. Murray-Darling Basin Commission, Canberra.

Boon, P.J., Calow, P. and Petts, G.E. (eds) (1992) *River Conservation and Management*. John Wiley, Chichester.

Boulton, A.J. and Lloyd, L.N. (1991) Macroinvertebrate assemblages in floodplain habitats of the lower River Murray, South Australia. *Regulated Rivers: Research and Management*, **6**, 183–201.

Briggs, S.V. and Maher, M.T. (1985) Limnological studies of waterfowl habitat in south-western New South Wales. II Aquatic macrophyte productivity. *Australian Journal of Marine and Freshwater Research*, **36**, 707–715.

Caldwell Connell Engineers (1981) *Katarapko Island Evaporation Basin Environmental Study*. For the South Australian Engineering and Water Supply Department, Adelaide.

Carter, J. and Nicolson, C. (1992) *Managing Wetlands of the River Murray in South Australia. A Guide for Communities of Common Concern*. South Australian Department of Environment and Planning, Adelaide.

Close, A. (1990) The impact of man on the natural flow regime. In McKay, N. and Eastburn, D. (eds) *The Murray*. Murray-Darling Basin Commission, Canberra.

Geddes, M. and Hall, D. (1990) The Murray Mouth and Coorong. In McKay, N. and Eastburn, D. (eds) *The Murray*. Murray-Darling Basin Commission, Canberra.

Jacobs, T. (1990) River regulation. In McKay, N. and Eastburn, D. (eds) *The Murray*. Murray-Darling Basin Commission, Canberra.

Jensen, A. (1983) *Lake Merreti. A Baseline Environmental Study*. South Australian Department of Environment and Planning, SADEP 61.

Jensen, A. (1993) Assigning Values to Wetlands and Natural Resources in the South East of South Australia. Background paper prepared for the Upper South East Dryland Salinity and Flood Management Plan Environment Impact Statement, South Australian Department of Environment and Land Management, Adelaide.

Jensen, A. (1998) The response to changed hydrology at Bulyong Island. In Jensen, A., Nichols, S., Nicholls, K. and Seaman, R. (eds) *Adding a Few Pieces to the 'Mysteries of the Murray' Jigsaw*. South Australian Department of Environment and Natural Resources, Adelaide.

Jensen, A. and Nicholls, K. (1996) Flow Management in the Lower River Murray. Working Paper, Wetlands Management Program, South Australian Department of Environment and Natural Resources, Adelaide.

Jensen, A., Lloyd, L. and Bennett, M. (1994a) Rehabilitation techniques for floodplain wetlands in

the Murray-Darling Basin. In *Managing Floodplain Wetlands in the Murray-Darling Basin*. Proceedings of Murray-Darling Basin Floodplain Wetlands Management Workshop, Albury, October 1992. Murray-Darling Basin Commission, Canberra.
Jensen, A., Nicolson, C. and Carter, J. (1994b) Restoration techniques for wetland systems: a case study in the South Australian River Murray Valley. Paper presented at the 1st International Lowland Stream Restoration Workshop, Lund, Sweden, 1991. South Australian Department of Environment and Natural Resources, Adelaide.
Jensen, A., Nicolson, C. and Carter, J. (1994c) Preservation and Management of Natural Wetlands in South Australian Murray Valley. *Wat. Sci. Tech.*, **29**(4), 325–333.
Mackay, N. and Eastburn, D. (1990) *The Murray*. Murray-Darling Basin Commission, Canberra.
Maltby, E. (1991). Wetlands and their values. In Finlayson, M. and Moser, M. (eds) *Wetlands*. International Waterfowl and Wetlands Research Bureau, Slimbridge, UK.
Margules and Partners, P. and J. Smith Ecological Consultants and Department of Conservation, Forests and Lands Victoria (1990) *River Murray Riparian Vegetation Study*. For the Murray-Darling Basin Commission.
Mitchell, D.S. and Frankenberg, J. (1993) Willows in Australia: weeds or beneficial plants. Discussion Paper, Murray-Darling Freshwater Research Centre, Albury.
Muller, K. and Nicolson, C. (1994) Abundance and viability study of *Eucalyptus camaldulensis* seed and *E. largiflorens* at Disher Creek, South Australia. South Australian Department of Environment and Natural Resources, Adelaide.
Murray-Darling Basin Ministerial Council (1987) *Murray-Darling Basin Environmental Resources Study*. State Pollution Control Commission, Sydney.
Murray-Darling Basin Ministerial Council (1989) *Murray-Darling Basin Natural Resources Management Strategy, Getting It Together*. Discussion Paper No. 2, The Council, Canberra.
Murray-Darling Basin Commission (1989). *Proceedings of the Workshop on Native Fish Management*. The Commission, Canberra.
Murray-Darling Basin Commission (1992) *Annual Report 1991–92*. The Commission, Canberra.
Murray-Darling Basin Commission (1994) *Wetlands Management Strategy*. The Commission, Canberra.
Nicolson, C. (1993) The development of a flow management strategy for the lower River Murray. Discussion paper for Flow Management Workshop, April 1993, South Australian Department of Environment and Land Management, Adelaide.
Ohlmeyer, R.G. (1991) *Report on Implications of Manipulating Water Levels in the River Murray Between Locks 1 and 10*. South Australian River Murray Wetlands Management Program, Murray-Darling Basin Natural Resources Management Strategy, South Australian Engineering and Water Supply Department, Adelaide.
Pierce, B. (1992) River Murray fish research: applying the results for Murray fish enhancement. Progress report for Murray-Darling Basin Natural Resources Management Strategy, South Australian Research and Development Institute, Adelaide.
Pressey, R.L. (1986) *Wetlands of the River Murray*. RMC Environmental Report 86/1, River Murray Commission, Canberra.
Pressey, R.L. (1987) *The Murray Wetlands in South Australia: Management Considerations and Research Needs*. Background Paper No. 5, Murray Valley Management Review, Department of Environment and Planning, Adelaide.
Sharley, T. (1992) Strategies for revegetating degraded floodplains. In *Catchments of Green*. Proceedings of a national conference on Vegetation and Water Management, Adelaide. Greening Australia Ltd.
Stone, A. (1991) Valuing wetlands. A contingent valuation approach. Paper presented at 35th Australian Agricultural Economics Society Conference, University of New England, Armidale, NSW.
Suter, P.J., Goonan, P.M., Beer, J.A. and Thompson, T.B. (1994) A biological and physico-chemical monitoring study of wetlands from the River Murray floodplain in South Australia. Report No. 7/93, Australian Centre for Water Quality Research, Adelaide.
Thompson, M.B. (1986) *River Murray Wetlands: Their Characteristics, Significance and Management*. For the Nature Conservation Society of South Australia and the Department of Environment and Planning, Adelaide.

Walker, K.F. (1986) The Murray-Darling River System. In Davies, B.R. and Walker, K.F. (eds) *The Ecology of River Systems*. Junk Publishers, the Netherlands.

Walker, K.F., Thoms, M.C. and Sheldon, F. (1991) The significance of the river-edge boundary for conservation of the River Murray. In Dendy, T. and Combe, M. (eds) *Conservation in Management of the River Murray System – Making Conservation Count*. Proceedings of the Third Fenner Conference on the Environment, Canberra, September 1989. South Australian Department of Environment and Planning in association with the Australian Academy of Science.

Walker, K.F., Thoms, M.C. and Sheldon, F. (1992) Effects of weirs on the littoral environment of the River Murray, South Australia. In Boon, P.J., Calow, P. and Petts, G.E. (eds) *River Conservation and Management*. John Wiley, Chichester, pp. 271–292.

Wetlands Working Party (1989) *Enhancing Wetlands*. South Australian River Murray Wetlands Working Party, Engineering and Water Supply Department and Department of Environment and Planning, Adelaide.

Part Three
IMPLEMENTATION: CASE STUDIES

14 Efforts for In-stream Fish Habitat Restoration within the River Iijoki, Finland – Goals, Methods and Test Results

TIMO YRJÄNÄ

North Ostrobothnia Regional Environment Centre, Oulu, Finland

INTRODUCTION

For some time, a high proportion of Finnish rivers have been used for floating timber for commercial purposes. In the 1950s, some 13 000 km of channels were being employed for this purpose, whilst the total length of the channels that could be employed for floating timber in the Finnish rivers and lakes is approximately 40 000 km (Lammassaari, 1990). The floating channels were being continually managed, with the most drastic modifications taking place in the late 1950s, when the channels were subjected to dredging by bulldozers. The most dramatic changes took place in small brooks and streams. Furthermore, 2000 dams were constructed in the running waters in Finland to store water for the floating of timber (Lammassaari, 1990). In addition to this, a great number of other guiding structures were installed to facilitate the transportation of timber. Lammassaari (1990) has evaluated that the value of the damages, caused by the dredging, to riverine and river spawning fisheries are several million ECUS. Ever since the end of the main floating period in the 1960s and 1970s, restoration of these channels has been ongoing, with the principal objectives being the improvement of fisheries, riverine habitat diversity and enhancement of recreational use.

THE IIJOKI RIVER SYSTEM AND MEASURES UNDERTAKEN TO FACILITATE THE FLOATING OF TIMBER

The catchment area of the River Iijoki (Figure 14.1) covers an area of 14 000 km^2. The length of the main channel is approximately 370 km, with a 250 m fall in altitude. A total of 70 km extending from the river mouth has been utilised for hydropower plants. The annual mean flow of the River Iijoki is 176 m^3 s^{-1}, with the maximum flow being 1429 m^3 s^{-1} (Ottavainen and Röpelinen, 1993). The target rivers for restoration measures within

---

Rehabilitation of Rivers: Principles and Implementation. Edited by L. C. de Waal, A. R. G. Large and P. M. Wade.
© 1998 John Wiley & Sons Ltd.

Figure 14.1 The catchment area of the River Iijoki, Finland

the Iijoki catchment are mainly rivers of stream order 2–3, the widths of which are 10–100 m, the mean flow 0.5–25 m^3 s^{-1}, and the maximum flow 10–500 m^3 s^{-1}. The gradient of these rivers ranges between 0.1 and 0.5%.

The River Iijoki runs from east to west stretching across the entire country, which has made it a very important route for the transportation of both products and people. At the height of timber floating activity, the floating channels within the Iijoki river system amounted to a total of 2614 km. As late as in the 1970s, the amount of timber floated on the river was 0.3–0.6 million m^3 per year.

There was an early need to dredge and clear the River Iijoki for the purposes of floating as the river system has an abundance of rapids. Approximately 200 km of Iijoki rapids previously suitable for smolt production have been dredged, this impact being magnified by the construction of 140 dams. The quantity of material removed during the dredging process amounted to over 1 million m^3, the tributaries of the river facing the greatest changes. A great majority of these instances of dredging took place illegally (Lammassaari, 1990), the result of which has been financial compensation to landowners (North Ostrobothnia Regional Environmental Centre, unpublished data). The river dredgings and construction of dams increased the area of commercial forestry radically in the 1950s, when the road network in northern and eastern Finland was insufficient (Lammassaari, 1990).

The dredging of the river led to a decrease in the area covered by water, since embankments were employed to isolate the shallow bays from the main channel, and the largest meanders were realigned. Another measure used, with a similar result, was the modification of the cross-section of the channel. These activities resulted in a situation where the riverine habitats were severly impaired and monotonised, and the diversity of the water velocity, water depth, and the substrate particle size were radically reduced.

RESTORATION OF THE RIVERS DREDGED FOR THE FLOATING OF TIMBER

Restoration of the rivers that had been dredged to facilitate timber floating began in Finland in the 1970s, and by 1995, some 180 restoration projects had been completed. Almost all these projects have been funded mainly by the Finnish Government. The projects have consisted of habitat restoration measures each covering an area of 950 ha of rapids, together with the enhancement of approximately 390 spawning grounds in the course of these projects. A majority of the projects dealt with entire river lengths. Only a few of these river restoration projects have included any monitoring or post-project evaluation (Yrjänä and Huusko, 1992; Yrjänä, unpublished data).

Restoration work to enhance sections of the catchment of the River Iijoki began in 1988, whereas field mapping of the area had already been going on for as long as 15 years before. The people involved in the Iijoki Project have been able to benefit from the experience gained in similar, earlier projects conducted elsewhere in Finland; personal encounters and visits to restoration sites having been the means through which the experience was delivered. Regular contacts have been maintained with organisations involved with restoration projects elsewhere in northern Finland.

By the end of 1996, restoration of the River Iijoki system had been completed in 9 of the tributaries (Pärjänjoki, Naamankajoki, Kouvanjoki, Livojoki, Loukusanjoki, Korpuanjoki, Kutinjoki, Siuruanjoki and Pushosjoki), with another 2 in the middle of the process (headwaters of the main stream and Korpijoki), and another 5 rivers or sections still unaltered. The time-scale of restoration has varied, with individual rivers taking between one and four years to restore. Costs of the whole restoration project in the Iijoki water system will be about 4.7 million ECU. In the period 1988–1996 costs have been 2.0 million ECU. The cost of habitat improvement work has varied between 3500 and 7000 ECU per hectare. The price of building a 100 m^2 spawning area has ranged between 850 and 1700 ECU.

All larger restoration measures conducted in Finland require a permit issued by the Water Court. The aim is to reconcile the needs of those owning and using the watercourse, and the restoration measures intended to repair the damage done to nature by the dredging activity, and to do this as early as possible in the planning phase of the project. The demands set by these factors often conflict with each other and, due to differences in local circumstances, various factors have different weights in different areas. The Water Court will ultimately decide in each individual case whether the plan has succeeded in combining the various uses of the water course in question and therefore whether to issue a permit for the river in question.

GOALS OF RESTORATION

The main goal of restoration has been to minimise the damage caused by the dredging and other measures undertaken to facilitate the floating of timber. The minimisation of this damage involves attempts (i) to increase the diversity of the riparian enviroment, (ii) to enhance fish habitats, and (iii) to rehabilitate the riparian landscape to a more natural, and less unaltered state. Since, however, the restoration work cannot take into account the habitat requirements of all species present in the area, an increase in the overall habitat diversity is an essential principle of the restoration attempts.

METHODS

Material removed from the channel during the dredging process for the floating of timber was originally placed in the embankments surrounding the floating channel and in the deflectors constructed to guide the timber. This meant that the material needed for the construction of various structures of the restoration projects was readily available at the sites and, in most places, the only thing needed was the dismantling of the embankments. Concrete structures were utilised solely for the construction of submerged weirs employed to restore the water levels in larger lakes and pools.

RESTORATION OF THE PATTERN OF FAST FLOWING–SLOW FLOWING SECTIONS

The mean length of the slow flowing sections in the tributaries of the River Iijoki range between 100 and 1500 m, with the slow flowing sections being followed by rapids or fast flowing areas 300–500 m in length. Dredging operations took place mainly in the rapids areas, which meant a decrease in the slow flowing–fast flowing section pattern and, consequently, a loss of most of the fish habitats (Swales, 1994). When the rapids areas are restored, the slow flowing–fast flowing section will be approximately the same as it was in

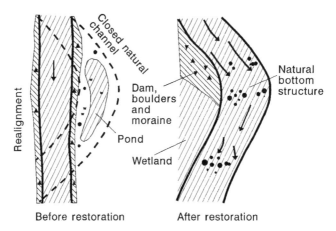

Figure 14.2 Restoration of a closed natural channel

the previous, more natural state of the river. This will mean a general increase in water depth and a decrease in water velocity in the slow flowing sections.

RECONNECTION OF DRAINED SECTIONS OF THE CHANNEL

Dredging and clearing of the channels also led to realignment of the rivers, and the cutting off of shallow bays and side channels (Figure 14.2), resulting in a situation where the reproduction sites and nursery areas of the migrating fish populations in the River Iijoki decreased from 193 ha in the 1950s to 109 ha in the 1980s. The reconnection of drained sections of the river back to the river system has been found to be an effective means of increasing the morphological diversity of a river (Huusko, 1995).

HABITAT RESTORATION IN A DREDGED CHANNEL

Boulder dams

The main in-stream method employed in the restoration of the River Iijoki to increase the diversity lost by dredging has been the construction of boulder dams (Figure 14.3). The boulder dams have varied in structure from a few large boulders set side by side, to the raising of the entire level of the river for a distance of 100 m.

A natural-looking boulder dam can be created by setting large boulders across the river so that the dam is 5–15 m wide, and by partly filling up the inside of the structure with cobble and rubble. Wesche (1985) states that a boulder dam is most useful in a river with a gradient of 0.5–2.0%, and in the River Iijoki boulder dams have been utilised in the river sections with a gradient of 0.3–1.5%. With gradients higher than 2.0%, boulder dams cannot be constructed without the aid of special supportive structures (Wesche, 1985).

Large boulder dams have turned out to be the most effective single restoration measure to increase smolt production of salmon fish in the dredged river sections (Näslund, 1987; Jutila et al., 1994). Wide boulder dams have the advantage of not only supplying plenty of hiding places for fish but also they are effective in diversifying the habitat available for fish in a larger area as far as water depth and water velocity are concerned. The studies

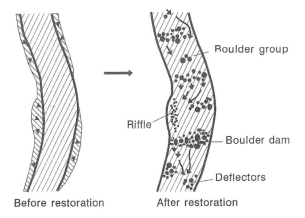

Figure 14.3 The main in-stream methods used in habitat restoration in dredged rivers in the Iijoki Basin, Finland

conducted in 5 tributaries of the River Iijoki, by Jutila et al. (1994) (Pärjänjoki, Naamankajoki, Kouvanjoki, Loukusanjoki and Livojoki) the test sites restored with the aid of large boulder dams showed that a year after overdense plantings, the mean density of 1-year-old trout present in the area was 16–23 individuals per 100 m². In the rapids restored with the aid of other types of measures and in the unrestored ones, the density of 1-year-old trout was less than 10 individuals per 100 m² (Jutila et al., 1994). The number of trout fry present in the vicinity of the boulder dams constructed during the restoration of the River Iijoki has been up to 40–80 individuals per 100 m² (Yrjänä, unpublished data).

Deflectors

Deflectors (Figure 14.3) can also be employed in river restoration in order to increase diversity in water velocities, deepen the channel, and assist in the development of erosion areas and pools trapping gravel. Since they are constructed of relatively coarse material, deflectors also offer cover for benthic macroinvertebrates and fishes.

A deflector is a simple and inexpensive structure to build. Wesche (1985) has found that the length of a deflector has to be at least half the channel width for the deflector to be successful. Brookes (1988) states that deflectors can only be effective in guiding the current if they are placed in sections where the water velocity exceeds 0.6–0.9 m s^{-1}. In the Iijoki Project, deflectors have been employed to assist other structures and in places where boulder dams cannot be utilised as a result of boat traffic.

Groups of boulders

Single boulders and groups of boulders have been utilised in the Iijoki with the specific purpose of increasing the number of feeding and cover areas for larger fish. The boulders have been used to provide better conditions for fish swimming through sections with a fast flow, such as areas in the vicinity of submerged weirs. In rapids with moderate water velocity, emplacing groups of boulders has been the main method of restoration.

The size of the individual single boulders placed in the stream should be relatively large. Wesche (1985) recommends the use of boulders 0.6–1.5 m in diameter. In a northern climate, it is vital that the boulders are large enough to rise above the water surface, which means that they will assist the formation of the ice cover and, consequently, shorten the period of anchor and frazil ice formation, which can be deleterious for aquatic organisms. The boulders will also prevent ice cover from descending to the bottom of the water column during the low flow period in the winter and offer hiding places during anchor ice formation. Calkins (1990) has described the formation of non-floating ice in small rivers.

Cobble ridges formed from former embankments

The former embankments lining the dredged rivers may be utilised by leaving behind some cobble material to create longitudinal ridges parallel to the channel. These ridges offer ideal cover for fingerlings and, furthermore, if some rubble and gravel is present, they may also provide suitable spawning grounds. Gore (1985) states that the production of benthic invertebrates is most abundant in places with a swift current when the substrate material is either gravel or cobble. Electrofishing experiments conducted in the River Iijoki have shown densities of brown trout (*Salmo trutta*) fry to be highest in the ridge

Figure 14.4 Construction of artificial spawning ground

areas formed from the former embankments. The density of 0-year-old trout in the best areas was 72 individuals per 100 m². In addition, the same area showed a 1- and 2-year-old density of 14 individuals per 100 m² (Jutila, unpublished data).

Enhancement of spawning habitat

Several of the river-spawning fish bury their eggs in gravel. In a river subjected to dredging, the amount of unsorted bottom material is definitely less than in a river in its natural state (Brookes, 1988). The dredging of the River Iijoki led to a situation where the gravel could only accumulate in the pool areas. Poorly sorted gravel could be found in the upper regions of pools, where its presence did little to assist the spawning of the river-spawning fish.

The spawning habitats of river-spawning fish have been improved by the positioning of natural gravel in suitable in-stream areas with a relatively swift current, and by the creation of artificial spawning grounds using sorted gravel (Figure 14.4). The work done in the larger rivers has focused on the positioning of natural gravel. The mean particle size of the gravel used in artificial spawning grounds was approximately 15 mm, with the smallest particles being 8 mm and the largest ones 45 mm in diameter. The mean thickness of the gravel layer was 20 cm. Between 1990 and 1991, 50 artificial spawning grounds were created in four of the tributaries of the River Iijoki. The work was planned using substrate size, water depth and water velocity criteria drawn from the experiments conducted in a restored rapids section of a river in Central Finland (Järvisalo et al., 1984). Results from the experiment with the spawning site selection and the evaluation of artificial spawning grounds (see below) were later taken into account in the planning of new spawning grounds.

EVALUATION AND DEVELOPMENT OF METHODS

MONITORING WATER QUALITY DURING RESTORATION

Water quality during the restoration activities was monitored at every restoration site within the Iijoki river system. Between 1988 and 1992, 363 water samples were analysed for turbidity, suspended solids content, pH, and iron content. The greatest amounts of suspended solids found were 100–160 mg l^{-1}, the highest values of turbidity being 70–700 FTU, and the greatest amounts of iron content being 6000–26 000 µg l^{-1}. Mean changes were much smaller than the changes mentioned above. pH showed no variance either up- or downstream from the restoration sites.

Since the effects of the restoration activities at each site were temporary, lasting in general for only a few days, none of the changes in water quality were found to be deleterious to the aquatic organisms involved. This was the case even when the results were compared with the tolerance criteria stated by Alabaster and Lloyd (1982), since one must take into account that the majority of the iron content found was probably attached to suspended solids.

THE EFFECT OF RESTORATION ON BENTHIC FAUNA

The short-term effects of the restoration activities during the Iijoki projects have been analysed, and results have shown that the restoration was immediately followed by a decrease in the number of benthic fauna. The recovery period needed by the macroinvertebrate fauna to return to levels observed prior to the disturbance was, however, very short and consisted only of a period of two weeks. The restoration practises used in the study site were rather moderate ones. The study site was rehabilitated by digging holes and inserting boulders (Laasonen *et al.*, 1993; Tikkanen *et al.*, 1994).

Information concerning the long-term effects of restoration was gained from the comparison between the state of benthic fauna in tributaries of three boreal drainage systems, the Rivers Iijoki, Oulankajoki and Oulujoki. Study sites restored 0, 1, 3, 8 and 16 years before sampling were compared to three unrestored dredged rivers and four rivers in their natural state. Samples were taken in October by the kick-sampling method. Abundance of macroinvertebrates were lowest in the +0 rivers and highest in the natural state rivers. Invertebrate abundances in all other restored streams were somewhat higher, but did not differ significantly from abundances in the dredged channels. There is no great demand to improve the habitat of benthic invertebrates in a typical Finnish boreal river dredged for timber floating although the restoration in most cases seems to increase the habitat heterogenity (Laasonen *et al.*, 1998).

The restoration activities increased the heterogenous nature of the river bottom, consequently raising the retention capacity of organic material. Changes in bryophyte cover is another criterion that may explain the differences in benthic fauna. Bryophyte cover even in the test sites restored eight years prior to the study was nowhere near to the cover present in the undredged test sites (Tikkanen *et al.*, 1994).

Effects of restoration to the population of the mussel *Margaritifera margaritifera* (L.) have been monitored by scuba-diving methods in the River Iijoki in 1990 and 1994. The interpolated number of specimens were almost the same before and after restoration (6645 and 7185 individuals) (Valovirta and Yrjänä, 1996).

MONITORING OF FISH DENSITIES AT RESTORATION SITES

Fish densities in the various sections of the River Iijoki, both in the sections restored in various ways and in the unrestored sections, were monitored by electrofishing between 1989 and 1992. The smolt production potential of the study areas was tested by the annual plantings of large numbers of 0-year-old and 1-year-old trout fry (Jutila et al., 1994).

The capacity of examined rapids showed no statistically significant difference to maintain stocked 0-year-old trout between the areas restored through a variety of methods and the unrestored areas, whereas the biomass and density of yearlings was significantly greater in the restored areas than in the unrestored ones. The greatest densities of yearlings were found in the test areas that had been restored by constructing large boulder dams (Jutila et al., 1994). A large boulder dam, in this context, means filling the channel with boulders in several layers for a longer distance. Since restoration did not seem to increase the habitat area of 0-year-old trout fry, their habitat demands have received more attention in the restoration operation conducted in the summer of 1994 at the River Iijoki rapids.

Population densities of other common running-water fish, such as minnow (*Phoxinus phoxinus*), bullhead (*Cottus gobio*) and Siberian bullhead (*Cottus poecilopus*), were considerably lower in the restored test areas than in the unrestored ones. Possible reasons for this could be changes in habitat structure which have resulted in changes in competition and predation system. The restoration seemed to have no effect on the habitat of grayling (*Thymallus thymallus*) (Jutila et al., 1994).

To improve the evaluation methods concerning the effect of restoration on the quality and quantity of the habitats of various fishes, the PHABSIM model, used with success on North American streams to estimate habitat quality (*cf.* Bovee, 1982), was applied to the Iijoki (Huusko and Yrjänä, 1997).

SPAWNING SITES SELECTION EXPERIMENT

The selection of the spawning sites for the brown trout was tested in a fenced area in the Livojoki, which is one of the biggest tributaries of the Iijoki. The test was arranged in the autumn of 1991. Ten female trout ready to spawn were placed in the area together with ten male trout. The area included two artificial spawning grounds made from sifted gravel. Natural gravel/sand and rubble were also available in the test area (Tähtinen and Yrjänä, unpublished data).

All the females laid eggs during the test, and nine of the spawning nests were located. All the nests were situated in the vicinity of a structure guiding the current and offering cover: five nests were protected by one or more boulders, two were situated in the vicinity of both the bank and a boulder, one was situated near the bank with no other protection, and one was situated next to the fence isolating the test area. The water velocity above the nests was usually greater than in the surrounding area. Four of the nests were situated on artificial spawning grounds. The results of the tests led to the conclusion that, besides gravel, it is necessary to place boulder groups on the spawning ground to offer cover and to increase local velocities. The results also indicate that the spawning grounds will be more successful if they are small, preferably only a few metres in diameter, since the central areas of a large gravel bed seem to remain unused (Tähtinen and Yrjänä, unpublished data).

Table 14.1 The operationability of artificial spawning grounds in the tributaries and headwaters of the River Iijoki. The spawning grounds were installed in 1990–1991 and were monitored in 1993 (Seija Pekkala, unpublished data)

Number of spawning grounds installed	47
Gravel *in situ*	9 sites
More than 50% of gravel *in situ*	13 sites
Less than 50% of gravel *in situ*	11 sites
No gravel present	14 sites
Secondary spawning grounds emerged	18 sites
Gravel deposited in the spawning grounds	1670 m^3
Gravel found *in situ*	900 m^3
Gravel found at new sites	350 m^3
Gravel no longer found	420 m^3

EVALUATION OF ARTIFICIAL SPAWNING GROUNDS

Forty-seven of the artificial spawning grounds created in 1990 and 1991 in the River Iijoki were tested in the summer of 1993. The results are shown in Table 14.1. The area covered with gravel and the thickness of the gravel layer were measured at each spawning site, together with water velocity. The factors describing the sites were recorded. If some of the gravel was no longer present at a site, the river channel was investigated 200–500 m downstream, which is to say to the site of the next pool. Gravel emanating from the artificial spawning site was easily detectable from natural gravel because of its unnaturally uniform particle size. After a period of 2 years, approximately 75% of the gravel beds were still operational (Pekkala and Pekkala, 1995).

During the two year period, all spawning grounds had collected a certain amount of sediment, with no correspondence to the water velocity. Spawning grounds with water velocities as low as 0.3–0.4 m s^{-1} had gathered mud on their surface. Only one of the 18 secondary spawning grounds had gathered mud, this site being the only one with periphyton present.

The analysis of the features of the artificial spawning grounds that remained *in situ* and the secondary ones showed that the best places for artificial spawning grounds are places with accumulations of natural gravel, widenings of the channel inside an area of rapids, the inner curves of bends, and large side-channels.

SUMMARY

Starting in 1988, the Iijoki river system has been the object of a comprehensive restoration project in a river previously subjected to dredging to facilitate floating of timber. The main objective at the macrohabitat level has been to restore the slow flowing–fast flowing pattern and to reclaim the original channel. The habitats in the dredged channels have been improved by utilising the boulders and the cobble and rubble originally removed from the channels, for the construction of boulder dams, deflectors, gravel beds and groups of boulders. The spawning opportunities of anadromous fish have been increased by the restoration of natural gravel beds and by the construction of artificial spawning grounds from sifted gravel. The effect of the restoration activities on the water quality in the River Iijoki has been monitored, and the results have shown that the changes in the water quality during the restoration operations are local and that their effect on the aquatic fauna has been fairly restricted.

The development of benthic fauna has been studied at the Iijoki restoration sites by Laasonen *et al.* (1993, 1998) and Tikkanen *et al.* (1994), and the development of fish by Jutila *et al.* (1994). The restoration has been shown to disturb the benthic fauna to some extent, but the recovery after the operation has been extremely swift. Bryophyte cover has been shown to grow very slowly after the restoration activities. Despite the fact that the project has been shown to increase the habitat of 1-year-old trout (*Salmo trutta*), it seemed to have no effect on the habitat of the youngest trout fry, which was also the case with the numbers of grayling (*Thymallys thymallus*) fry in the rapids sections. Other typical running-water fish species, such as minnow (*Phoxinus phoxinus*) and bullheads (*Cottus gobio* and *Cottus poecipolus*), were considerably less abundant in the restored areas than in the unrestored ones. Possible reasons for this could be the changes in habitat structure, which have resulted in changes in competition and in the predation system.

Results of experiments with artificial spawning grounds showed that the gravel beds remain best *in situ* when situated in the inner curves of the channel and in channel widenings. Other potential places for artificial spawning grounds are the side-channels and sites with accumulations of natural gravel. Tests at the trout spawning sites led to the construction of smaller spawning grounds and the addition of boulder groups at the gravel beds. The restoration of the running waters within the Iijoki river system will continue for at least another five years, with the methods under constant revision.

ACKNOWLEDGEMENTS

The English translation of the chapter was carried out by Ms Eeva-Kaisu Kemppainen. Thanks are also due to several colleagues at the North Ostrobothnia Regional Environment Centre, who have kindly co-operated with the publication of this chapter by rendering a considerable amount of unpublished material at my disposal.

REFERENCES

Alabaster, J.S. and Lloyd, R. (1982) *Water Quality Criteria for Freshwater Fishes*. Butterworths, London.
Bovee, K.D. (1982) A guide to stream habitat analysis using instream flow incremental methodology. Instream Flow incremental Paper No. 12. FW/OBS 82/26. US Fish and Wildlife Service, Fort Collins, Colorado.
Brookes, A. (1988) *Chanelized Rivers: Perspectives for Environmental Management*. John Wiley, Chichester.
Calkins, D.J. (1990) Winter habitats of Atlantic salmon and brook trout in small ice-covered streams. *Proceedings of the 10th Symposium of Ice*, Volume 3, IAHR, pp. 113–126.
Gore, J.A. (1985) Mechanisms for colonization and habitat enhancement for benthic macroinvertebrates in restored river channels. In Gore, J.A. (ed.) *The Restoration of Streams and Rivers*. Butterworth, Stoneham, pp. 103–164.
Huusko, A. (1994) The effects of restoration to the amount and quality of fish habitats at certain sites in the River Iijoki. Finnish Game and Fisheries Research Institute, Kalaraportheja 14 (in Finnish).

Huusko, A. and Yrjänä, T. (1997) Effects of in-stream enhancement structures on brown trout habitat availability in a channelized boreal river: a PHABSIM approach. *Fisheries Management and Ecology*, **4**, 453–466

Järvisalo, O., Heikkilä, T. and Kärkkäinen, P. (1984) *The Spawning Habitat for Brown Trout in the Restored Rapids of Äyskoski*. Publications of the Water and Environment Administration, **225** (in Finnish).

Jutila, E., Karttunen, V. and Niemitalo, V. (1994) Better one stone in the rapid than ten on the bank – Influence of various restoring methods on the parr densities of brown trout in the rapids of the tributaries flowing into the Iijoki River. Kalatutkimuksia 87. Finnish Game and Fisheries Research Institute, Helsinki (in Finnish with Englih summary).

Laasonen, P., Muotka, T., Tikkanen, P. and Kuusela, K. (1993) In Tuomisto, J. and Ruuskanen, J. (eds) *Proceedings of the 1st Finnish Conference of Environmental Sciences*. Kuopio University Publications C, Natural and Environmental Sciences 14, pp. 151–154.

Lammassaari, V. (1990) *Floating and Its Effects on Watercourses*. Publications of the Water and Environment Administration, Series A 54 (in Finnish with English summary).

Laasonen, P., Muotka, T. and Kirijärvi, I. (1998) Recovery of macroinvertebrate communities from stream habitat restoration. *Aquatic conservation: Marine Freshwater Ecosystems*, **8**, 101–113.

Näslund, I. (1987) Effects of habitat improvement on the brown trout (*Salmo trutta* L.) population of a North Swedish River. Information från Sötvattenslaboratoriet, Drottningholm 1987 (3) (in Swedish with English summary).

Ottavainen, P. and Röpelinen, J. (1993) Flood control in the middle course of the river Iijoki. *Vesitalous*, **34**(6), 8–12 (in Finnish).

Pekkala, J. and Pekkala, S. (1995) Evaluation of artificial spawning sites. In Yrjänä, T. (ed.) *Restoration of Former Log Floating Rivers using Iijoki Water System as an Example*. Publications of the Water and Environment Administration, Series A (in Finnish with English summary).

Swales, S. (1994) Strategies for conservation of stream fish populations. In Cowx (ed.) *Rehabilitation of Freshwater Fisheries*. Blackwell Scientific, Oxford, pp. 34–47.

Tikkanen, P., Laasonen P., Muotka, T., Huhta, A. and Kuusela, K. (1994) Short term recovery of benthos following disturbance from stream habitat rehabilitation. *Hydrobiologia*, **273**, 121–130.

Valovirta, J. and Yrjänä, T. (1996) Effects of restoration of salmon rivers on the mussel *Margaritifera margaritifera* (L.) in Finland. Council of Europe, Convention on Conservation of European Wildlife and Natural Habitats T-PVS (96)51, pp. 31–48.

Wesche, T.A. (1985) Stream channel modifications and reclamation structures to enhance fish habitat. In Gore, J.A. (ed.) *The Restoration of Streams and Rivers*. Butterworth, Stoneham, pp. 103–164.

Yrjänä, T. and Huusko, A. (1992) Enhancement of fish habitat in Finnish rivers. *Suomen Kalastuslehti*, **99**(6), 26–27 (in Finnish).

15 Rehabilitation of the Acque Alte Drainage Canal on the River Po Alluvial Plain, Italy

BRUNA GUMIERO,[1] GIANPAOLO SALMOIRAGHI,[1]
MARCO RIZZOLI[2] and ROBERTA SANTINI[3]
[1] Department of Biology, University of Bologna, Italy
[2] Provincia di Bologna, Bologna, Italy
[3] Unità Sanitaria Locale 26, San Giovanni in Persiceto (BO), Italy

INTRODUCTION

BACKGROUND

Systems of parallel ditches or dikes on lowlands are useful for enhancing agricultural production. However, the same ditches and dikes may cause severe problems when they drain rainfall away too quickly. This rapid drainage of agricultural lands through ditch systems significantly increases pollution of watercourses. Beside this the drainage can significantly lower the local water table where wetlands serve to replenish groundwater (Heckman, 1990).

Some studies (Petts, 1990; Gardiner, 1991; Kajak, 1992) have indicated the basic necessity of including bufferstrips in any rehabilitation of such canals. The terrestrial–aquatic ecotone is a region of hydrological and biological control of nutrient flux. Because riparian zones link the stream with its terrestrial catchment, they can modify, incorporate, dilute or concentrate substances before they enter a lotic system. Forested and grass vegetated buffer strips have been shown to reduce shallow subsurface inputs of nutrients from cultivated lands to streams (Peterjohn and Correll, 1984). The retention and transformation processes, vegetation uptake and microbial denitrification, work together to provide an effective buffer zone that can protect the aquatic environment (Haycock et al., 1993). This chapter examines an approach to rehabilitation on the Acque Alte Canal to the north-east of Bologna in the Po Basin.

EVOLUTION OF THE CANAL SYSTEM ON THE PO RIVER PLAIN

In the original Po River Plain environment the riparian forests were dominated by *Quercus robur*, *Carpinus betulus* and *Ulmus minor*, except in the wetlands where *Fraxinus oxycarpa* and *Populus alba* were more prevalent. Along the dikes and rivers the riparian vegetation was composed of *Alnus glutinosa*, *Populus nigra*, *Populus alba*, *Salix eleagnos*, *Salix triandra*, *Salix alba* and *Salix cinerea*.

Rehabilitation of Rivers: Principles and Implementation. Edited by L. C. de Waal, A. R. G. Large and P. M. Wade.
© 1998 John Wiley & Sons Ltd.

In the past, the Po River Plain was subjected to marked modifications essentially caused by drainage to increase the land area for agriculture. In more recent times other uses have been added on to the original and sole function of drainage. Because of increased urbanisation, as well as industrial and zootechnical activities, the dikes still offer the most economic and consequently the most used way of removing waste waters. At the same time they are utilised to impound water for irrigation needs by agriculture.

The relevant reduction of the wet surface in the Po Plain from $42\,000\,km^2$ in the last century to $7\,km^2$ today has caused a significant reduction in biodiversity. The wetlands are necessary to protect the biodiversity of the surrounding areas, which have been substantially altered. In this situation, the artificial canals represent suitable environments for the conservation and reproduction of many species that find shelter in them. Unfortunately there are many factors which contribute to the reduction of this environment quality: (i) the hydraulic management exclusively for the conveyance of waste water (in winter) and for irrigation (in summer); (ii) the extremely linear channel shape (without vegetation) required to reduce the cost of maintenance, and (iii) the maintenance of maximum hydraulic efficiency for water speed. Increased interest in the environment and the consequent need to utilise these ecosystems for recreational objectives brought about initial ecological studies to widen the limited knowledge of the drainage canal systems in Italy. The Po is one of the first to be studied in this way.

ACQUE ALTE CANAL

The development of a wide drainage system began in the 10th century on the Po River Plain, north-east of Bologna, affecting an area of 29 000 ha. Today 24 000 ha of this are used for agriculture and there are about 470 km of dikes. The Acque Alte Canal represents the main hydraulic artery of this network of dikes (Figure 15.1) and its initial plan and construction date back to 1487. It is an artificial waterway, with banks from 0.80 to 5.5 m high, built to achieve maximum hydraulic efficiency. This dike is the primary means of drainage and conveyance of the sewage effluent via seven ditches from an area of 23 000 ha. Besides this, it contributes to flooding downstream after a period of heavy rainfall, therefore the safety of the area depends on its efficiency. The Acque Alte Canal is also the largest surface water reserve for 16 000 ha of agricultural land.

Today the Acque Alte Canal is managed through a particular hydraulic regime: for eight months of the year it becomes a lotic system for drainage needs, whilst during the summer months it becomes a lentic environment for irrigation needs. In the summer period about 17 km of the Acque Alte Canal are dammed at three different points and become a reservoir of $565\,000\,m^3$ capacity (Figure 15.1). During the three months of summer it receives a total hydraulic load of about $8.3 \times 10^6\,m^3$: $4.0 \times 10^6\,m^3$ coming from the natural down flow of the catchment basin, $1.0 \times 10^6\,m^3$ from the urban effluences and $3.3 \times 10^6\,m^3$ are collected and pumped from a border dike called CER (Canale Emiliano-Romagnolo) which is fed by the Po River. The exit water volumes are also divided: irrigation accounts for $2.0 \times 10^6\,m^3$, evaporation for $1.0 \times 10^6\,m^3$, dispersion for $0.8 \times 10^6\,m^3$, downstream flow for $1.0 \times 10^6\,m^3$ and drain overflow for $3.5 \times 10^6\,m^3$. Today the quality of the water and surrounding area substantially reduces the historic and naturalistic value of the Acque Alte Canal.

The main aim of ecological investigations, which began in 1991, is to back up a rehabilitation programme at the catchment scale. It was felt that quasi-pristine

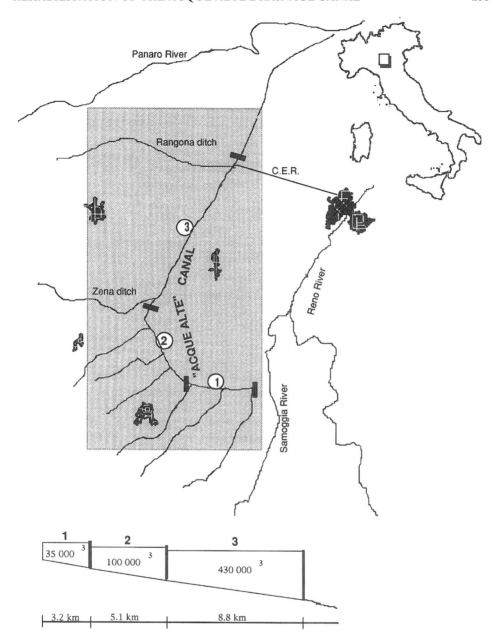

Figure 15.1 Study site (shaded) and surrounding areas. At the bottom is the scheme of the Acque Alte Canal during reservoir conditions. The figure indicates the three reaches (1, 2 and 3). C.E.R. = Canale Emiliano Romagnolo

conditions should only serve as a point of reference and not as a goal for stream restoration because a truly pristine state is not realistic in artificial canals and their catchment basins where humans have modified the land use and cover to a large extent. The restoration programme aims to create a system with a dike in dynamic equilibrium that supports a self-sustaining and functionally diverse community assemblage (Osborne et al., 1993). An interdisciplinary study group was established in order to examine the various aspects of the three reaches and the two different kinds of hydrological management (lentic/lotic) of the Acque Alte Canal by several investigations (Figure 15.2).

MAIN RESULTS

THE POLLUTANT EFFLUENTS

The pollutant sources have been divided into three different categories: urban, industrial and zootechnical, all of which have been transformed into 'population equivalents' (as proposed by the European Union, i.e. the organic degradable load having a BOD_5 of 60 g of oxygen per day) (Figure 15.3 and Table 15.1). The population equivalent could be reduced from 528 000 to 138 000 considering the case of the best available technologies to manage the three different wastes (Table 15.1). The primary source of pollutant effluent in the catchment basin is livestock effluents (81% of the total) but in the medium and final reach basin, domestic effluents are also important. It is evident that there is a great difference in spatial distribution for both the quality and quantity of pollutants. If the

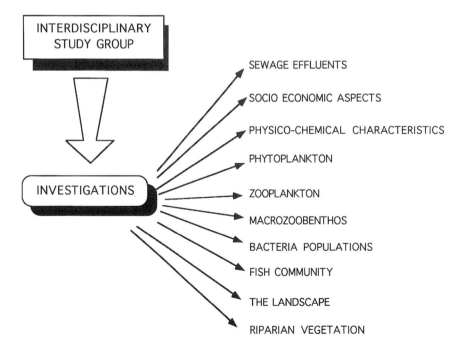

Figure 15.2 The investigations made by the interdisciplinary study group

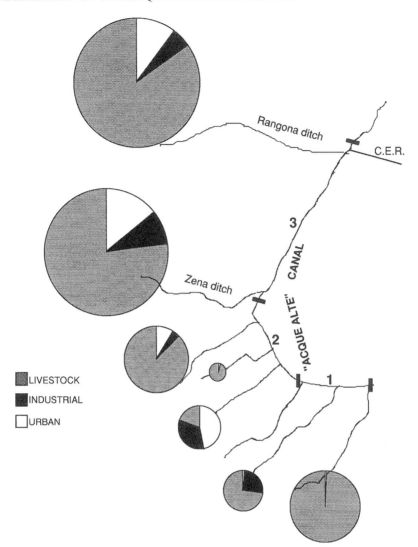

Figure 15.3 Load of organic pollutants originating in the catchment basin of the most important tributaries of the Acque Alte Canal in terms of population equivalent

data referred to in the three reaches are analysed, it can be seen that, during the summer, the pollutants are principally present in the third reach (71% of the total).

WATER QUALITY

The geographical location and high solar radiation intensity, due to a lack of riparian vegetation, cause high summer water temperatures with a maximum of 26 °C. In the study no stratification was observed in the first and second reaches, but occasionally a low thermic gradient (1.5 °C per 3 m of depth) was recorded in the third reach. Dissolved

Table 15.1 Organic pollutants originating from domestic sewage, industrial and livestock wastes in the catchment basin

Sources	Urban	Industrial	Livestock	Total
Raw				
Population equivalent	64 250	36 027	427 914	528 191
	12%	7%	81%	100%
Treated				
Population equivalent	44 390	10 770	82 550	137 710
	32%	8%	60%	100%

oxygen concentrations show an evident difference between superficial supersaturation (177%) and a strong deficit (6%) near the bottom (Table 15.2). BOD and COD values are rather high and reveal a considerable quantity of organic matter.

As far as the assessment of N and P concentrations is concerned, it must be pointed out that they were well above the allowable limit for eutrophic water. Consequently the quantity of macronutrients was high enough to support algal blooms and a high growth of macrophytes (Chiaudiani and Vighi, 1974). The nitrogen and phosphorus concentration rate was rather variable both in space (over the three different reaches) and in time (the two different hydraulic managements) (Figure 15.4).

THE BIOLOGICAL COMPONENTS

In the Acque Alte system, 33 species of phytoplankton were found and separated into five different groups (Table 15.3). The specific variability on each reach did not exceed 27 species, of which at least 18 were common. The phytoplankton community was partly composed of species with high body size, typical of environments with turbulent waters (Harris, 1986). Besides this, there were many species with exponential growth strategies suitable to a new recolonisation every year. In all three reaches (Figure 15.1) the community was represented mostly by Diatomeae and Chlorophyceae when the lotic system was in operation, while during impoundment Peridineae and Cyanophyceae became dominant (Figure 15.5).

In terms of water storage, the greater part of the phytoplankton community was considered edible (Burns, 1968; Vanni, 1987) and there was available food resource for grazing zooplankton (Gliwicz, 1980; Knisely and Geller, 1986). There was a higher exploitation of the phytoplankton resource in the second and third reaches than in the first (Figure 15.6).

However, the temporal trend of edible phytoplankton and zooplankton biomass did not always account fully for the survival of grazing organisms. This was caused by the high quantity of organic matter found. In these kinds of environments the fine and ultrafine particulate organic matter entered into a grazing process utilising the zooplankton. The zooplankton diversity was extremely low (Table 15.4). Only a few species of Rotatoria were found, although because of their parthenogenetic ability, a higher diversity could have been expected.

No significant differences were found on the macrozoobenthic communities among the three reaches (Table 15.5). Most of the species were gathering-collectors with only one species of predator, *Ischnura elegans* (Insecta, Odonata), and all were very well adapted to

Table 15.2 Results of physico-chemical analyses in two different types of hydrological management in the three reaches during 1991

	8 April	7 June	21 June	5 July	19 July	2 Aug.	20 Aug.	5 Sept.	25 Oct.
Temperature (°C)		18	25	26	26	25	25	21	8
pH	7.8	7.6	7.9	7.5	7.6	7.5	7.5	7.9	6.1
Conductivity (μS cm^{-1})	735	1008	618	600	727	583	407	689	824
Oxygen									
surface (mg l^{-1})	9.9	8.1	11.4	13.2	10.9	11.2	10.9	7.3	9.6
bottom (mg l^{-1})	–	1.5	6.2	5.8	5.7	9.2	8.8	4.0	–
N-NH$_3$ (μg l^{-1})	–	3720	444	600	3100	930	130	3625	1676
N-NO$_3$ (μg l^{-1})	–	650	260	210	408	130	150	94	4040
N-NO$_2$ (μg l^{-1})	–	53	33	88	212	99	565	415	22
Inorganic-N (μg l^{-1})	–	4423	737	898	3720	1159	845	4134	5738
Total-N (μg l^{-1})	–	4795	3436	1150	4362	3186	3108	4282	8020
P-PO$_4$ (μg l^{-1})	–	500	420	524	860	614	555	695	523
Total-P (μg l^{-1})	–	512	595	572	876	632	584	701	581

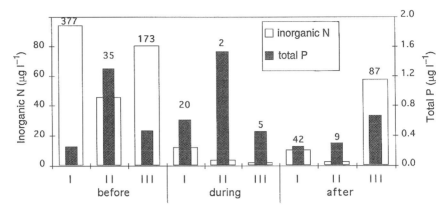

Figure 15.4 Temporal variations of average concentrations of inorganic nitrogen and phosphorous, before, during and after reservoir conditions, in the three reaches (I, II and III). The upper values indicate the N/P rate

Table 15.3 Phytoplankton composition with the relative cell biomass

Diatomeae	μm^3 cell^{-1}	Chlorophyceae	μm^3 cell^{-1}
Cyclotella comensis	400	*Ankistrodesmus falcatus*	200
Cyclotella comta	2500	*Coelastrum microporum*	200
Melosira sp.	590	*Closterium aciculare*	4000
Navicula sp.1	480	*Dictyospharium pulchellum*	150
Navicula sp.2	12 000	*Euglena* sp.	11 200
Nitzschia sp.	950	*Micractinium pusillum*	200
Synedra actinastroides	250	*Phacus lenticularis*	10 000
Synedra acus	2000	*Pediastrum duplex*	300
Synedra acus radians	800	*Scenedesmus quadricauda*	100
Synedra minuscola	500	*Scenedesmus acuminatus*	100
		Scenedesmus armatus	100
Cryptophyceae		*Sphaerocystis schroeteri*	75
Cryptomonas erosa reflexa	1800	*Staurastum gracile*	515
Rhodomonas minuta	200	*Tetraedron caudatum*	60
Cyanophyccac			
Anabaena sp.	33	**Peridineae**	
Chroococcus dispersus	200	*Gymnodinium* sp.	15 000
Mycrocistis robusta	30	*Peridinium* sp.	16 000
Oscillatoria rubescens	75	*Peridinium aciculiferum*	8500

polluted water. Furthermore, many shells of Gasteropoda, some of which were species able to live only in unpolluted water, were identified. Extremely few organisms were found alive however. Further consideration should be given to the very low abundance of the collected organisms. Besides the organic effluents which would have reduced primarily the richness, it is believed that toxic pollutants probably caused the heavy decrease of the abundance.

The study of the fish community of the Acque Alte Canal also showed a low species richness (Figure 15.7), with only 10 species being present. The diversity of the fish

REHABILITATION OF THE ACQUE ALTE DRAINAGE CANAL

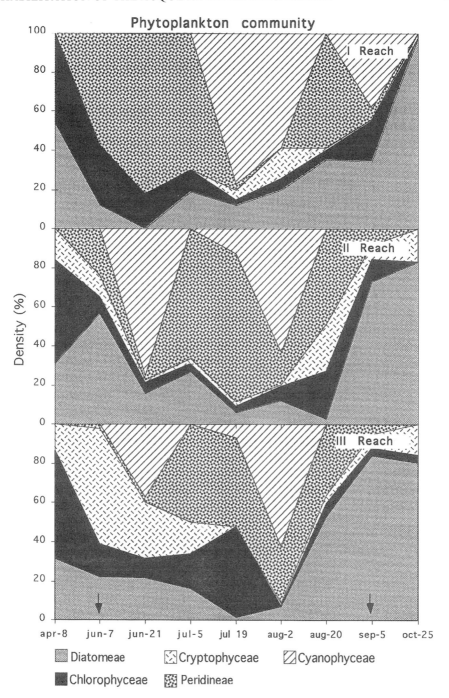

Figure 15.5 Temporal composition in percentage of phytoplankton density in the three reaches of the 'Acque Alte' Canal

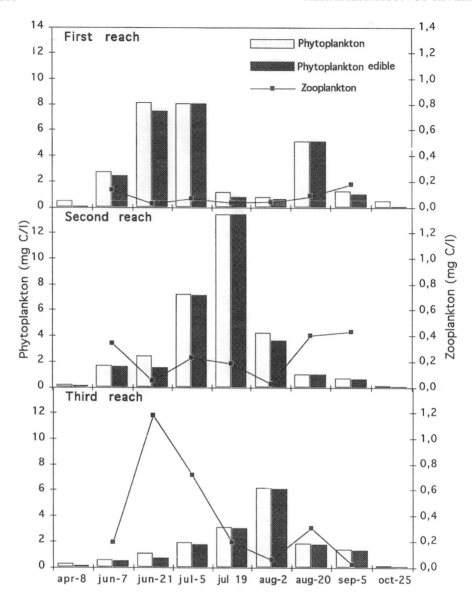

Figure 15.6 Temporal variations of phytoplanktonic (total and edible) and zooplanktonic biomasses during the season of 1991.

community increased in abundance from upstream to downstream and from lotic to lentic flow. This latter difference was due to the remarkable natural fish migration from the Panaro River (see Figure 15.1) that takes place during the spring months, before the closing of the floodgates. Almost all of these species grow inside the Acque Alte Canal, to complete their reproductive cycle. However, during the coldest months they migrate downstream. The presence of *Pseudorasbora parva* (stone moroko) is very interesting: this

Table 15.4 Zooplankton composition with the relative biomass

Cladocera	mm^3 org^{-1}	Rotatoria	mm^3 org^{-1}
Bosmina longirostris	0.017	Asplanchna priodonta	0.165
Daphnia hyalina	0.115	Brachionus calyciflorus	0.006
Simocephalus vetulus	0.560	Filinia longiseta	0.001
Ceriodaphnia pulchella	0.030	Keratella quadrata	0.001
Copepoda		Lepadella patella	0.001
Cyclops sp. (adulti)	0.500	Polyarthra sp.	0.006
Cyclops sp. (copepoditi)	0.200	Trichotria sp.	0.002
naupli	0.002		

Table 15.5 Macrobenthos community with their relative functional feeding group (R = gathering-collectors, S = scrapers, T = shredders, P = predators)

MACROZOOBENTHOS COMMUNITY			
Ephemeroptera		**Gasteropoda**	
Cleon sp.	R	Bithynia sp. *	S
Caenis sp.	T	Gyraulus sp.	S
Diptera		Physa sp.	S
Chironomidae	R	Theodoxus sp.	S
Eteroptera		Valvata sp.*	S
Naucoris sp.	P	**Oligochaeta**	
Odonata		Tubifex sp.	R
Ischnura elegans	P		
Acheta			
Dina sp.	P		

*Only empty test.

is an allochthonous species of Asiatic origin which has recently found very suitable conditions for its development in the downstream watercourse and dike environments of the Po River Plain.

In the Acque Alte Canal the macrophytic component did not underline any qualitative difference among the three reaches. The more frequent species were *Phragmites australis* (Cav.), *Bolboschoenus maritimus* (L.), *Ceratophyllum demersum* L., *Polygonum amphibium* L., *Typha angustifolia* L., *Typha latifolia* L., *Carex riparia* Curtis, *Carex eleata* All. and *Iris pseudocorus* L. The occurrence of a macrophytic community was extremely discontinuous. This was due not only to its natural development, but also to the aquatic vegetation which is cut every year to guarantee a suitable fast flow crosssection of the canal. Riparian vegetation was almost totally absent along the banks. This is the reason for the total lack of shaded areas and the very low input, throughout the year, of coarse particulate organic matter (CPOM). This situation could be one of the causes of the limited quality–quantity of the macrobenthic fauna (Cummins *et al.*, 1989).

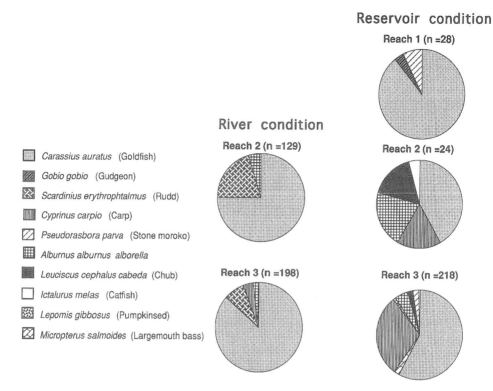

Figure 15.7 Composition in percentage of fish community, in the two different types of hydrological management (n = total organisms)

RESTORATION PROPOSALS

CANAL

The drainage dikes were well engineered for removing water. This water-removal function was appropriate a century ago when the groundwater and the marine environment were not overloaded, and more land was needed for agriculture. This management, however, is a clear case of disregard of sustainable development, and biological degradation took place because of habitat loss (Petersen *et al.*, 1992). For this reason there is a pressing need to improve the biodiversity which has been adversely affected in this environment by the extreme linear channel shape. In this condition water is transported faster to the sea, the self-cleaning capacity is reduced, and nutrient transport to the sea is increased.

Firstly the following actions can be suggested, which are also supported by a large number of international case studies (e.g. Gore, 1985; Brookes, 1988; Statzner *et al.*, 1989; Gardiner, 1991; Petersen *et al.*, 1992).

1. Restoration of the sinuosity and the regrading of a pool–riffle sequence so as to have a range of current velocities between pools and riffles.

2. The production of asymmetrical cross-sections at bends and symmetrical cross-sections in straight reaches, and the avoidance of uniform trapezoidal sections and steep slopes. The combined effect of the above points is to diversify the distribution of erosion and sedimentation zones.
3. The re-introduction of large-scale roughness elements. These include boulders and woody debris which change the distribution of the hydraulic forces over a stream bed and thereby cause scouring and sorting of fine sediment and gravel. The different bed load materials provide good habitat for a variety of benthic organisms. In turn these organisms provide essential food supplies to other animals.
4. The creation of a marsh which facilitates the dynamic flow between water and sediment, water and water, water and vegetation and the planting of common reeds (*Phragmites australis*) which will aid the removal of excess nutrients.
5. The reforestation of banks to provide cover and to insulate water from extremely high temperatures. This will also increase the input of coarse particulate organic matter and thus food for macroinvertebrate shredders throughout the year.
6. The biomanipulation of the food chain to control the growth of fish plankton feeders by the introduction of fish predators.

All the above suggestions can be used effectively within the lotic environment and at least some of them are suitable for use where the environment is impounded. Investigations verified that the stored water helped self-cleaning and, as a result, the general quality of the water improved. During the impoundment period, the natural processes of particulate sedimentation increased, the phytoplankton community made greater use of macronutrients and the zooplankton community also fed on allochthonous particulate organic matter. On the other hand, the phytoplankton and zooplankton communities found suitable conditions to grow and to maintain the equilibrium between the processes of production and degradation. Also, under these hydraulic conditions, the water system received relatively good water quality in the third reach from the CER. It would therefore be advisable to reduce the water velocity as much as possible, by modifying the Acque Alte Canal longitudinally with a series of lentic environments. A further benefit for the water quality would be to install a down–up mechanism in order to pump and distribute some of the CER water, to the upper part of the Canal, theoretically throughout the year. It is imperative, however, that the pollutants in the waste waters are reduced where possible; for example, by biological uptake by plants.

BUFFER STRIPS AND CORRIDOR

In many cases, along the dikes and rivers of the Po Plain, the original riparian vegetation has disappeared and arable land borders the dikes and rivers directly. Once the buffer strip is set aside, it will revegetate naturally or be replanted with cover crops or trees. The EC regulations on set-aside (Reg. EC 1765/92) and on environmentally friendly production methods (Reg. EC 2078/92) offer good opportunities for the realisation of buffer strips.

Under Reg. EC 1765/92:

- set-aside is compulsory on the surface cultivated with cereals and oilseeds (0–15%);
- voluntary set-aside could be extended to up to 50% of the surface cultivated with cereals and oilseeds;

- exchange of compulsory set-aside between farmers is possible;
- farmers that set aside arable lands in the Bologna lowland are supported by compensatory aid of 358 ECU ha^{-1}.

Under Reg. EC 2078/92:

- set-aside for 20 years is proposed in the Bologna lowland, with reforestation supported by grant aid of 600 ECU ha^{-1};
- restoration projects could be supported to the value of 2000 ECU ha^{-1}.

By the integration of agricultural and environment policies the realisation of buffer strips could be possible without any additional costs.

In conclusion, recommendations include (i) promoting set-aside of arable lands in the corridor for environmental aims; (ii) planting indigenous species of trees (e.g. *Quercus robur*) on the outer banks; (iii) planting of *Populus alba* within the floodplain in the lower part of the dike where the floodplain is wider; and (iv) the selective removal of water plants to permit recolonisation.

Preserving the terrestrial–aquatic interface by setting aside riparian land corridors is critical to all stages of restoration. In this case, concerns about the hydraulic safety of the dike prevents the creation of a continuous corridor. Consequently, the most suitable areas were chosen in order to create zones of high biological diversity.

LANDSCAPE

Landscape architecture guidelines are currently available and useful for restoration projects. Aesthetic considerations are particularly important in urban and highly cultivated areas, where the return of the riparian corridor to leisure or recreational uses is a primary goal and pristine goals are elusive.

Further to those interventions outlined above for water quality, buffer strips could be added to improve the landscape in order to make the environment suitable for other activities, including tourism, sport and leisure.

To the main infrastructure components of the region – such as the watercourses, the dikes and urban centres and the 'links' which may be established among them – can be added features of naturalistic or historical–archaeological interest ('sub-components'), which enrich both main components and links.

The Acque Alte Canal represents at the same time one of the main landscape elements, a means of connection and a concentrated line of various 'sub-components'. Therefore, the rehabilitation and improvement of the canal may produce an important process of holistic development of the territory.

The rehabilitation plan started with the identification of the main components of the landscape. These are the Samoggia River, the Reno River, the Zena Ditch, and the Acque Alte Canal, and also some towns. The second step was to establish connections between these components, e.g. biking tracks, streets, dikes and so on, to link the urban and recreational areas. The recreational areas which will be located along the dikes or nearby have already been identified. Some are to be rehabilitated, others created. In addition, the unused lagoons (lagooning industrial wastes), although completely artificial, could be suitable sites for avifauna, substituting the reduced natural wetlands.

REHABILITATION OF THE ACQUE ALTE DRAINAGE CANAL

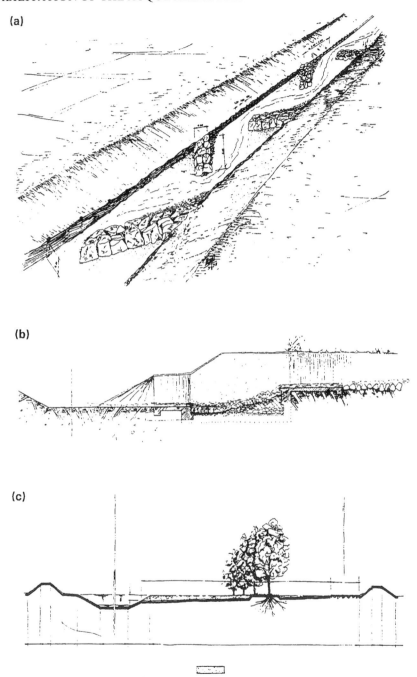

Figure 15.8 (a) Meandering of Romita ditch; (b) shallowing banks; (c) lowering of the floodplain level

COMMENTS

The economics of stream restoration is an important consideration when deciding to restore lowland streams. For the buffer strips the value of the farmland in terms of the market price and its profitability has to be compared to the financial benefits that the corridor can provide. The regulation of set-aside offers a good opportunity to promote buffer strips. The poor conditions described and the limited economic resources do not allow predictions to be made about the time-scale of the Acque Alte Canal rehabilitation. However, at least, the local authorities are now more sensitive to the issue of the conservation of nature. In fact, some of the suggestions outlined here have been financed by them.

The funding includes the following:

- meandering for a 250 m length of the Romita ditch (Figure 15.8(a));
- diversifying the ditch bed for a 50 m length by roughness elements;
- shallowing the banks close to each check dam (Figure 15.8(b));
- lowering of the floodplain in a part of the dike (Figure 15.8(c));
- some planting interventions in different parts of the area.

ACKNOWLEDGEMENTS

We thank the others involved in the study group: N. Agostini, P. Altobelli, M. Felicori, D. M. Gioia, L. Locchi, A. Morisi and G. Soverini. Comments by G. Pinay, B. Maiolini and U. Peruch substantially improved the manuscript.

This research was supported by regional authorities (Provincia di Bologna, USL No. 26, Comune di San Giovanni in Persiceto, Comune di Crevalcore, Comune di Sant'Agata Bolognese, Consorzio di Bonifica Reno-Palata).

REFERENCES

Brookes, A. (1988) *Channelized Rivers. Perspectives for Environmental Management.* John Wiley, Chichester.

Burns, C.W. (1968) The relationship between body size of filtering Cladocera and the maximum particle ingested. *Limnology and Oceanography*, **13**, 675–678.

Chiaudiani, G. and Vighi, M. (1974) The N:P ratio and tests with *Selenastrum capricornutum* to predict eutrophication in lakes. *Water Research*, **8**, 1063–1069.

Cummins, K.W., Wilzbach, M.A., Gates, D.M., Perry, J.B. and Tagliaferro, W.B. (1989) Shredders and riparian vegetation. *Bioscence*, **39**(1), 24–30.

Gardiner, L. (1991) *River Projects and Conservation. A Manual for Holistic Appraisal.* John Wiley, Chichester.

Gliwicz, Z.M. (1980) Filtering rates, food size selection and feeding rates in cladocerans – another aspect of interspecific competition in filter-feeding zooplankton. In Kerfoot, W.C. (ed.) *Evolution and Ecology of Zooplankton Communities*. University Press of New England, Hanover, New Hampshire.

Gore, J.A. (1985) *The Restoration of Rivers and Streams.* Butterworth, London.

Harris, G.P. (1986) *Phytoplankton Ecology. Structure, Function and Fluctuation.* Chapman and Hall, London.

Haycock, N.E., Pinay, G. and Walker C. (1993) Nitrogen retention in river corridors: a European perspective. *Ambio*, **22**(6), 340–346.

Heckman, C.W. (1990) Agricultural reclamation. In Patten, B.C. (ed.) *Wetlands and Shallow Continental Water Bodies*. SPB Academic, Netherlands.

Kajak, Z. (1992) The River Vistula and its floodplain valley (Poland): its ecology and importance for conservation. In Boon, P.J., Calow, P. and Petts, G.E. (eds) *River Conservation and Management*. John Wiley, Chichester, pp. 35–50.

Knisely, K. and Geller, W. (1986) Selective feeding of four zooplankton species on natural lake phytoplankton. *Oecologia*, **69**, 86–94.

Osborne, L.L., Bayley, P.B., Higler, L.W.G., Statzner, B., Triska, F. and Moth Iversen, T. (1993) Restoration of lowland streams: an introduction. *Freshwater Biology*, **29**, 187–194.

Peterjohn, W.T. and Correll, D.L. (1984) Nutrient dynamics in an agricultural watershed: observations of the role of riparian forest. *Ecology*, **65**, 1466–1475.

Petersen, R.C., Petersen, L.B.-M. and Lacoursiére, J. (1992) A building-block model for stream restoration. In Boon, P.J., Calow, P., Petts, G.E. (eds) *River Conservation and Management*. John Wiley, Chichester, pp. 293–310.

Petts, G.E. (1990) The role of ecotones in aquatic landscape management. In Naiman, R.J. and Décamps, H. (eds) *The Ecology and Management of Aquatic and Terrestrial Ecotones*. MAB-Series, **4**. Unesco and the Parthenon Publishing Group, Paris, 227–262.

Statzner, B., Bernhart, H.H. and Becker, C. (1989) Hydraulic stream ecology. Fundamentals, applications and future perspectives. In *Limnology in the Federal Republic of Germany*. Lampert, W. and Rothhaupt, K.O. (eds). 24th Congress of the International Association of Theoretical and Applied Limnology, Munich, pp. 148–151.

Vanni, M.J. (1987) Effects of food availability and fish predation on a zooplankton community. *Ecological Monographs*, **57**, 61–88.

16 Degradation and Rehabilitation of Waterways in Australia and New Zealand

CHRISTOPHER J. GIPPEL[1] and KEVIN J. COLLIER[2]
[1] Centre for Environmental Applied Hydrology, Department of Civil and Environmental Engineering, The University of Melbourne, Parkville, Victoria, Australia
[2] National Institute of Water and Atmosphere, Hamilton, New Zealand

INTRODUCTION

The geographical proximity of Australia and New Zealand belie their physiographic differences. The geologically ancient, low relief, climatically dry and erratic, and vast continent of Australia contrasts with the geologically active and young, high relief, climatically humid islands of New Zealand. The rivers and streams in these countries therefore have different geomorphological and hydrological characteristics. These differences provide for an interesting contrast in some of the problems facing management of waterways in the two countries. However, the countries have experienced a similar duration of European settlement, and share many modern cultural similarities, so they have some common waterway management problems. This chapter describes the main waterway degradation problems in the two countries, assesses the capacity of the administrative framework to encourage rehabilitation, illustrates past and current attempts to rehabilitate riverine environments, and discusses the prospects for future rehabilitation and management programmes.

Restoration implies returning a degraded system to a previous (usually pristine, or less human-modified) condition (Brink et al., 1988), which in Australasia would normally be interpreted as the condition existing just before European settlement. Such a vision may be useful for planning, but is unattainable in most situations (Osborne et al., 1993). More practical is rehabilitation, which the Murray-Darling Basin Commission (1993, p. 92) loosely defined as 'making a degraded system useful'. Utility might be defined solely in terms of satisfaction of direct human needs, such as provision of acceptable water quality or a recreation resource, or in terms of the ability of a waterway to support a self-sustaining and functionally diverse assemblage of aquatic life. It is recognised that the ecological communities of rehabilitated stream systems may be quite different to those that existed prior to disturbance. This chapter is concerned mainly with ecological rehabilitation of waterways, but often this is a secondary expectation or outcome of projects that are driven by utilitarian needs.

Rehabilitation of Rivers: Principles and Implementation. Edited by L. C. de Waal, A. R. G. Large and P. M. Wade.
© 1998 John Wiley & Sons Ltd.

WATERWAY MANAGEMENT IN AUSTRALIA

In presenting an overview of waterway management in Australia it is impossible to avoid focusing on the south-eastern part of the country (Figure 16.1). This is where the majority of the population lives, so it is also the area where land and water resources development (and degradation) have been most intense, where most of the research on river systems has been conducted, and where most of the recent initiatives for rehabilitation have emerged. The legislative, planning and administrative frameworks for river management vary throughout Australia's States and Territories. To simplify the discussion of these issues, special emphasis has been placed on the situation in the State of Victoria.

HISTORY AND EXTENT OF WATERWAY DEGRADATION

For around 50 000 years, Aboriginal people actively managed waterways for the purpose of food gathering (Government of Victoria, 1988; Lloyd, 1988). The most impressive Aboriginal-built hydrological structure appears to be a 100 m long embankment on a

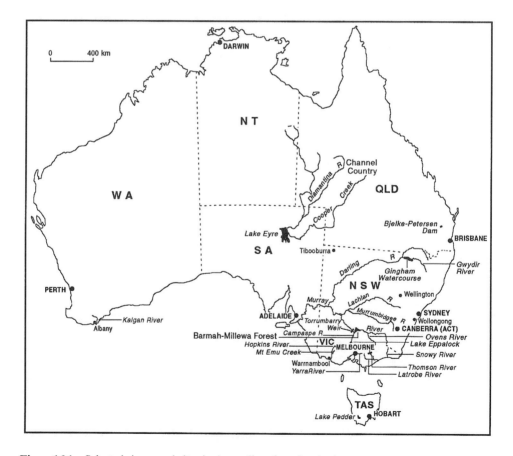

Figure 16.1 Selected rivers and sites in Australia referred to in the text

stream near Tibooburra, in the north-east of New South Wales (Figure 16.1), which would have created an impoundment of 0.7 Ml (Lloyd, 1988, p. 17). Aboriginal use of fire is believed to have had an impact on the distribution of forest vegetation in Australia, and although the details of this impact are largely unknown (Gifford et al., 1992), it could have indirectly affected waterways through increased catchment erosion (Hughes and Sullivan, 1981). It is known that the unique riparian forests of river red gum (*Eucalyptus camaldulensis*) along the River Murray developed after European settlement, when they replaced the savanna woodland vegetation that was maintained by regular Aboriginal firing (Jacobs, 1955). Aboriginal impacts can be regarded as local and minor in comparison with the scale and intensity of catchment and channel disturbances that have taken place since European settlement began around 200 years ago.

In modern times, the solution to the country's natural aridity and variable distribution of water resources was seen to be the imposition of a high degree of river regulation, including construction of impoundments, water diversion schemes, and channel modification such as levee construction, de-snagging, straightening and channelisation (Pigram, 1986; McMahon et al., 1992). The purposes of these works were to facilitate large-scale irrigation schemes, allow channel navigation, and mitigate the negative impacts of floods and droughts on agricultural activities and urban settlements. Rivers have also been adversely affected by sand and gravel dredging, widespread clearing of riparian vegetation, and damage to banks by stock due to unrestricted access (Warner, 1984; Vollebergh, 1992). During the early part of the century, most stream systems in Victoria and many in eastern Australia were disturbed by gold mining activities.

Catchment changes have also impacted on the hydrology of waterways. Approximately one-third of Australia's vegetation has undergone major structural change since European settlement, with the greatest impact being the replacement of native forests and woodlands by alien pasture and crop species (Gifford et al., 1992). Although not well documented, it is reported (I. Rutherfurd, Monash University, pers. comm.) that much of the extensive waterway network in the vast channel country of western Queensland (Figure 16.1) has been badly degraded by grazing stock and feral animals. Other catchment changes that have impacted on waterways are urbanisation, construction of farm dams (Warner, 1984; Warner and Bird, 1988; Smith and Finlayson, 1988; McMahon et al., 1992) and degradation of the soil resource (Chartres et al., 1992).

The inherently variable hydrology of Australian catchments (Finlayson and McMahon, 1988) means that it is difficult to discern the nature and severity of anthropogenic impacts on riverine environments. Also, the more recent deliberate channel modifications were superimposed on a system that was still adjusting to the major catchment alterations which began only about 150 years ago. Another difficulty is the usual lack of suitable pre-disturbance data for comparison (Warner, 1984; Warner and Bird, 1988). However, some general conclusions about the degradation of Australia's waterways can be drawn from the literature. Impoundments and diversions have reversed the seasonal pattern of flows, reduced the frequency of inundation of wetlands, altered water quality and temperature, created periods of very low or cease-to-flow conditions, and caused channel contraction (Sherrard and Erskine, 1991; Gippel and Finlayson, 1993; Finlayson et al., 1994). On inland draining streams, such as those in the Murray-Darling system (Figure 16.1), the drier lower catchment areas make relatively small contributions to total runoff, and the impact of a headwater impoundment can persist for a considerable distance downstream (Walker, 1985).

Although stream flow can be affected by forestry practices, including the conversion to softwood plantations which now cover 900 000 ha, the impact is temporary (McMahon et al., 1992). Urbanisation, land clearing, and changes in soil structure associated with clearing tend to increase runoff and increase the magnitude of flood peaks (Smith and Finlayson, 1988; Chartres et al., 1992; McMahon et al., 1992). Land clearing has led to widespread gullying, often in association with soil piping, and other forms of soil erosion (Government of Victoria, 1988; Smith and Finlayson, 1988; McMahon et al., 1992). In 1975 it was estimated that one-third of agricultural land in Victoria was affected by water erosion (Woods, 1984). Although land degradation is widespread in Australia, the sediment delivery system is relatively inefficient because of low gradients and the variable hydrological regime (Olive and Rieger, 1986). In many cases, much of the detached coarse sediment is stored within the catchment close to its source (Rutherfurd and Smith, 1992; Erskine et al., 1993), but there are cases where trunk streams have suffered siltation (Ian Drummond and Associates Pty Ltd, 1992). Deep chemical weathering is common to most of Australia's soils, and when they are disturbed, clays and fine silts are available for entrainment (Rieger and Olive, 1988). Once entrained they tend to remain in suspension, because of their low settling velocity, their dispersive nature, and the sodium-dominated ionic composition of runoff waters which promotes dispersion (Hart and McKelvie, 1986). Low concentrations of clay give streams a turbid appearance, especially in deep flows (Gippel, 1995). Although many Australian streams would have been naturally turbid, it is widely believed that the turbidity of Australian waterways has increased in response to catchment disturbance, even though data which demonstrate this are generally lacking (Government of Victoria, 1988).

Because of the low fertility of many Australian soils, agricultural production benefits from the application of fertilisers. Nutrients are transported to waterways attached to fine eroded soil particles, and also via sewerage treatment discharges. It has been postulated that stream phosphorus loads have increased by an order of magnitude since European settlement (Government of Victoria, 1988). The geologically old landscape of Australia contains large stores of salt, and the natural salinity levels of water would have generally been higher than those for other continents (Smith and Finlayson, 1988). Clearing of deep-rooted vegetation from catchments, and irrigation of poorly drained terrace and floodplain areas, has more than doubled the median salinity of some rivers, but probably reduced the range of salt concentration (McMahon et al., 1992).

Construction of levees, channelisation, straightening and de-snagging have been carried out in the belief that this 'improvement' mitigates flooding by increasing channel capacity, increasing channel slope and reducing hydraulic resistance. In practice it is difficult to isolate the hydrological effects of these channel works and demonstrate a flood mitigation benefit. However, some channels have undergone incision and enlargement following channel works (Erskine, 1992; Gippel et al., 1992). This response is generally less severe than in other parts of the world, because many of Australia's lowland rivers meander through wide, low-gradient, low-energy floodplains and there is little relief for incision (Warner and Bird, 1988). However, in south-east Victoria, reclamation of swamps by drainage caused rejuvenation of natural watercourses upstream, creating deep gorge-like cross-sections, and the mobilised sediment has been deposited in channels downstream (Warner and Bird, 1988).

Riverine flora and fauna have undergone profound changes as a consequence of physical habitat alteration. Clearing of native floodplain vegetation has been extensive, with 39% lost to agricultural and residential uses on the River Murray floodplain (Margules and Partners Pty Ltd, 1989). The lignum forests of the Lowbidgee wetlands on the Murrumbidgee River, NSW (Figure 16.1) are recognised as one of the most important waterbird habitats in eastern Australia, but their area has been reduced by 40% in the last 40 years due to clearing (Roberts, 1990).

The introduction of alien plant and animal species has impacted waterways. Willows (*Salix* spp.) were extensively planted to 'beautify' river verges and stabilise eroding banks. They have spread profusely throughout many of Australia's lowland river systems and require regular and expensive maintenance to prevent channel blockage. Water hyacinth (*Eichornea crassipes*) was first noticed on the Gwydir River, NSW (Figure 16.1) in the 1950s, and by the mid-1970s it covered an area in excess of 7000 ha (McCosker and Duggin, 1993). Many disturbed riparian areas are infested with large alien shrubs like blackberry (*Rubus fruiticosus*), which provide cover for feral animals such as rabbits and foxes. While it appears that trout (*Salmo trutta* and *Oncorhynchus mykiss*) can coexist with some native fish, Koehn and O'Connor (1990) are of the view that alien fish species have had an overall negative impact on native fish populations through predation, competition, habitat alteration and introduction of disease.

Cadwallader (1978) was critical of the practice of removing large woody debris because of its importance as fish habitat, particularly for Murray cod (*Maccullochella peeli*) and freshwater blackfish (*Gadopsis marmoratus*) which seem to prefer hollow logs for spawning. Channelisation has reduced fish populations (Hortle and Lake, 1983), and the concrete-lined urban channels found in the large cities were not designed with aquatic fauna in mind.

In the Murray-Darling Basin alone there are around 150 dams and weirs plus numerous causeways and small barriers that impede the passage of migratory fish. It has been estimated that the amount of habitat available to native freshwater fish in the coastal rivers of south-eastern New South Wales has been halved as a result of the construction of dams and weirs (Mallen-Cooper, 1994). Australian native freshwater fish generally have a weaker swimming ability than their northern hemisphere counterparts, and therefore, the steep ramp or pool-type fishways designed in the northern hemisphere for salmonid fish have proved unsuitable for Australian conditions (Mallen-Cooper, 1994).

Aquatic species whose behaviour is cued by floods have been impacted by the changed flooding regimes downstream of dams. Reduced flooding also affects those species that depend on regular wetland inundation for habitat (Walker, 1979; Lloyd *et al.*, 1991). Water temperatures below deep release dams are too low for spawning of some native fish species (Gippel and Finlayson, 1993). Excess sediment supply and reduced flushing flows have resulted in smothering of gravel spawning beds (Walker, 1983; Davey *et al.*, 1987).

Salinisation has caused declining productivity from affected agricultural land, but native vegetation has also been impacted. For example, on the River Murray floodplain, nearly 9000 ha of native woody floodplain vegetation have been severely degraded by land salinisation (Margules and Partners Pty Ltd, 1989). Available data suggest that macroinvertebrates and riparian and in-stream plants suffer direct adverse effects when stream water and wetland salinity reaches 1000 mg l^{-1} (Hart *et al.*, 1990). With the exception of naturally saline streams, most waterways still have salinities below this level (e.g. for Victoria, see Department of Water Resources Victoria, 1989). Hart *et al.* (1990) warned that the environmental impact of schemes proposed to control land salinisation by pumping groundwater and discharging the effluent to natural waterways should be carefully monitored.

Turbidity is considered mainly an aesthetic, rather than an environmental problem, but reduced light availability can limit flora and fauna, and the nutrients attached to fine suspended matter have been implicated in the proliferation of cyanobacterial blooms, some strains of which are highly toxic (Murray-Darling Basin Commission, 1993). Algal blooms acutely restrict the use of rivers and lakes for recreation and water supply. However, in Australia, there have been no deaths of animals (other than domestic stock and pets), birds or fish that have been unequivocally attributed to cyanobacterial poisoning (Murray-Darling Basin Commission, 1993, p. 21).

Because of extreme hydrological conditions and long geological isolation, Australian rivers naturally support a relatively small number of native fish species, so habitat degradation can have severe consequences (Lake, 1967; Bayly and Williams, 1973). For example, of the seven species of fish endemic to the Murray-Darling system, all but one is classified as being endangered or threatened in conservation status. Of the total 29 native species present, five are in danger of extinction in the near future, and a further nine will become endangered without intervention or changes in water management (Lloyd et al., 1991). Migratory fish species have been particularly affected, with Australian grayling (*Prototroctes maraena*) and Australian bass (*Macquaria novemaculeata*) no longer found in some coastal rivers of south-eastern Australia. In inland rivers, silver perch (*Bidyanus bidyanus*) numbers have declined by up to 93% over the past 50 years (Mallen-Cooper, 1994).

A recent survey of the environmental condition of Victorian streams (Mitchell, P., 1990) revealed that 27% of all streams were in poor or very poor condition, and for streams flowing through cleared land the figure was 65%. In summary, very little of Australia's freshwater aquatic environment remains unaltered by the effects of European settlement.

LEGISLATIVE, PLANNING AND ADMINISTRATIVE FRAMEWORK

Many of the developments that have caused degradation of Australia's waterways have been government sponsored. Irrigation consumes 70% of the total applied water on the continent, and 60% of this is supplied from government schemes (Smith and Finlayson, 1988). A system of transferable (freely tradable) water rights has been instituted in South Australia and is in the process of being introduced in Victoria, and it is hoped that this will lead to more efficient water use. However, in some catchments nearly all available water is already committed for agricultural, industrial and domestic water supply (80% is utilised in the Murray-Darling Basin in an average climatic year), and difficulties may arise in acquiring and retaining water entitlements for less directly market-driven purposes such as provision of environmental flows. There is currently a shift towards catchment-based waterway management (Dovers and Day, 1988), and a multiple objective planning approach might help integrate land and water management and promote consideration of environmental issues (Pigram, 1986).

In Australia, catchment-based management is complicated by the relatively large, and physically, economically and culturally heterogeneous catchments (the Murray-Darling Basin contains 16 000 km of major river channels draining an area of 1 060 000 km^2) (Figure 16.1). Involvement by the local community in planning is also a current government priority, although given the vested economic interests of some stakeholders, this will not necessarily ensure an equitable distribution of water, or an outcome which halts or reverses environmental degradation. The more drastic option of land use restraint is available in many parts of Australia, because 88% of land is directly held by governments, leased, or held on licence from them (Smith and Finlayson, 1988). However, there would be practical difficulties in enforcing controls over land use in such large and often remote areas.

The statutory framework of Australian water law has been described as a legislative sprawl which suppresses rather than enhances the natural attributes of rivers (Dovers and Day, 1988). Fortunately this has been recognised and in most States, the legislative, planning and administrative framework for waterway management has recently undergone, or is currently undergoing, a period of review and reform. An example of this is the major overhaul of Victoria's waterway management structure which began in the 1980s after it was recognised that, apart from environmental concerns, if water development continues at past rates, all available water will be developed within the next 35 years (Department of Water Resources Victoria, 1992).

About three-quarters of the funding for rural river management works in Victoria is provided by direct State government grants to local river management authorities (about A$3 million in the 1993/1994 financial year), and this is supplemented by local ratepayer contributions. River management authorities have recently been restructured and expanded, but they still cover only 17% of the area of the State. From the environmental perspective, much of their work has been parochial, short-sighted and ill-conceived, and has not addressed high-priority catchment-scale issues like salinity, land degradation and algal blooms (Department of Water Resources Victoria, 1992). The emphasis has clearly been on channel stabilisation, flood mitigation and protection of private and State-owned assets and infrastructure, with minimal concern for the aquatic environment. Such work makes little or no contribution to ecological rehabilitation, because environmental values are not a high priority and its aims are to minimise flooding and channel instability, both of which are ecologically important processes.

In Victoria, the revised Water Act (1989) states among its purposes, 'to provide formal means for the protection and enhancement of the environmental qualities of waterways and their instream uses; [and] to provide for the protection of catchment conditions' (Department of Water Resources Victoria, 1992, p. 9). New policies recognise the need for environmental flow allocations, but only for new diversion and abstraction schemes. Conservation of high-value riverine environments is facilitated through special legislation or programmes (Department of Water Resources Victoria, 1989). New catchment and land protection legislation has seen the creation of larger, more locally funded, and more professionally managed boards, which have the capacity to tackle catchment-scale environmental issues.

About 60–70% of Victoria's river and stream frontages are publicly owned for a width of 20 m or 40 m (25 000 km covering about 1000 km^2), and it is within the power of the Government to manage these frontages for the benefit of the natural environment. However, the sheer area of land involved, and the lack of access through adjoining freehold land, makes this difficult to implement and enforce. About half of the land is managed by adjacent landholders virtually as private property, and it is currently much cheaper for them to pay an annual licence fee that allows stock to graze on the frontage than to fence and revegetate. This is despite revegetation and fencing being advocated by the Government as the principal management solution to physical channel degradation (Vollebergh, 1992). With appropriate reforms, incentives, awareness raising, and management, these Crown frontages offer enormous potential for rehabilitation of the riparian condition of many of the State's waterways.

Urban waterways in Australia's large cities are managed by metropolitan boards. Such boards traditionally have been responsible for water supply and sewerage, and held little or no concern for maintaining ecological values in natural waterways. The ecologically degraded state of many urban waterways has recently become a major management issue. In response to growing public pressure, in 1985 the Victorian State Government announced a programme to promote development of urban riverine parks and waterways with the multiple objectives of conservation, recreation and flood mitigation (Senior, 1992). Parts of Melbourne's waterways have already been successfully rehabilitated as part of a vision for linear parks, and while much of the work has been driven by recreational requirements, many riparian habitats have benefited from revegetation, erosion control and bank reshaping (Senior, 1992). It is fortunate that most of the land required for the creation of a system of linear parks along Melbourne's stream corridors is already in public ownership.

A bloom of toxic cyanobacteria that occurred simultaneously over a 1000 km stretch of the Darling River (Figure 16.1) in November 1991 was considered a serious disaster, and the event was reported internationally. Shortly afterwards the Federal Government provided funding for catchment-based programmes aimed at improving water quality (mainly phosphorus) by reducing suspended solids loads (Department of Primary Industries and Energy, 1993). Ten projects were funded in Victoria in 1993–1994 through the Healthy Rivers and Better Catchments component of the National Landcare Program (total partnership funding is A$1.3 million). The proposal documents (unpublished documents, Department of Conservation and Natural Resources) make, at best, only vague references to pre- and post-project appraisal. This continues a long tradition in Australian waterway management for giving scant regard to the rigorous appraisal of publicly funded work which is ostensibly performed to 'improve' some environmental conditions for the public good. To be fair, this is partially in deference to the problem of inherent hydrological variability which makes scientific detection of change a difficult, expensive and long-term procedure.

An important Federal Government initiative was the National River Health Program (1993), funded with A$12 million over five years. The main goals of the programme are to monitor and assess the health of Australia's rivers and to manage flows for ecological sustainability (Anonymous, 1994a).

Government policy changes have created a framework that in principle should encourage slowing or halting the degradation of Australia's waterways. Improved ecological functioning might be an outcome in some areas, but intentional rehabilitation of natural physical and biological processes in degraded rivers currently relies on special, short-term initiatives, rather than a long-term planning strategy. Many of the initiatives appear to be driven by the current emphasis on nutrient, particularly phosphorus, control.

CASE STUDIES OF AUSTRALIAN WATERWAY REHABILITATION

It is apparent that the majority of Australia's waterways remain in an ecologically degraded condition. While government instrumentalities in Australia have the legislative power to rehabilitate rivers, it has rarely been exercised. Therefore, most examples of river rehabilitation in action are at the local scale. A rigorous survey of projects has not been conducted, but the range of approaches can be illustrated with some case studies.

Rural stream revegetation

Most large-scale improvement work on rural streams is carried out by river management authorities under the direction of engineers. The work is funded largely by the State governments, sometimes supplemented by a contribution from local ratepayers. The main goals of this work are to stabilise channels and mitigate flooding, but where revegetation of the riparian area is used as a stabilisation technique, the aquatic habitat is also improved. Revegetation is becoming a favoured technique, but local, native species are not always used, and many landholders are unenthusiastic about fencing riparian land because of reduced access for stock watering, maintenance costs and loss of productive land.

Recently there have been some modest attempts by local interest groups and concerned landholders to rehabilitate degraded rural streams by revegetating riparian land. For example, in south-west Western Australia, funding was provided from the National Landcare Program to plant and fence a total of 40 ha of riparian land along the Kalgan River (Figure 16.1). Landcare programmes promote community participation and most of the labour is voluntary. Further plantings will be funded by the local river management authorities and Green Skills, a community-based group that carries out federally funded Landcare programmes and at the same time provides opportunities for people seeking employment training. Some landholders have fenced off sections of river without waiting for financial assistance, and the educational opportunities provided by such projects have been recognised by some local schools (e.g. Yakamia Primary School, WA).

Degradation of rural waterways has impacted on the recreational angling industry. Freshwater angling groups focus their attention on trout rather than native fish species (Anonymous, 1994b). Fly fishing clubs in Victoria have been revegetating some streams in order to improve shade and physical habitat structure, and to improve water quality by reducing bank erosion. The groups claim that their work may have 'spin-off' benefits for native fish species, platypus and koala (Brown *et al.*, 1994). Warrnambool Fly Fishers Club obtained government funding to revegetate and fence sections of the Mt Emu Creek, Merri River and the Hopkins River (Figure 16.1). Voluntary labour enables rehabilitation to be done at a cost of A$1500–2000 km^{-1}. After fencing an area, weeds are poisoned or physically removed and then native tree and understorey species are planted. Access for stock watering is provided at suitable points. The projects are supported by local landholders because the fencing improves stock control and the bank revegetation helps prevent loss of land by erosion.

Gully erosion control

The upper catchments of the Campaspe and Coliban Rivers (2000 km^2), Victoria (Figure 16.1), were severely degraded during the mid- and late 1800s by land clearing, inappropriate agricultural practice, and rabbit plagues. Altered runoff response made flows even more variable than they were naturally. Gully development was widespread and because of this the channels were affected by siltation, and streams transported excessive sediment loads. In 1960 a scheme, reported in Department of Conservation, Forests and Lands (1985), was devised to rehabilitate the catchment in order to overcome declining agricultural productivity. An important part of the scheme was to regulate flows by constructing an impoundment at Eppalock (with a 312 000 Ml capacity) (Figure 16.1).

Fears that erosion would lead to rapid sedimentation of the Eppalock reservoir led to demands that channel and gully stabilisation works be carried out immediately. This involved construction of weirs, grade control structures, and battering, revegetation and fencing of active gully banks. Pasture improvement by contour chisel seeding reduced runoff and increased the effectiveness of, and need for further development of, channel erosion control works.

By the mid-1970s, the project was justified in economic terms, with a benefit to cost ratio of 2:1, but improvements in the aquatic environment have also been claimed, although poorly documented. Stream sediment loads were reduced, and 22 bird species had newly occupied the revegetated gully areas. This project had a strongly utilitarian origin, but a secondary outcome was rehabilitation of headwater streams. However, it must be remembered that these ecological improvements in the upper part of the catchment were achieved at the expense of the erection of a major and several minor stream barriers.

Fish passage

Experimental research by the Fisheries Research Institute of New South Wales on selected native fish species (Mallen-Cooper, 1994) has found that successful fishways require low velocities (< 1.8 m s^{-1}), low slopes (< 1 in 15) and low turbulence (41–100 W m^{-3}). Since the 1980s, six vertical slot-type fishways with these characteristics have been constructed on rivers in New South Wales (NSW). Monitoring of fish passage through a fishway constructed in 1991 on the 6 m high Torrumbarry Weir (Figure 16.1) has demonstrated its effectiveness (Mallen-Cooper, 1994). Because of its gentle slope, this type of fishway is expensive to build (A$250 000–500 000).

Navigation locks on the River Murray (Figure 16.1) are normally opened only for the short period of time it requires for boats to enter the lock. Mallen-Cooper (1994) reported that a trial 16-hour opening of a navigation lock on the River Murray during the main migration season resulted in a forty-fold increase in the numbers of fish passing through. The Murray-Darling Basin Commission now plans to partially open 13 locks on the River Murray during spring, making 1300 km of river available for migration.

Grade control (or drop) structures are often used to help manage the problem of stream bed degradation. The Department of Water Resources, NSW, have recently modified the design of these structures so that they do not impede fish passage. The current fishway designs used by the Department have a maximum height of 1 m and a maximum slope of 1 in 20 (unpublished information provided by W. Hader, Department of Water Resources, NSW). Three grade control structures have recently been constructed on the Bell River at Wellington, NSW (Figure 16.1). The Schauberger ramps are constructed of machine-placed boulders, with a fishway in the low flow section. The fishway consists of a series of boulder ridges, spaced two boulder widths apart. The Department of Water Resources reported (unpublished documents) that the ramps have experienced floods without disruption, and fish (trout) have been observed in the pools of the fishways under low flow conditions. A similar ramp has been constructed on Minamurra Creek near Wollongong, and urban drains in the Eastern Creek area of Sydney (Figure 16.1) have also been similarly rehabilitated.

A fishpass has recently been constructed on the Lower Yarra River over a tidal barrier at Dight's Falls, Melbourne, Victoria (Figure 16.1). While in principle, migratory species can now access the upper parts of the river system, this possibility raises other rehabilitation issues. Fish entering the river system will have to cope with degraded water quality, a modified flow regime (abstraction and urban runoff effects), altered channel conditions (affecting the hydraulic environment, the physical habitat, and riparian vegetation) and alien species (the river is periodically stocked with trout). Fish will also have difficulty negotiating the many hard-lined channel sections, drop structures, culverts, concrete fords, water offtake weirs, and flow gauging weirs that are scattered throughout the catchment.

Meander reinstatement

The lower Latrobe River, Victoria (catchment area of 5200 km^2) (Figure 16.1), is one of the most disturbed river systems in Australia, having long been subjected to river 'improvement', including 66 artificial meander neck cut-offs which reduced the length of the lower part of the river by 25%. With the exception of the seven lowest cut-offs they were planned as shallow floodways, but most of them progressively eroded to the full river depth. In the 1970s river erosion was dealt with using the expensive techniques of bank battering and rock protection. By the 1980s it was hoped that the alternative engineering approach of fencing and planting would provide a better long-term solution, but the work was often destroyed by continuing erosion.

In an attempt to reduce velocity and therefore alleviate erosion of the lower Latrobe River, some of the cut-off meanders are now being reinstated at a cost of A$15 000–25 000 per cut-off (Reinfelds et al., 1995). Although the environmental impacts of the works are not being rigorously monitored, the authority responsible for river management and the local community are hoping that this physical restoration will improve habitat conditions. However, the sediment stored in the reinstated cut-offs is being allowed to re-enter the river, and there are some questions about its toxicity, and the possible undesirable geomorphic consequences of this sediment input (Erskine et al., 1990).

Provision of environmental flows

The idea of rehabilitating habitats in rivers degraded by impoundment is being widely discussed (Australian Water and Wastewater Association, 1994), but little practical progress can yet be reported. However, new dams have required investigation of likely downstream impacts, and development of management plans which will mitigate adverse effects. Two examples are described below.

Construction of the Bjelke-Peterson Dam on the Barker Creek, Queensland (Figure 16.1), commenced in 1982 and was completed in 1988. It is a medium-sized impoundment with a capacity of 125 000 Ml and a wall 32 m high. Recommendations for management of the impoundment have been made on the basis of integrating irrigation and other off-stream water demands with the requirements for conservation of the stream ecosystem (Arthington et al., 1992). In-stream incremental flow methodology (IFIM) was used to help specify environmental flow requirements for fish. As well as requirements for maintaining water temperature and quality within tolerable limits, the strategy specifies boundary flow conditions for achieving optimum in-stream protection, minimum permissible flows for maintenance of the in-stream ecosystem during drought years, and high flows which act to flush pools, maintain the channel and provide biological cues for fish and invertebrates. The strategy also recommends additional artificial flood flows for specific purposes such as salinity control and removal of weed infestations (Arthington et al., 1992).

An integrated approach was taken to developing guidelines for maintaining in-stream habitats in the Thomson River, Victoria (Figure 16.1), downstream of a major dam (1 200 000 Ml capacity) completed in 1983, but only recently filled to capacity (Gippel and Stewardson, 1995). The river has a high conservation value due to the presence of a threatened species of native fish, the freshwater blackfish (*Gadopsis marmoratus*). The dam is large enough to significantly alter flood hydrology, and it has been proposed to annually release an artificial flood if the dam does not naturally spill (Gippel and Stewardson, 1995). Experience suggests that without these floods, channel contraction, and therefore loss of habitat, is likely. Minimum flows have also been set for each month. These flows were formulated to provide at least the same amount of blackfish habitat that existed prior to regulation. The inverted U-shaped relationship between habitat availability and discharge allowed velocity-limited high discharges to be traded for depth-limited low discharges (Gippel and Stewardson, 1995).

Wetland rehabilitation

The Gingham Watercourse is a 2000 km^2 area of riverine wetland on the Gwydir River, northern NSW (Figure 16.1) (McCosker and Duggin, 1993; Bennett and McCosker, 1994). The flat terrain and fertile soils have long been exploited for agricultural purposes, but this area has also long been renowned for the richness and abundance of its bird fauna. Completion of the Copeton Dam in the upper part of the catchment in 1976 enabled establishment of an irrigation industry on the lower Gwydir River floodplain. Since then, all regulated flows, and a large proportion of unregulated flows, have been diverted away from the wetlands to irrigate 850 km^2 of cotton. Landholders in the Gingham Watercourse have suffered an estimated 30% loss of productivity, but the loss of wetland flora and fauna has been even more dramatic. Only 1000 ha of the former 13 500 ha of core semi-permanent wetland remains undisturbed. The 42 000 ha of low floodplain has not been flooded since 1984 and almost all of the wetland vegetation has been decimated. Waterbirds have also disappeared, with the last significant breeding event occurring in 1984.

Although recent regulation has severely disturbed the hydrology of the Gingham Watercourse, the prospects for rehabilitating the wetlands are good. Flow records suggest that the wetlands developed under a highly irregular flooding regime. During the 85 years prior to regulation there were four separate occasions when floods did not occur for five or more consecutive years. Seedbank germination trials demonstrated that propagules will regenerate given inundation for 30 days or more during the warmer months. Based on an analysis of 100 years of flow records, and known water requirements of vegetation and bird communities, McCosker and Duggin (1993) and Bennett and McCosker (1994) have recommended a flooding regime to rehabilitate the wetlands. Water can be supplied from unregulated inflows to the Gwydir River below the dam, or allocated from water stored in the dam.

Similar plans have been proposed and trialed for other large wetland areas affected by regulation, including wetlands on the Lachlan River, NSW (Wettin and Bennett, 1986), the Lower Latrobe River, Victoria (E. Keogh, Department of Conservation and Environment, Central Gippsland Region, pers. comm.), the Snowy River, Victoria (Department of Conservation and Environment, 1992) and the Barmah-Millewa Forests on the River Murray (McPhail and Young, 1992; Atkins, 1993) (Figure 16.1). Some other wetland rehabilitation schemes planned or under way in NSW and South Australia are outlined in McPhail and Young (1992).

Artificial habitat enhancement

The physical habitat of a short reach of the Ovens River, Victoria (Figure 16.1) was enhanced in 1983 by the addition of rocks and a low, V-shaped weir constructed of logs (Koehn, 1987). The fish fauna was found to be dominated by the two-spined blackfish (*Gadopsis bispinosis*), a little known species not described until 1984. The rocks provided hydraulic diversity and a variety of shelter areas, and the log structure created a deep scour pool, thereby greatly increasing the area of habitat suitable for fish shelter. By 1987 surveys demonstrated a nine-fold increase in the blackfish population, and a three-fold increase in the trout population.

The Upper Yarra Reservoir was constructed in 1958 and functions as the major storage for distribution of water to the Melbourne metropolitan area (Figure 16.1). Environmental flows were never required and no water has been intentionally released from the dam since its construction. In response to regulation, the channel downstream of the dam has contracted and the in-stream habitat has become degraded. In an effort to increase the population of freshwater blackfish, a trial was conducted in 1993 to determine the effect of enhancing the habitat by emplacing artificial spawning tubes and releasing a small environmental flow. Blackfish are known to prefer hollow logs for spawning, but the surrogate plastic tubes also proved successful, with evidence of spawning being observed at all of the trial sites (Doeg and Saddlier, 1993).

At their own expense, the Victorian Fly Fishing Association have recently introduced gravel redds into the Emu Creek, a stream near Melbourne degraded by bank erosion. The banks were stabilised by fenced vegetation. A large flood has since removed much of the silt, and the gravel, although redistributed, stayed in the rehabilitated section. Around 6000 trout eggs were introduced to the redds and they had a hatch rate of 97% (J. Beavis, VFFA, pers. comm.). The Warrnambool Fly Fishers Club decided to overcome the natural lack of gravel spawning habitat in rivers in the Western District (these low gradient streams naturally have fine bed material), by importing river gravel. With financial assistance from a fishing gear retailer, voluntary labour, and heavy machinery loaned by the local Framlingham Aboriginal Trust, they have introduced 120 tonnes of gravel for a cost of A$1400. The WFFC advise that the depth of gravel required is 200–300 mm and about 1 m^2 is required per trout per redd (Anonymous, 1994b).

Water quality improvement

Dandenong Creek flows for 53 km through the south-eastern suburbs of Melbourne (Figure 16.1). Water quality in the Creek has suffered from the impacts of urban and industrial development. However, physico-chemical and microbiological surveys indicated an overall improvement in water quality over the past 20 years. The improvements were attributed to improved pollution surveillance, pollution control, and progressive sewering of the catchment (Nuttall, 1983; Ferdinands and D'Santos, 1994).

Despite the improvement in water quality, Dandenong Creek remains in relatively poor biological condition. With the exception of the headwaters, the diversity of macroinvertebrates in the stream system has been described as poor (Nuttall, 1983; Ferdinands and D'Santos, 1994). A survey of fishes indicated low diversity and population size, compared with other lowland streams (Koehn, 1986). It is likely that in this case, the biotic community did not respond to the improved water quality because of poor habitat conditions. Some 60% of the stream length has been piped or channelised, much of the riparian vegetation has been removed, the channel has been extensively de-snagged, and several weirs and grade control structures form barriers to fish passage. Urbanisation has almost certainly altered the hydrology of the stream. Despite its enormous potential for recreational fishing, Koehn (1986) could not recommend stocking with salmonids until the physical habitat is rehabilitated.

Urban stream rehabilitation

Mullum Mullum Creek is a 16 km long tributary of the Yarra River, Melbourne (Figure 16.1), with a catchment area of 44 km^2. The floodplain is relatively undisturbed and the stream corridor still contains important areas of remnant native vegetation, but urbanisation of the whole catchment changed the stream's hydrology which resulted in channel destabilisation. The initial phase of massive bed erosion has now passed, and in many places the stream bed now contacts bedrock and a hard clay base. Bank erosion is still occurring, leading to loss of land, degraded water quality, and loss of important remnant trees and riparian habitats. In 1992 a Linear Park Concept Plan was developed by the local government in association with the metropolitan water authority and various consultants (Biosis Research Pty Ltd, 1992). The Plan is a proposal to manage the entire waterway for recreation, protection of remnant vegetation, and physical and ecological rehabilitation, so that the stream will eventually provide a continuous ecological and recreational link from the headwater area to the Yarra River.

In-stream bed and bank stabilisation works began in mid-1993. The works involved construction of low graded rock weirs designed with a 10% slope to allow passage of wildlife. The structures are intended to act as catalysts for natural sedimentation and regeneration of vegetation. The structures have survived repeated overbank floods without damage. Community groups are removing alien weeds and revegetating eroded banks, public access pathways have been constructed, and a system of off-stream wetlands was constructed in 1995. Waterbirds, aquatic insects and fish have apparently returned to the rehabilitated section (Craigie and Eldridge, 1994).

Proposed decommission of a major dam

Lake Pedder, formerly a small glacial lake (9.7 km^2 in area) on the Serpentine River, Tasmania (Figure 16.1), was the site of one of the Australian environmental movement's most bitter defeats. In 1972, the famous pink quartzite sand beach and unique faunal community was submerged by an impoundment (240 km^2 in area) as part of the Gordon hydroelectric scheme (Anderson, 1994). The environmental group Pedder 2000 has recently called for the dam to be drained and the lake to be rehabilitated. This was supported by a motion by 1000 delegates at the World Conservation Union meeting in Buenos Aires in January 1993 (Anonymous, 1994c).

Only one animal endemic to the former Lake Pedder is known to have survived the inundation, and it appears unlikely that the previous unique faunal assemblage can be re-established. Waves have heavily eroded the current shoreline, presenting a difficult challenge for rehabilitation (Anderson, 1994). However, Tyler et al. (1993) confirmed that the geomorphological features of the former lake are largely intact and the soil is still bound by root systems. The feasibility of the scheme is reinforced by the current surplus of electricity generating capacity in Tasmania. Despite the worldwide support for the rehabilitation scheme, there is local political resistance, with the State energy minister quoted as saying, 'I cannot understand why anyone would want to demolish such valuable assets' (Anonymous, 1994c). Given the undisputed environmental value of the former lake, the irony of this statement is disturbing.

FUTURE PROSPECTS FOR REHABILITATION OF AUSTRALIA'S WATERWAYS

Recent water law reforms and government initiatives have created a framework which should encourage reduced degradation of the aquatic environment, and reduced risk of new developments causing detrimental impacts. Most of the river rehabilitation efforts being driven by Federal and State Governments appear to be directed towards reducing the salinity and nutrient levels of stream water, even though aquatic communities are probably not chronically impacted by existing conditions. Thus far, this chapter has highlighted flow regulation and channel modifications as the major causes of habitat degradation. The Australian case studies suggest that these problems have so far been addressed only at the relatively local scale.

Flow regulation is widespread, and rehabilitation of some impacted streams should be possible with the adoption of a policy of allocating environmental flows. Difficulties will arise in situations where all water is already committed for off-stream use, and where structural limitations prevent the release of artificial environmental floods of a suitable magnitude and temperature. Under conditions of increasing water demand, allocation of environmental flows will reduce the length of time over which an existing impoundment can provide a guaranteed security of supply of water for consumptive use. Future impoundments will create unavoidable, undesirable environmental impacts, so while environmental flows must be specified for sustainability under conditions of scientific uncertainty, a liberal margin of safety is a luxury that cannot be afforded. It is crucial that the uncertainty be reduced by appropriate multidisciplinary research. Much of the research conducted in the northern hemisphere is of limited relevance in the Australian context because of the highly variable flow regimes, the common occurrence of cease-to-flow and extreme flood conditions, and quite different behaviour and habitat requirements of the native flora and fauna (Arthington and Pusey, 1993). It appears that it will be necessary to preserve a period of high flows in the regulated regime in order to stimulate fish migration and spawning (Swales, 1994). Arthington and Pusey (1994) have also argued for the preservation of flow variability in regulated regimes. In many cases, re-establishment of fish passage past major barriers will have to be complemented with other rehabilitation efforts. In particular, techniques will have to be devised to overcome the numerous minor barriers that are scattered throughout most catchments.

A challenge for urban stream rehabilitation will be to devise ways of offsetting the impacts of altered hydrological conditions. A major step towards rehabilitating the physical habitat of urban and rural streams is revegetation of riparian areas. This should be encouraged, but it must also be recognised that in most situations ongoing maintenance will be required in order to maintain hydraulic performance and allow access for stock and recreation. The channel country of western Queensland and the semi-arid rangelands are so vast that stock exclusion over very large areas may be the only feasible option for rehabilitating the waterways in these areas.

In the foreseeable future, most wetland rehabilitation will be localised and require engineered watering strategies. Catchment-wide reinstatement of near-to-natural flooding by manipulation of dam releases and/or removal of levees is unlikely unless there is a change in the community attitude to one where regular flooding is accepted as a manageable risk. Similarly, it may be some time before morphological channel instability (a degree of which is probably necessary for ecological sustainability) can be accepted and managed by a community that traditionally has pursued a vision of absolute stability, despite its proven expense and ultimate futility.

There is evidence of a general decline in erosion rates in south-eastern Australia since the 1950s (Rutherfurd *et al.*, 1993). For a considerable time now, streams and rivers have been adjusting to the severe disturbances imposed during the phase of land clearing, gold mining, channelisation and flow regulation. Erskine *et al.* (1993) found that in the Goulburn River catchment in Victoria, most gully heads have now either eroded to bedrock or been controlled by structural works, and many incised reaches of channels have developed new, stable vegetated floodplains. Relatively geomorphologically stable channels such as these present the best opportunity for successful ecological rehabilitation programmes.

WATERWAY MANAGEMENT IN NEW ZEALAND

HISTORY AND EXTENT OF WATERWAY DEGRADATION

The historical factors which have led to the degradation of New Zealand's waterways are similar in many ways to those that operated in Australia. This section provides a brief synopsis of the main factors and highlights important differences in the waterway degradation problems of the two countries.

Compared with Australia, river systems are short (no point of land is further than 130 km from the sea) (Figure 16.2), but their physical diversity is high due to the geologically young and active nature of the landscape. The presence of extensive and

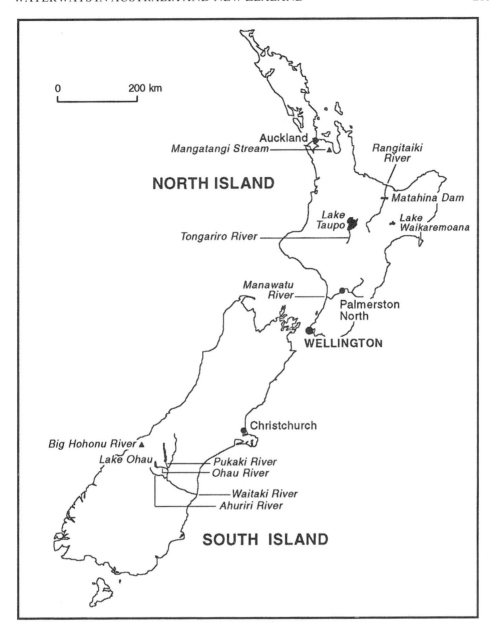

Figure 16.2 Selected rivers and sites in New Zealand referred to in the text

highly dissected mountain ranges coupled with frequent and sometimes heavy rainfall in many areas of New Zealand provides an overall abundance of flowing water in most regions. When the first Polynesian migrants arrived in Aotearoa (New Zealand), forest covered three-quarters of the land area (Fahey and Rowe, 1992). Subsequent land clearance reduced this to around 57% by the time European settlers arrived, and today only about 23% of the country remains in native forest. The impacts of land development have been most severe on the lowlands where only 15% of forest is intact and over 90% of wetlands have been drained.

Approximately half of New Zealand's land area is now pastorally farmed (Department of Statistics, 1990), and almost half the length of rivers in the North Island flow through catchments with modified vegetation cover (Collier, 1993). Few other places in the world have intensive pastoral farming extending onto such steep land with high rainfall as in some parts of New Zealand (Poole, 1983). Associated soil erosion and altered flow regimes along with river channelisation works and nutrient enrichment have greatly modified habitat and water quality in many rivers (Hickey and Rutherford, 1986; Wilcock, 1986; McColl and Ward, 1987). Dairy factories, wool scours, tanneries and pulp and paper mills have historically used rivers (as well as coastal waters) for organic waste disposal, and agricultural runoff makes a significant contribution to nutrient levels in receiving waters. Ward et al. (1985) estimated that the value of phosphorus fertiliser lost in runoff was in the range NZ$3–6 million. Growing urbanisation, particularly in the North Island, has also led to increased demands on local water resources (McColl and Ward, 1987).

Large areas of land (1.3 million ha by 1990) in many parts of New Zealand have been converted to *Pinus radiata* forestry, and this has caused stream flows to decline due to higher evapotranspiration rates from forest canopies (Fahey and Rowe, 1992). Logging has been associated with increased sediment loads, higher water temperatures, and changes in invertebrate and fish community composition (Graynoth, 1979). Recently, the Logging Industry Research Association has developed an environmental code of practice which should reduce such impacts.

Despite these catchment alterations, many upland rivers are still of a high chemical quality, support diverse invertebrate and periphyton communities, and are extensively colonised by trout (Biggs et al., 1990). However, most lowland streams and rivers are in poor ecological condition due to the effects of catchment and riparian clearance and associated chemical and habitat changes.

The discovery of gold in the South Island in the 1800s led to the dredging of substrates in many rivers, a practice which continues today and impacts upon aquatic life (Quinn et al., 1992). In the catchment of one major South Island river alone, it was estimated that 23 million m^3 of tailings were produced, with about 20% of this remaining in the river (Acheson, 1968). In northern New Zealand, 'splash dams' that were constructed on many small forested streams to store surges of water for the transport of logs must have had a major impact on aquatic life (McDowall, 1990). In addition, the construction of hydropower dams and abstraction of water for irrigation and water supply has altered the flows of many rivers and led to changes in river morphology, substrate armouring, proliferations of periphyton, reduced habitat for aquatic life and encroachment of marginal vegetation (e.g. Biggs, 1982).

As in Australia, dam, weir and culvert construction has caused problems for various species including many native fish and alien sports fish which require access up and down rivers to complete their natural life cycles (McDowall, 1990). Seventeen of the 27 described native fish species are diadromous. Some galaxiid and bully species spawn in freshwater, and young move to the sea either as developing eggs, larvae or juveniles before migrating back upstream in spring and early summer. In contrast, mature eels swim downstream and out to sea to spawn, and the young return to freshwater in large migrations during spring. Trout and salmon also require access up rivers and streams to spawn, and this has been disrupted on some rivers by the construction of dams. In the Taranaki and Rangitikei regions of the North Island alone, fish access was restricted by dams and weirs on at least 32 streams and rivers (Hicks, 1984).

More than 228 species of alien aquatic plants and animals have been introduced into New Zealand (Collier, 1993). Trout and salmon were first introduced in the late 19th and early 20th centuries, and brown and rainbow trout are now widespread largely due to the stocking efforts of Acclimatisation Societies. These species form the basis of an important recreational fishery, the welfare of which is often used to advocate against the degradation of aquatic habitats (Scott, 1987). Recent work has shown that trout can adversely affect some native galaxiid fish in New Zealand through predation (Townsend and Crowl, 1991). The relatively high number of alien species relative to the native flora and fauna (for example, there are 20 alien fish species), and the widespread distribution of many, may limit the level of ecological restoration that can be achieved in New Zealand rivers.

THE LEGISLATIVE FRAMEWORK

The development of the New Zealand lowlands for agriculture required river control and land drainage, and this was facilitated by the Land Drainage Acts of 1893 and 1908. However, agricultural development and flood control resulted in severe erosion problems in some places, leading to loss of productive farmland and the overloading of rivers with silt and shingle. These problems led to the passing of legislation (Soil Conservation and Rivers Control Act, 1941) designed to link river and land control by whole catchment management. Under this Act, comprehensive catchment control schemes were undertaken, and large areas (approximately 600 000 ha) of floodplain soils were upgraded to support intensive horticulture and agriculture (Poole, 1983).

The poor water quality of many rivers as a result of pollution inputs had become clear by the 1960s when water was recognised as a national asset to be managed for the benefit of the whole community (McColl and Ward, 1987). In 1963, the addition of Water Pollution Regulations to the Water Pollution Act 1953 gave the first effective powers to control pollution. Although this legislation was effective in restoring good water quality to some rivers, there was insufficient administrative support for adequate planning, monitoring and enforcement at the regional level. In 1983, tougher penalties for pollution and a means of recovering restoration costs were introduced to combat situations where it was cheaper to pollute than to institute proper treatment.

A more comprehensive system of water and soil resource management was achieved with the passing of the Water and Soil Conservation Act 1967 which vested the sole rights to dam, divert, take, discharge to, or use natural water in the Crown (McColl and Ward, 1987). Regional management was achieved through water boards which made water quality assessments and allocated resources to users through a system of water rights. By the 1980s, catchment management plans were being developed, and there was a trend towards matching water demand with availability, emphasising efficiency and multiple use, and keeping people away from flooded areas (Fenemor, 1992). An amendment to the Water and Soil Conservation Act in 1981 established a means for the retention of water flows and quality in rivers having outstanding wild, scenic, recreational, scientific or other values, enabling for the first time the consideration of in-stream uses and values.

New Zealand's resource laws recently underwent a major change with the passing of the Resource Management Act 1991 (RMA). The purpose of the RMA is to promote sustainable management of natural and physical resources; it provides for 'safeguarding the life-supporting capacity of air, water, soil and ecosystems [as well as] avoiding, remedying, or mitigating any adverse effects of activities on the environment'. The RMA stresses an integrated approach to the management of all natural resources (including water), and places more emphasis on monitoring (Fenemor, 1992).

The preservation of the natural character of rivers and their margins is recognised as a matter of national importance by the RMA, and discharges are not permitted if they are likely to cause the production of surface films, scums, a conspicuous change in colour or clarity, an objectionable smell, or significant adverse effects on aquatic life. Eleven national water quality classes are established under the third schedule of the RMA based on whether the water is managed for aquatic ecosystem purposes, fishery purposes, fish spawning, the gathering or cultivating of shellfish, contact recreation, water supply, irrigation, industrial abstraction, the natural state, aesthetic purposes, or for cultural purposes.

CASE STUDIES OF NEW ZEALAND WATERWAY REHABILITATION

The widespread degradation of river environments, particularly on the lowlands, means that there is considerable scope for ecological rehabilitation in New Zealand. Given the historical emphasis of New Zealand's resource laws on soil conservation and the maintenance of water quality, and the recreational significance attached to the salmonid sports fishery, it is not surprising that most examples of river rehabilitation have focused on these issues. Urban stream rehabilitation is also beginning to gather momentum in some New Zealand cities because of the potential amenity and recreational values such streams offer (Couling, 1992). In general terms, ecological rehabilitation efforts carried out on New Zealand rivers can be grouped under the six broad headings described below.

Improving bank stability

Removal of native forest from floodplains and river banks, and accelerated runoff from cleared land (Fahey and Rowe, 1992) have commonly had the unwanted side-effect of bank erosion leading to loss of productive land and the degradation of in-stream habitat. In response to widespread bank erosion in the agricultural landscape, thousands of kilometres of willow (*Salix* spp.) trees have been planted to stabilise banks, and where these do not give sufficiently rapid or effective protection, other means such as rock rip-rap and groynes have been used (Mosely, 1992). Similar works have also been undertaken on some urban streams where bank erosion has been a problem (Roper-Lindsay, 1994).

The ecological effects of these bank rehabilitation works have not been widely investigated, but are likely to be generally favourable due to the reduction in fine sediment inputs. However, dense planting of willows alongside agricultural streams and rivers can have adverse ecological effects. These may be caused by the growth of willow roots into the channel constricting flow and increasing fine particle retention, by excessive shading reducing the production of aquatic plants and low bankside vegetation, and by decreasing suitable habitat for birds feeding on open riverbeds or ecotone areas (Collier, 1994).

Several regional authorities have undertaken willow clearance operations to mitigate flooding problems associated with channel invasion by willows, although Green *et al.* (1989) recommend willow management rather than removal to minimise habitat damage.

Rehabilitation of channel morphology

Deliberate modification of river channels has been undertaken on an extensive scale in New Zealand to mitigate the effects of bank erosion, bed aggradation and flooding (Mosely, 1992). Channel rehabilitation work has also been carried out to mitigate the effects of alluvial mining. Studies on the West Coast of the South Island have suggested that alluvially mined channels should not be restored to a hydraulically efficient shape because these do not provide suitable cover for larger fish (Eldon *et al.*, 1993). Where cover for fish is absent, rehabilitation efforts should focus on introducing boulders and logs, and planting suitable overhanging riparian vegetation. However, high daily flow variability and flood discharges in many parts of New Zealand (Parde, 1966; Jowett and Duncan, 1990) may limit the success of in-stream structures in river rehabilitation. For example, 'boulder drops' installed in Big Hohonu River (Figure 16.2) to increase habitat heterogeneity for aquatic biota were wiped out by a single large flood (Eldon *et al.*, 1989).

An example of channel rehabilitation in an urban setting is provided by the Heathcote River in Christchurch (Roper-Lindsay, 1994) (Figure 16.2). The back-up of floodwaters caused by a long meander was alleviated by the construction of a canal (effectively transforming the meander into a large island) to improve the transmission of floodwaters to the estuary. However, this resulted in salt-water penetrating almost 2 km further upstream than it had previously, leading to a reduction in the extent of freshwater habitat, the death of riparian plants with low salt tolerance, and destabilisation of the banks. The local council insisted that the solution to this problem be environmentally driven, and this led to a decision to restore the channel pattern to its previous condition while maintaining flood control by building a set of gates at the upstream end of the canal. The gates will remain closed during normal flow conditions, but will be opened during floods.

Water quality improvement

Most stream and river rehabilitation work in New Zealand has focused on restoring water quality to conditions that are suitable for aquatic life (primarily trout), contact recreation, and/or water supply. There are numerous examples of 'clean-ups' involving the treatment of point-source discharges from sewage and industrial outfalls, dairy sheds, and mining operations. Examples of rehabilitation of poor water quality caused by diffuse inputs are more difficult to find, however, and non-point-source inputs to rivers (often associated with grazing animals) are still perceived as a common problem in many parts of New Zealand (Hoare and Rowe, 1992).

The lower Manawatu River in the southern North Island (Figure 16.2) receives inputs from domestic sewage works, a dairy company, meat works and a pharmaceutical company which collectively discharged a total load of up to 23 500 kg BOD per day into the river (Currie, 1980). In 1978, several hundred trout died as a result of severe night-time dissolved oxygen depletion caused by prolific growths of sewage fungus and algae (Quinn and Gilliland, 1988). Since then, over NZ$4 million has been spent on waste-water treatment, and BOD loads of individual discharges have been reduced by 60–90%. This has greatly improved river appearance and the conditions for river biota, but prolific algal growths can still cause stressful dissolved oxygen depletions and pH fluctuations in summer (Quinn and Gilliland, 1988).

Water quality can also be improved by appropriate riparian management. A 20-year study of the effects of riparian protection on water quality was conducted on a second-order agricultural stream draining to Lake Taupo in the central North Island (Figure 16.2) (Howard-Williams and Pickmere, 1994). This study revealed that nutrient removal capacity was enhanced in the first 13 years of protection due to the growth of plants with high nutrient stripping capacity. However, stream nutrient levels later increased, following a decline in plant abundance due to increased shading, thereby demonstrating that a high light environment is necessary for the continued effectiveness of macrophytes in removing nutrients. Alternatively, nutrients may be removed before they enter the stream, and work in New Zealand has shown that the protection of near-stream seepage areas is effective in reducing nitrate inputs to agricultural watercourses (Cooke and Cooper, 1988).

Provision of environmental flows

In recent years, in-stream flow incremental methodology (IFIM) has become increasingly popular in New Zealand as a basis for resolving minimum flow issues, particularly in setting flows to maintain or restore recreational salmonid fisheries. An excellent example of the use of this technique to rehabilitate salmonid habitat is found in the Ohau River, South Island (Figure 16.2). A weir constructed across the outlet of Lake Ohau for hydroelectric purposes resulted in the dewatering of the outlet river for a distance of 11.5 km. A study using IFIM to estimate the potential trout stocks if a residual flow was maintained indicated that adult trout habitat would be optimised at between $8 \, m^3 \, s^{-1}$ and $12 \, m^3 \, s^{-1}$ (Jowett, 1993). Modifications to the lake outlet were completed in 1992 so that a summer flow of $12 \, m^3 \, s^{-1}$ and a winter flow of $8 \, m^3 \, s^{-1}$ were maintained. Although surveys to monitor ecological changes following the implementation of residual flows have not been carried out, the river is proving to be a good fishery and other aquatic biota are also likely to have benefited. It should be remembered that increasing flows below dams and impoundments is not always necessary for ecological rehabilitation. A study on the Mangatangi Stream, near Auckland, North Island (Figure 16.2), which has been impounded for water supply, found that existing residual flows were adequate and that most benefit to the stream would accrue from restoring riparian shade to reduce water temperatures and excessive periphyton growths (McBride et al., 1994).

Rehabilitation of flow regimes for native fish species and riverine birds (e.g. the blue duck, *Hymenolaimus malacorhynchos*) in regulated rivers has been hampered in New Zealand by poor knowledge of their flow preferences, although studies are addressing this issue. Jowett and Richardson (1995) identified four flow habitat guilds of native fish in a range of New Zealand rivers: a fast-water guild found in central portions of riffles ($> 0.75 \, m \, s^{-1}$), an edge-dwelling guild preferring shallow water ($< 0.1 \, m$) at velocities less than $0.25 \, m \, s^{-1}$, an intermediate guild preferring velocities in the range 0.27–$0.44 \, m \, s^{-1}$, and a ubiquitous guild with broad habitat preferences. Adult blue duck appear to prefer shallow water with slow-moderate velocities (0.1–0.6 m s^{-1}) for feeding in Tongariro River, central North Island (Figure 16.2) (K. J. Collier, unpublished data).

Fish passage

The upstream climbing abilities of the different native fish species vary, but some native galaxiids are apparently able to negotiate damp, vertical rocky faces as high as 60 m (McDowall, 1990). Nevertheless, passage of even the most skilled climbers is disrupted by vertical drops of water such as below culverts, lack of flow refugia as found on smooth culvert bottoms, and by excessively high or dry structures. The 1947 Fishpass Regulations gave fisheries authorities in New Zealand the right to require a fishpass on any dam or weir on rivers where trout or salmon do or could exist, and by 1983 authorities had the right to require fishpass facilities on any dam or diversion structure. Restoration of fish passage on New Zealand rivers has typically involved the use of pool and weir or weir and orifice systems designed primarily for salmonid fishes (Jowett, 1987).

Recently, success in restoring passage has been achieved for native eels and some galaxiids using wetted PVC pipes with a large 'bottle-brush' or gravel lining (J. Boubee, pers. comm.; Mitchell et al., 1984). For example, between January and March 1992, 15 000 elvers used one of these devices to pass over the 63 m high Matahina Dam located on the Rangitaiki River in the eastern North Island (Figure 16.2) (Boubee and Mora, 1993). Mature eels require passage downstream to spawn, and many are thought to die in power station turbines. Electrical screens have been used successfully to divert eels away from the intake at Piripaua power-station, Lake Waikaremoana, in the central North Island (Figure 16.2) (Jowett, 1987). Diversion by light offers another possible means of deterring eels from power station intakes.

Provision of wildlife habitat

Several native galaxiid fishes spawn in riparian grasses close to the upstream limit of salt-water intrusion (McDowall, 1990). The larvae of these species travel downstream after hatching and swim back upstream as juveniles; collectively these are known as 'whitebait' and they form the basis of a recreationally important fishery. The hatching success of whitebait eggs laid on riparian grasses is believed to be adversely affected by grazing cattle due to trampling, exposure and dehydration. Fencing of streambanks to exclude cattle is a recommended tool for rehabilitating whitebait spawning grounds, although some grazing outside the spawning season may be necessary to maintain suitable riparian grass cover (Mitchell, 1990a, 1991). Replanting with appropriate spawning vegetation is also recommended at heavily modified sites (Mitchell, 1990b).

A large-scale habitat rehabilitation project named Project River Recovery is currently under way in the Waitaki River Basin, southern South Island (Figure 16.2). Here, flow regulation and the invasion of alien plants, particularly willows and lupins, have degraded the riverbed habitat (Rawlings, 1993). As a result, a number of threatened bird species such as the black stilt (*Himantopus novaezelandiae*) and wrybill (*Anarhynchus frontalis*) have lost feeding and nesting grounds, and are subject to greater predation risk from animals like stoats and ferrets which shelter in the increased vegetation cover.

Habitat rehabilitation is being coupled with a captive breeding programme for the black stilt which is raising juveniles for release into the restored habitat.

FUTURE PROSPECTS FOR REHABILITATION OF NEW ZEALAND'S WATERWAYS

The emphasis of current resource management laws on remedying or mitigating adverse impacts of development provides considerable future scope for river rehabilitation in New Zealand, particularly in lowland areas where there has been widespread degradation of aquatic habitats. The recently completed '100 Rivers Project' characterised and classified New Zealand's rivers in terms of flow, water quality, and faunal characteristics, and should greatly assist the development of rehabilitation goals for water quality and biological communities (Biggs et al., 1990 and references therein).

The remedial effects associated with appropriate riparian management have increased awareness of this as a tool in stream rehabilitation. In-stream benefits (such as reducing maximum summer water temperatures, decreasing nutrient inputs, and increasing allochthonous carbon inputs) are generally most pronounced on small watercourses, and guidelines have recently been developed to assist resource managers in rehabilitating agricultural streams using riparian management (Collier et al., 1995).

Where possible, guidelines for river rehabilitation have used modelling techniques to assist in predicting the likely outcomes of different management strategies or to determine optimum riparian zone widths under different conditions. The modified CREAMS model of Cooper et al. (1992) was used to determine optimum zone widths for managing inputs of contaminants to streams in overland flow. In addition, the temperature model SEGMENT (Blackett, 1993) was used to predict maximum summer water temperatures on different sized streams under a variety of riparian shade scenarios so that recommendations could be made on the extent of planting required to achieve temperature rehabilitation goals.

Temperature modelling used in conjunction with IFIM and a dissolved oxygen model identified riparian restoration as the best option for optimising habitat in a recent ecological assessment of a stream impounded for water supply (McBride et al., 1994). Given the prohibitively high costs of detailed site-specific studies, it is likely that such techniques will play an increasingly important role in determining management options in the future. Scientifically based monitoring of the ecological outcomes of rehabilitation works is necessary for the continuing development of successful techniques and the refinement of models, and this should be built in to the costs of projects. Project River Recovery, for example, has placed considerable emphasis on monitoring the effects of vegetation removal and herbicide application on water quality, birds, trout and aquatic and terrestrial invertebrates as part of its rehabilitation programme.

Because of the complex nature of river systems and the processes that affect them, rehabilitation projects need a multidisciplinary approach that includes the expertise of aquatic and terrestrial ecologists, geomorphologists, hydrologists and engineers. In the urban environment, input from landscape architects will also be necessary to design systems that provide aesthetic appeal as well as habitat for aquatic life (Roper-Lindsay, 1994). The long-term success of rehabilitation works may depend on the support and involvement of the local community and river users, and sociologists are necessary to help distil their feelings into ecological goals. Recent research into community attitudes to riparian restoration in the northern South Island has highlighted the need for public education and appropriate focusing of rehabilitation goals, and the importance of resolving conflicts to bring about change (O'Brien, 1994).

In New Zealand there has been a move towards the formation of working parties to address potential conflicts in river rehabilitation proposals. Project River Recovery provides an excellent example of this approach. The impending expiry of the Electricity Corporation of New Zealand's (ECNZ) right to take water for electricity generation from the Waitaki Basin (Figure 16.2) led to the convening of a public forum in 1988 to discuss the issues in light of the RMA which was pending at the time (Rawlings, 1993). Concerns of local Maori, farmers, fishers, recreationalists, conservation groups and local bodies were represented by a working party which sought to produce a compromise that was relatively free of conflict. ECNZ was not prepared to restore flows to the Pukaki River (Figure 16.2) because of the millions of dollars it would cost in lost power generation. This led to a proposal to rehabilitate habitat elsewhere in the catchment equivalent to or greater than the net loss of habitat and conditions attributable to the Waitaki hydro-development. The programme is funded wholly by ECNZ for 21 years and will rehabilitate 31 km of habitat on the Ahuriri River (Figure 16.2) alone, compared with the 17 km lost in the Pukaki River.

CONCLUSION

Turbidity, salinity and nutrient enrichment are the major current concerns in Australian waterway management, even though the biota seem to have been more chronically impacted by flow regulation and modification of the channel and riparian zone. Water quality is relatively good in many parts of New Zealand, but catchment land use, riparian vegetation clearance, channel modification and the introduction of alien species of flora and fauna have caused major problems, particularly in the lowlands. In Australia there is a shortage of water and impoundments have been constructed mainly to provide a reliable water supply to irrigation areas. By comparison, New Zealand has a relative abundance of water, but rivers have been impounded to regulate seasonal flow variations and to take advantage of the relief to generate electricity. Both countries occasionally experience catastrophic floods, so mitigation of flooding of lowland rivers has been a major objective of river management. Thus, while New Zealand and Australia are physiographically very different, flow regulation, channel modification and introduction of alien species are major factors compromising the ecological integrity of rivers in both countries. Lowland streams are worst affected and it is these habitats that are likely to be a major focus of future rehabilitation work.

In recent years there has been growing awareness by the Australasian community of the importance of the health of the aquatic environment. Recent water law reform is likely to encourage the rehabilitation of physical habitat structure, and provision of environmental flow regimes downstream of regulating impoundments. The behaviour and habitat requirements of the native aquatic fauna, while known to be quite different to those of northern hemisphere species, are as yet poorly understood. Thus, authorities have had difficulty agreeing on the minimum flows that are required to protect the habitats of native species; a problem that requires immediate research attention. Recreational angling groups have demonstrated a willingness to rehabilitate streams to increase trout numbers, and although the habitat for many native species should also be improved by these efforts, there are doubts about the ability of some native fish to coexist successfully with trout.

The concept of rehabilitating waterways (urban and rural) for ecological sustainability is relatively new in Australasia. The efforts so far have been mostly isolated, localised and small scale. Few, if any, projects have attempted to scientifically appraise the outcomes of rehabilitation works. Such an undertaking would probably be too expensive for small community-based groups, but should be incorporated into larger scale projects. In addition, Australasian rivers are probably too hydrologically and biologically different from those of the northern hemisphere to allow simple transfer of rehabilitation methodologies. It is recommended that a few larger-scale experimental/demonstration projects be supported by waterway management authorities in order to establish general methodologies that can be successfully and expediently applied, and to encourage the concept of river rehabilitation throughout the wider community. Rehabilitation is more likely to succeed in rivers that are relatively geomorphologically stable. Fortunately, it appears that many rivers in Australasia have by now adjusted to the major catchment and channel disturbances inflicted in the early part of the century.

Opportunities for rehabilitating rivers purely for ecological purposes are currently limited, but with some imagination, ecological rehabilitation can be integrated with rehabilitation for other utilitarian purposes. A vision of ideal, fixed channels, rehabilitated by expensive structures, is inappropriate in Australasia because the risk of damage by extreme floods is too great. It would be more feasible to remove the causes of degradation, revegetate riparian areas, and encourage as far as possible the re-establishment of natural geomorphic and hydrologic processes. It has to be accepted that even if natural fluvial processes are re-established, in most places, the presence of alien species precludes restoration of the pre-European ecology.

Unappraised, parochial, single-interest river management has, in the past, led to environmental degradation and conflict. It is vital that river rehabilitation programmes involve communication between stakeholders, have well-defined goals, use multidisciplinary planning and management teams, use available modelling tools, and where possible make use of scientifically based monitoring.

ACKNOWLEDGEMENTS

The contribution of Chris Gippel was partly funded by an Australian Vice-Chancellors Committee Bicentennial Fellowship and a British Council Academic Links and Interchange Scheme Grant. The chapter was improved on the basis of reviews by Ian Rutherfurd and Brian Finlayson.

REFERENCES

Acheson, A.R. (1968) *River Control and Drainage in New Zealand*. Ministry of Works, Wellington, New Zealand.
Anderson, I. (1994) Can Lake Pedder resurface? *New Scientist*, **143** (1941), 14–15.

Anonymous (1994a) *Australian River Research Newsletter 1*, National River Health Program. Published by CRC for Freshwater Ecology, Monash University, Caulfield East, Victoria, Australia, pp. 1–4.
Anonymous (1994b) President's Report, *The Australian Trout 1*, Published by the Australian Trout Foundation Inc., Moonee Ponds, Victoria, Australia, pp. 5–6.
Anonymous (1994c) Demand for Lake Pedder to be restored. *Civil Engineers Australia*, **66**(5), 9.
Arthington, A.H., Bycroft, B.M. and Conrick, D.L. (1992) Environmental Study Barker-Barambah Creek Volume 1: Executive Summary. Report to Water Resources Commission, Department of Primary Industries by the Centre for Catchment and In-Stream Research, Griffith University, Nathan, Queensland, Australia.
Arthington, A.H. and Pusey, B.J. (1993) In-stream flow management in Australia: methods, deficiencies and future directions. *Australian Biologist*, **61**, 52–60.
Arthington, A.H. and Pusey, B.J. (1994) Essential flow requirements of river fish communities. In *Proceedings of Environmental Flows Seminar*, Canberra, 25–26 August, Australian Water and Wastewater Association Inc., Artarmon, NSW, Australia, pp. 1–8.
Atkins, B. (1993) Final project report: the Integrated Watering Strategy for the mid-Murray wetlands, Integrated Watering Strategy Report No. 2, Department of Conservation and Natural Resources, Shepparton, Victoria, Australia.
Australian Water and Wastewater Association (1994) *Proceedings of Environmental Flows Seminar*, Canberra, 25–26 August, Australian Water and Wastewater Association Inc., Artarmon, NSW, Australia.
Bayly, I.A.E. and Williams, W.D. (1973) *Inland Waters and Their Ecology*. Longman Australia, Melbourne, pp. 151–283.
Bennett, M. and McCosker, R.O. (1994) Estimating environmental flows requirements of wetlands. In *Proceedings of Environmental Flows Seminar*, Canberra, 25–26 August, Australian Water and Wastewater Association Inc., Artarmon, NSW, Australia, pp. 9–16.
Biggs, B.J.F. (1982) The effects of low flow regulation on the biology and water quality of rivers with particular reference to an impoundment-regulated flow. In *River Low Flows: Conflicts and Water Use*. Proceedings of a seminar, Lincoln 1982. New Zealand Ministry of Works and Development Water and Soil Miscellaneous Publication 47, pp. 28–42.
Biggs, B.J.F., Duncan. M.J., Jowett, I.G., Quinn, J.M., Hickey, C.W., Davies-Colley, R.J. and Close, M.E. (1990) Ecological characterisation, classification, and modelling of New Zealand rivers: an introduction and synthesis. *New Zealand Journal of Marine and Freshwater Research*, **24**, 277–304.
Biosis Research Pty Ltd (1992) Mullum Mullum Creek Linear Park, Concept Plan, Waterway Management Plan, and Stage One Management Plan. Report to City of Doncaster and Templestowe and Melbourne Water by Biosis Research Pty Ltd in conjunction with Ratio Consultants Pty Ltd and Neil Craigie and Associates Pty Ltd, Port Melbourne, Victoria, Australia.
Blackett, S. (1993) SEGMENT: an IBM PC model of temperature in a small stream. Unpublished report, NIWA Ecosystems, Hamilton, New Zealand.
Boubee, J. and Mora, A. (1993) Restoring fish passage at hydro dams (Abstract)., *New Zealand Limnological Society Newsletter*, **29**, 23.
Brink, P., Nilsson, L.M. and Svedin, U. (1988) Ecosystem redevelopment. *Ambio*, **17**, 84–89.
Brown, A., Butterworth, R., Saunders, G. and Mill, P. (1994) No fishing for an answer in this project. *Freshwater Fishing Australia*, **25**, 22.
Cadwallader, P.L. (1978) Some causes of the decline in range and abundance of native fish in the Murray-Darling River system. *Proceedings of the Royal Society of Victoria*, **90**, 211–224.
Chartres, C.J., Helyer, K.R., Fitzpatrick, R.W. and Williams, J. (1992) Land degradation as a result of European settlement of Australia and its influence on soil properties. In Gifford, R.M. and Barson, M.M. (eds) *Australia's Renewable Resources: Sustainability and Global Change*. Bureau of Rural Resources Proceedings No. 14, AGPS, Canberra, ACT, Australia, pp. 3–33.
Collier, K.J. (1993) Linking the mountains to the sea – conserving New Zealand's rivers. *Forest and Bird*, November 1992, 26–32.
Collier, K.J. (1994) Effects of willows on waterways and wetlands – an aquatic habitat perspective. In West, C.J. (comp.) *Wild Willows in New Zealand*. Proceedings of a Willow Control Workshop

hosted by Waikato Conservancy, Hamilton, Department of Conservation, Wellington, New Zealand, pp. 33–41.

Collier, K.J., Cooper, A.B., Davies-Colley, R., Rutherford, J.C., Smith, C. and Williamson, R.B. (1995) *Managing Riparian Zones along Watercourses in Modified Catchments*. NIWA Ecosystems Report, National Institute of Water and Atmosphere, Hamilton, New Zealand.

Cooke, J.G. and Cooper, A.B. (1988) Sources and sinks of nutrients in a New Zealand hill pasture catchment III. Nitrogen. *Hydrological Processes* **2**, 135–149.

Cooper, A.B., Smith, C.M. and Bottcher, A.B. (1992) Predicting runoff of water, sediment and nutrients from a New Zealand grazed pasture using CREAMS. *Transactions of the American Society of Agricultural Engineers*, **35**, 105–112.

Couling, K.C. (1992) Recent changes in management philosophy, policies and practices for Christchurch's waterways. Unpublished paper presented to Association of Local Government, Engineers of New Zealand Conference, June 1992, Wellington, New Zealand.

Craigie, N. and Eldridge, C. (1994) Mullum Mullum Creek – transforming a drain into a valued public asset. *River Basin News*, **49**, 6. Published by the River Basin Management Society, Hawthorn East, Victoria, Australia.

Currie, K. (1980) Co-operation cleans up the Manawatu. *Soil and Water* **16**, 5–8.

Davey, G.W., Doeg, T.J. and Blyth, J.D. (1987) Changes in benthic sediment in the Thomson River, southeastern Australia, during construction of the Thomson dam. *Regulated Rivers: Research and Management*, **1**, 71–84.

Department of Conservation and Environment (1992) Lower Snowy River wetlands proposed management plan. Orbost Region and National Parks and Public Land Division, DCNR, East Melbourne, Victoria, Australia.

Department of Conservation, Forests and Lands (1985) *Eppalock Catchment Project*. Published by the DCFL in collaboration with the Soil Conservation Authority of Victoria, Melbourne, Victoria, Australia.

Department of Primary Industries and Energy (1993) 1994/95 Guide to project funding under the Commonwealth/State component of the National Landcare Program. Unpublished document by the Department of Primary Industries and Energy, National Landcare Program, Canberra, ACT.

Department of Statistics (1990) *New Zealand Official 1990 Year Book*. Department of Statistics, Auckland, New Zealand.

Department of Water Resources Victoria (1989) *A Resource Handbook*. Water Victoria, Victorian Government Printing Office, North Melbourne, Victoria, Australia.

Department of Water Resources Victoria (1992) *A Scarce Resource*. Water Victoria, DWRV, Melbourne, Victoria, Australia.

Doeg, T.J. and Saddlier, S.R. (1993) The impact of water released from the Upper Yarra Reservoir, and the introduction of artificial blackfish spawning habitats on the fish and macroinvertebrate fauna and habitats of the Yarra River, final report. Unpublished report to Melbourne Water Corporation by Freshwater Ecology Section, Flora and Fauna Division, Department of Conservation and Natural Resources, Heidelberg, Victoria, Australia.

Dovers, S.R. and Day, D. (1988) Australian rivers and statute law. *Environmental and Planning Law Journal*, **5**, 98–108.

Eldon, G.A., Sagar, P.M., Taylor, M.J., Bonnett, M.J. and Kelly, G.R. (1989) *Recolonisation of a Stream Diversion by Aquatic Invertebrates and Fish, Big Hohonu River, Westland*. New Zealand Freshwater Fisheries Report No. *115*, MAF Fisheries, Christchurch, New Zealand.

Eldon, G.A., Taylor, M.J. and Jellyman, D.J. (1993) *Aquatic Fauna Survey of Duffers Creek Catchment (Grey River), February 1990, Following Cessation of Alluvial Gold Mining and Channelling*. New Zealand Freshwater Research Report No. *3*, National Institute of Water and Atmospheric Research, Christchurch, New Zealand.

Erskine, W.D. (1992) Channel response to large-scale river training works: Hunter River, Australia. *Regulated Rivers: Research and Management*, **7**, 261–278.

Erskine, W.D., Rutherfurd, I.D. and Tilleard, J.W. (1990) Fluvial Geomorphology of the Tributaries to the Gippsland Lakes. Report to Department of Conservation and Environment, Ian Drummond and Associates Pty Ltd, Wangaratta, Victoria, Australia.

Erskine, W.D., Rutherfurd, I.D., Ladson, A.R. and Tilleard, J.W. (1993) Fluvial Geomorphology of the Goulburn River Basin. Report to the Mid Goulburn Catchment Co-Ordinating Group Inc., Ian Drummond and Assoc Pty Ltd, Wangaratta, Victoria, Australia.

Fahey, B.D. and Rowe, L.K. (1992) Landuse impacts. In Mosely M.P. (ed.) *Waters of New Zealand*. New Zealand Hydrological Society, pp. 265–284.

Fenemor, A. (1992) Water resource management in New Zealand. In Mosely, M.P. (ed.) *Waters of New Zealand*. New Zealand Hydrological Society.

Ferdinands, K.B. and D'Santos, P. (1994) A biological assessment of Dandenong Creek using benthic macroinvertebrates: stage 1. Melbourne Water Labs Scientific Report SR93/04, Melbourne Water Laboratories, Melbourne, Victoria.

Finlayson, B.L. and McMahon, T.A. (1988) Australia v the world: a comparative analysis of streamflow characteristics. In Warner, R.F. (ed.) *Fluvial Geomorphology of Australia*. Academic Press, Sydney, Australia, pp. 17–40.

Finlayson, B.L., Gippel, C.J. and Brizga, S.O. (1994) The effects of reservoirs on downstream aquatic habitat. *Journal Australian Water Works Association*, **21**(4), 15–20.

Gifford, R.M., Cheney, N.P., Noble, J.C., Russell, J.S., Wellington, A.B. and Zammit, C. (1992) Australian land use, primary production of vegetation and carbon pools in relation to atmospheric carbon dioxide concentration. In Gifford, R.M. and Barson, M.M. (eds) *Australia's Renewable Resources: Sustainability and Global Change*. Bureau of Rural Resources Proceedings No. 14, AGPS, Canberra, ACT, Australia, pp. 151–187.

Gippel, C.J. (1995) Potential use of turbidity monitoring for measuring the transport of suspended solids in streams. *Hydrological Processes*, **9**, 83–97.

Gippel, C.J. and Finlayson, B.L. (1993) Downstream environmental impacts of regulation of the Goulburn River, Victoria. In *Hydrology and Water Resources Symposium*, 30 June–2 July 1993, Newcastle, Institution of Engineers Australia, Canberra, ACT, Australia, pp. 33–38.

Gippel, C.J. and Stewardson, M. J. (1995) Development of an environmental flow management strategy for the Thomson River, Victoria, Australia. *Regulated Rivers, Research and Management*, **10**, 121–135.

Gippel, C.J., O'Neill, I.C. and Finlayson, B.L. (1992) *The Hydraulic Basis of Snag Management*. Centre for Environmental Applied Hydrology, Department of Civil and Environmental Engineering, The University of Melbourne, Parkville, Victoria, Australia.

Government of Victoria (1988) *Victoria's Inland Waters, State of the Environment Report 1988*, Office of the Commissioner for the Environment, Melbourne, Victoria, Australia.

Graynoth, E. (1979) Effects of logging on stream environments and faunas in Nelson. *New Zealand Journal of Marine and Freshwater Research*, **13**, 79–109.

Green, W.R., McPherson, B.R. and Scott, D. (1989) Vegetation Aspects of River Management. NWPC Technical Publication No. 7, New Zealand Acclimatisation Societies.

Hart, B.T. and McKelvie, I.D. (1986) Chemical limnology in Australia. In deDekker, and Williams, W.D. (eds) *Limnology in Australia*. CSIRO, Canberra, ACT, Australia, pp. 3–31.

Hart, B.T., Bailey, P., Edwards, R., Hortle, K., James, K., McMahon, A., Meredith, C. and Swadling, K. (1990) Effects of salinity on river, stream and wetland ecosystems in Victoria, Australia. *Water Research*, **24**, 1103–1117.

Hickey, C.W. and Rutherford, J.C. (1986) Agricultural point source discharges and their effects on rivers. *New Zealand Agricultural Science*, **20**, 104–110.

Hicks, B. (1984) Regional modification to waterways. Part 2. Taranaki and Rangitikei. *Freshwater Catch*, **22**, 26–30.

Hoare, R.A. and Rowe, L.K. (1992) Water quality. In Mosely M.P. (ed.) *Waters of New Zealand*. New Zealand Hydrological Society, pp. 365–380.

Hortle, K.G. and Lake, P.S. (1983) Fish of channelized and unchannelized sections of the Bunyip River, Victoria. *Australian Journal of Marine and Freshwater Research*, **34**, 441–450.

Howard-Williams, C. and Pickmere, S. (1994) Long-term vegetation and water quality changes associated with the restoration of a pasture stream. In Collier, K.J. (ed.) *The Restoration of Aquatic Habitats*. Department of Conservation, Wellington, New Zealand, pp. 93–109.

Hughes, P.J. and Sullivan, M.E. (1981) Aboriginal burning and Late Holocene geomorphic events in eastern N.S.W. *Search*, **12**, 277–278.

Ian Drummond and Associates Pty Ltd (1992) *Glenelg Catchment Waterway Management Study: Final Report*. Report to Shire of Dundas on behalf of the Steering Committee for the Glenelg

Catchment. In association with Erskine, W.D., Frankenberg, J., Sharp, B., Fisher Stewart Pty Ltd and Read Sturgess and Associates, Wangaratta, Victoria, Australia.

Jacobs, M.R. (1955) *Growth Habits of the Eucalypts*. Australian Government Printing Service, Canberra, ACT, Australia.

Jowett, I.G. (1987) Fish passage, control devices and spawning channels. In Henriques, P.R. (ed.) *Aquatic Biology and Hydroelectric Power Development in New Zealand*. Oxford University Press, Auckland, New Zealand, pp. 138–155.

Jowett, I.G. (1993) Setting flows for the restoration of the brown trout fishery in the Ohau River (Abstract). *New Zealand Limnological Society Newsletter*, **29**, 27.

Jowett, I.G. and Duncan, M.J. (1990) Flow variability in New Zealand rivers and its relationship to in-stream habitat and biota. *New Zealand Journal of Marine and Freshwater Research*, **24**, 305–318.

Jowett, I.G. and Richardson, J. 1995. Habitat preferences of common riverine New Zealand native fishes and implications for flow management. *New Zealand Journal of Marine and Freshwater Research*, **29**, 13–23.

Koehn, J.D. (1986) *Dandenong Creek: Fishes, Their Habitats and Management Recommendations*. Arthur Rylah Institute for Environmental Research Technical Report Series No. 51, Department of Conservation Forests and Lands, Fisheries Division, Heidelberg, Victoria, Australia.

Koehn, J.D. (1987) *Artificial Habitat Increases Abundance of Two-Spined Blackfish* (Gadopsis bispinosis) *in Ovens, River, Victoria*. Arthur Rylah Institute for Environmental Research Technical Report Series No. 56, Department of Conservation Forests and Lands, Fisheries Division, Heidelberg, Victoria, Australia.

Koehn, J.D. and O'Connor, W.G. 1990. *Biological Information for Management of Native Freshwater Fish in Victoria*. Department of Conservation and Environment, Freshwater Fish Management Branch, Arthur Rylah Institute for Environmental Research, Victorian Government Printing Office, North Melbourne, Victoria, Australia.

Lake, J.S. (1967) Principal fishes of the Murray-Darling River system. In Weatherly, E.H. (ed.) *Australian Inland Waters and Their Fauna*. ANU Press, Canberra, ACT, Australia, pp. 192–213.

Lloyd, C.J. (1988) *Either Drought or Plenty, Water Development and Management in New South Wales*. Department of Water Resources New South Wales, Parramatta, NSW, Australia.

Lloyd, L.N., Puckridge, J.T. and Walker, K.F. (1991) The significance of fish populations in the Murray-Darling system and their requirements for survival. In Dendy, T. and Coombe, M. (eds) *Conservation in Management of the River Murray System – Making Conservation Count*. Proceedings of the 3rd Fenner Conference on the Environment, Canberra, 1989. South Australia Department Environment and Planning and the Australian Academy of Science, Adelaide, Australia, pp. 86–99.

Mallen-Cooper, M. (1994) How high can a fish jump? *New Scientist*, **142**(1921), 32–36.

Margules and Partners Pty Ltd (1989) River Murray riparian vegetation study. Report to the Murray-Darling Basin Commission in association with P. and J. Smith Ecological Consultants, Department of Conservation, Forests and Lands, Canberra, ACT, Australia.

McBride, G.B., Cooke, J.G., Cooper, A.B. and Smith, C.M. (1994) Optimising habitat in a stream impounded for water supply. In Collier, K.J. (ed.) *The Restoration of Aquatic Habitats*. Department of Conservation, Wellington, New Zealand, pp. 111–123.

McColl, R.H.S. and Ward, J.C. (1987) The use of water resources. In Viner A.B. (ed.) *Inland Waters of New Zealand*. DSIR Bulletin 241, Wellington, New Zealand, pp. 441–459.

McCosker, R.O. and Duggin, J.A. (1993) *Gingham Watercourse Management Plan Final Report*. Report to Gingham Watercourse Association and Department of Water Resources NSW, Department of Ecosystem Management, University of New England, Armidale, NSW, Australia.

McDowall, R.M. (1990) *New Zealand Freshwater Fishes – A Natural History Guide*. Heinemann Reed MAF Publishing Group, Auckland, New Zealand.

McMahon, T.A., Gan, K.C. and Finlayson, B.L. (1992) Anthropogenic changes to the hydrological cycle in Australia. In Gifford, R.M. and Barson, M.M. (eds) *Australia's Renewable Resources: Sustainability and Global Change*. Bureau of Rural Resources Proceedings No. 14, Australian Government Printing Service, Canberra, ACT, Australia, pp. 35–66.

McPhail, I.R. and Young, E.M. (1992) Water for the environment in the Murray-Darling basin. In Pigram, J.J. and Hooper, B.P. (eds) *Water Allocation for the Environment*. Proceedings of an

International Seminar and Workshop, 27–29 November 1991, The Centre for Water Policy Research, The University of New England, Armidale, NSW, Australia, pp. 191–210.
Mitchell, C.P. (1990a) *Whitebait Spawning Grounds on the Bay of Plenty*. New Zealand Freshwater Fisheries Miscellaneous Report No. 40, MAF Fisheries, Rotorua, New Zealand.
Mitchell, C.P. (1990b) *Whitebait Spawning Grounds on the Lower Waikato River*. New Zealand Freshwater Fisheries Miscellaneous Report No. 42, MAF Fisheries, Rotorua, New Zealand.
Mitchell, C.P. (1991) *Whitebait Spawning Ground Management: Interim Report*. New Zealand Freshwater Fisheries Report No. 131, MAF Fisheries, Rotorua, New Zealand.
Mitchell, C.P., Huggard, G. and Davenport, M. (1984) Home grown fish pass helps eels get by. *Freshwater Catch*, **24**, 21–22.
Mitchell, P. (1990) *The Environmental Condition of Victoria's Streams*. Report to the Department of Water Resources Victoria, Melbourne, Victoria, Australia.
Mosely, M.P. (1992) River morphology. In Mosely, M.P. (ed.) *Waters of New Zealand*. New Zealand Hydrological Society, pp. 285–304.
Murray-Darling Basin Commission (1993) Technical Advisory Group Report, Algal Management Strategy, MDBC with Land and Water Resources Research and Development Corporation, Canberra, ACT, Australia.
Nuttall, P.M. (1983) *Temporal Changes in Water Quality in the Dandenong Creek Catchment*. Technical Report 24, Dandenong Valley Authority, Dandenong, Victoria, Australia.
O'Brien, M. (1994) Community perspectives of riparian management and restoration: a case study in Marlborough. In Collier, K.J. (ed.) *The Restoration of Aquatic Habitats*. Department of Conservation, Wellington, New Zealand, pp. 145–171.
Olive, L.J. and Rieger, W.A. (1986) Low Australian Sediment Yields – A Question of Inefficient Sediment Delivery. IAHS Publication No. 159, pp. 355–366.
Osborne, L.L., Bayley, P.B., Higler, L.W.G., Statzner, B., Triska, F. and Iversen, T.M. (1993) Restoration of lowland streams: an introduction. *Freshwater Biology*, **29**, 187–194.
Parde, M. (1966) Flow regimes of New Zealand rivers. *Journal of Hydrology (N.Z.)*, **5**, 2–19.
Pigram, J.J. (1986) *Issues in the Management of Australia's Water Resources*. Longman Cheshire, Melbourne, Victoria, Australia.
Poole, A.L. (1983) *Catchment Control in New Zealand*. Water and Soil Miscellaneous Publication 48, New Zealand Ministry of Works and Development, Wellington, New Zealand.
Quinn, J.M. and Gilliland, B.W. (1988) The Manawatu River cleanup – has it worked? *Transactions of the Institute of Professional Engineers New Zealand*, **16**, 22–26.
Quinn, J.M., Davies-Colley, R.J., Hickey, C.W., Vickers, M.L. and Ryan, P.A. (1992) Effects of clay discharges on streams. *Hydrobiologia*, **248**, 235–247.
Rawlings, M. (1993) Lupins, willows, and river recovery: restoration in the upper Waitaki. *Forest and Bird*, November 1993, 10–15.
Reinfelds, I., Rutherfurd, I. and Bishop, P. (1995) History of channelisation and its effects on the Latrobe River, Victoria. *Australian Geographical Studies*, **33**, 60–76.
Rieger, W.A. and Olive, L.J. (1988) Channel sediment loads: comparisons and estimation. In Warner, R.F. (ed.) *Fluvial Geomorphology of Australia*. Academic Press, Marrickville, NSW, Australia, pp. 69–85.
Roberts, J. (1990) Flood plain ecosystems and their protection. In *Wetlands, Their Ecology, Function, Restoration and Management*. Proceedings of the Applied Ecology and Conservation Seminar Series, October–December 1989, Wildlife Reserves, La Trobe University, Bundoora, Victoria, Australia, pp. 75–81.
Roper-Lindsay, J. (1994) Tales of the riverbank – examples of bank restoration on urban rivers. In Collier, K.J. (ed.) *The Restoration of Aquatic Habitats*. Department of Conservation, Wellington, New Zealand, pp. 125–143.
Rutherfurd, I.D. and Smith, N. (1992) *Sediment Sources and Sinks in the Catchment of the Avoca River, North Western Victoria*. Water Division, Department of Conservation and Natural Resources Report No. 83, Monash University, Department of Geography and Environmental Science, Department of Conservation and Natural Resources, Melbourne, Victoria, Australia.
Rutherfurd, I.D., Reinfelds, I.V., Bishop, P. and Grayson, R.B. (1993) Historical sediment storage and declining sediment yields in Victorian catchments: implications for river management. In Yu, B. and Fielding, C.R. (eds) *Proceedings of the 5th International Conference on Fluvial*

Sedimentology, Modern and Ancient Rivers, Their Importance to Mankind, 5–9 July, The University of Queensland, Brisbane, Queensland, Australia, p. 115.

Scott, D. (1987) Sports fisheries. In Viner, A.B. (ed.) *Inland Waters of New Zealand*. Department of Scientific and Industrial Research Bulletin 241, Wellington, New Zealand, pp. 413–440.

Senior, J.G. (1992) Melbourne's waterways enhancement. In *International Symposium on Urban Stormwater Management*, Sydney, 4–7 February 1992, Institution of Engineers Australia, Canberra, ACT, Australia, pp. 413–418.

Sherrard, J.J. and Erskine, W.D. (1991) Complex response of a sand-bed stream to upstream impoundment. *Regulated Rivers: Research and Management*, **6**, 53–70.

Smith, D.I. and Finlayson, B.L. (1988) Water in Australia: its role in environmental degradation. In Heathcote, R.L. and Mabbutt, J.A. (eds) *Land Water and People, Geographical Essays in Australian Resource Management*. Academy of the Social Sciences in Australia, Allen and Unwin, Sydney, Australia, pp. 7–71.

Swales, S. (1994) Streamflow requirements of native fish in NSW rivers – the role and importance of flushing flows. In *Proceedings of Environmental Flows Seminar*, Canberra, 25–26 August. Australian Water and Wastewater Association Inc., Artarmon, NSW, Australia, pp. 178–183.

Townsend, C.R. and Crowl, T.A. (1991) Fragmented population structure in a native New Zealand fish: an effect of introduced brown trout. *Oikos*, **61**, 347–354.

Tyler, P.A., Sherwood, J., Magilton, C. and Hodgson, D. (1993) A geophysical survey of Lake Pedder and region. Unpublished report from the School of Aquatic Science and Natural Resources Development, Deakin University, to the Lake Pedder Study Group, for the Lake Pedder Restoration Committee, Ferntree, Tasmania, Australia.

Vollebergh, P. (1992) *River Frontage Management in Gippsland*. Report No. 78 Water Resource Management Report Series, Gippsland Water Strategy, Water Victoria, Department of Water Resources Victoria, Melbourne, Victoria, Australia.

Walker, K.F. (1979) Regulated streams in Australia: the Murray-Darling River system. In Ward, J.V. and Stanford, J.A. (eds), *The Ecology of Regulated Streams*. Plenum Press, New York, pp. 143–163.

Walker, K.F. (1983) Impact of Murray-Darling basin development on fish and fisheries. In Petr, P. (ed.) *Summary Report and Selected Papers Presented at the IPFC Workshop on Inland Fisheries for Planners*, Manila, The Philippines, FAO Fisheries Report 288, pp. 139–149.

Walker, K.F. (1985) A review of the ecological effects of river regulation in Australia. *Hydrobiologia*, **125**, 111–129.

Ward, J.C., Talbot, J.M., Denne, T. and Abrahamson, M. (1985) Phosphorus losses through transfer, soil erosion and runoff: processes and implications. Centre for Resource Management Information Paper No. 3, Lincoln College, Canterbury, New Zealand.

Warner, R.F. (1984) Man's impact on Australian drainage systems. *Australian Geographer*, **16**, 133–141.

Warner, R.F. and Bird, J.F. (1988) Human impacts on river channels in New South Wales and Victoria. In Warner, R.F. (ed.) *Fluvial Geomorphology of Australia*. Academic Press, Marrickville, NSW, Australia, pp. 343–363.

Wettin, P.D. and Bennett, M.W.A. (1986) Wetlands of the Lachlan Valley: location, characteristics and potential for water management. In *Proceedings of the International Symposium on Wetlands*, 5–8 June 1986, Shortland Wetlands Centre, Shortland, Newcastle, NSW, Australia, pp. 445–457.

Wilcock, R.J. (1986) Agricultural run-off: a source of water pollution in New Zealand. *Agricultural Science*, **20**, 98–103.

Woods, L.E. (1984) *Land Degradation in Australia*. Department of Home Affairs, AGPS, Canberra, ACT, Australia.

17 Rehabilitation of Japan's Waterways

CHRISTOPHER J. GIPPEL[1] and SHUBUN FUKUTOME[2]
[1]*Centre for Environmental Applied Hydrology, Department of Civil and Environmental Engineering, The University of Melbourne, Parkville, Victoria, Australia*
[2]*Nishinihon Institute of Technology, Kochi City, Japan*

INTRODUCTION

In Japan, the combination of steep topography, tectonically active geology, and highly seasonal and typhoon-prone climate, produces waterways which are geomorphologically and hydrologically highly dynamic and unstable. The archipelago has a high relief but its maximum width is only about 300 km (Figure 17.1), so the rivers tend to be short, with high gradients. Much of the country is too rugged and steep for urban development. Three-quarters of the total land area of 377 800 km^2 has a slope greater than 15° (Kornhauser, 1982), and two-thirds of the country is forested (Iyama, 1993).

Floodplain land occupies only about one-tenth of the total land area of Japan, but these areas are of great cultural and economic significance. Communities developed on floodplains where water was readily available for agriculture (particularly rice cropping), and works to control water for irrigation supply have a long history. Development of water resources for irrigation expanded considerably during the phase of rapid urbanisation in the early Tokugawa (Edo) Period (1603–1868), but flood mitigation also became a priority issue in the more densely settled areas. During this period the main course of the Tone River, which used to drain into Tokyo Bay, was diverted to Choshi, and a new channel named the Edo River was directed into Tokyo Bay (Figure 17.1). The main objectives of the scheme were to mitigate flooding, open up new land for rice cultivation, and develop navigable channels (Uzuka and Tomita, 1993). The modernisation of Japan during the Meiji Era (1868–1912) involved widespread industrial and urban development. A series of damaging floods in the 1890s prompted enactment of the River Law in 1896, and this marked a shift in the focus of river management from low-flow control (for navigation and irrigation) to flood control (Uzuka and Tomita, 1993; Ikebuchi, 1994). The works involved channelisation, channel straightening by meander cut-off, and the construction of embankments, floodways and dams.

Japan began to experience a period of remarkable economic growth in the 1960s. The small area of available floodplain land now contains half of the country's population of 125 million people and three-quarters of the nation's property and infrastructure (Iyama, 1993; Seki and Takazawa, 1993; Ikebuchi, 1994; Tamai *et al.*, 1994). Despite flood control works, flood damage has actually increased in recent times (Iyama, 1993). This can be attributed partly to the repeated occurrence of major typhoons, but channelisation has

Rehabilitation of Rivers: Principles and Implementation. Edited by L. C. de Waal, A. R. G. Large and P. M. Wade.
© 1998 John Wiley & Sons Ltd.

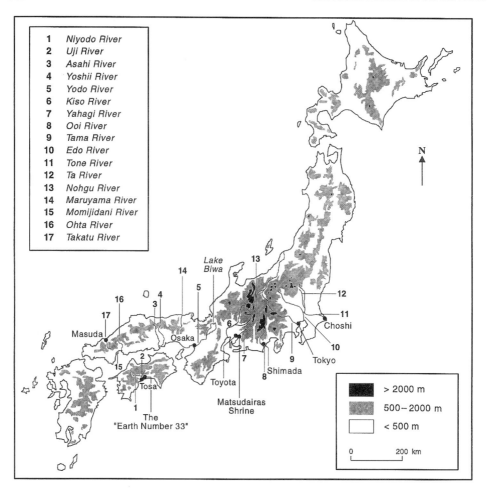

Figure 17.1 Sites of river rehabilitation referred to in the text

increased the magnitude of flood peaks (Kikkawa et al., 1975; Hirano, 1993), and the concentration of human activity in flood-prone areas has exacerbated the problem (Ikebuchi, 1994). Since 1945 there have been 12 major river flood disasters in Tokyo that have claimed a total of 374 lives, and in each event, thousands of residences suffered inundation damage (River Division, Bureau of Construction, 1988, p. 50).

Under conditions of intense rainfall, the highly erodible soils in the steep upland areas can produce debris flows which damage infrastructure and cause loss of life. In June 1961 a debris flow in the Tenryu River basin resulted in 136 casualties, and as recently as June 1985, three people were killed by a debris flow on the Yotagiri River (Imai, 1993). Sabo dams have been constructed on many of Japan's rivers to control coarse sediment transport.

The high level of flow and channel regulation required for flood mitigation and sediment control, plus pollution and hydrological changes associated with industrialisation and urbanisation, have led to a severe deterioration in the ecological condition of

Japan's waterways. Although freshwater fishing in Japan has significant economic and recreational value, artificial stocking technology has allowed the industry to develop without concern for conservation of spawning habitat and migration routes in the middle and upper reaches.

Japan's rivers are highly regulated, and future water resources development and hazard reduction will see the level of flow and channel regulation increase. Past regulation has caused severe degradation of natural biophysical processes. It has also had the effect of alienating river environments from people. From the earliest times the Japanese have held a close cultural association with nature, later reinforced by the Chinese philosophies of Taoism, Confucianism and Buddhism. The respect for natural phenomena is a balance of gratitude for provision of resources and awe of nature's probable dominance (Kornhauser, 1982, p. 144). This suggests an underlying ambivalence in the Japanese attitude towards environmental management. Not surprisingly, there is currently a high level of enthusiasm for developing and adopting practices that achieve engineering goals, but which also conserve or rehabilitate the natural riverine environment and make it accessible.

Over the past 20 years there has been a growing demand for urban recreation spaces, which has resulted in development of river foreshore improvement projects aimed at creating visually appealing areas with high amenity value. Ecological rehabilitation of waterways is a more recent phenomenon, beginning early in the Heisei Era (1989–present) in response to growing awareness by the community of environmental issues (Aso, 1991). This chapter examines the main environmental problems associated with management of Japan's waterways, charts the development of the legislative and planning framework for waterway management, and reviews recent progress in waterway rehabilitation.

WATERWAY MANAGEMENT PROBLEMS

Rivers in Japan convey large volumes of flood water. Average annual rainfall in Tokyo is 1460 mm, but it is highly seasonal. The average rainfall in June, September and October is 180–190 mm month^{-1}, but the winter months receive an average of only 50–60 mm month^{-1}. Typhoons can produce rainfall events of extreme intensity and duration. For example, rainfall intensity during a typhoon in July 1985 peaked at 91 mm over a one hour period, and 1000 mm fell over six days during an event in April 1978 (River Division, Bureau of Construction, 1988, p. 50). Flood runoff from the small, high-relief catchments can be rapid and of a high magnitude. Relative to basin area, peak flows are much larger than rivers in most other countries and hydrographs are generally very steep (Iyama, 1993). Geologically, Japan is a young country with active mountain building, earthquakes and volcanoes. Under conditions of steep relief and intense rainfall the soils have a high potential for erosion. Streams and rivers therefore tend to carry large volumes of bedload, sometimes in the form of catastrophic debris flows. This combination of geomorphological and hydrological conditions gives rise to rivers which are naturally highly flood-prone and morphologically unstable. Despite these hazards, the scarcity of floodplain land means that lowland areas are economically valuable and heavily developed. The main, and formidable, task of waterway management in Japan is to protect these areas against flooding and channel instability. The perceived importance of this task to the future of the nation's economy cannot be understated (Iyama, 1993; Ikebuchi, 1994).

Waterway regulation in Japan has involved channel straightening, hard-lining banks, grade control structures, ground sills on beds, embankments, reservoirs, retarding basins, sabo dams to control sediment transport, and removal of riparian vegetation. While channel works are generally aimed at flood mitigation, they are also conducted in association with disaster rehabilitation and road construction, so regulated channels can be found in middle and upper reaches (Ishida et al., 1994). Many streams in Tokyo were canalised with stone revetment in the early Showa Era (1926–1989). Most urban, and many rural streams, now have hard-lined banks and beds. Over 20% of the riparian edges of major streams are constructed of artificial material (Tamai et al., 1994). The shortage of land has resulted in removal of riparian habitats and alienation of channels from their floodplains. Japan's waterways have been engineered to function hydraulically, not ecologically.

Shimatani et al. (1993) found that the morphological changes associated with a meander cut-off on the Ta River (Figure 17.1) changed the hydraulic conditions from a heterogeneous pool and riffle environment to a uniform flat riffle environment. This tended to favour swimming fish, but benthic fish declined markedly or disappeared. Also, the decrease in tractive force of the wider, modified channel means that the possibility of bed material forming alternate bars is limited, so the uniform bed morphology is expected to persist.

Hatchery technology for salmon was introduced into Japan from the USA in the late 1870s, and since then the overwhelming emphasis has been on stocking rather than natural spawning. Ayu (*Plecoglossus altivelis*) is an anadromous fish that is very important in Japan for recreational angling. In the 1930s the technology for artificially incubating fry was developed, and the practice of stocking rivers with landlocked ayu from Lake Biwa (Figure 17.1) began. Thus, maintenance of fish passage and spawning beds in rivers was not an important issue to be considered in connection with channelisation and construction of barriers such as weirs, sabo dams, and grade control structures (Ishida et al., 1994). Recently it has been recognised that despite stocking, regulation has impacted on fish populations through declining availability of cover, shade, refuge and hydraulically suitable habitat (Ishida et al., 1994).

The hydrology of Japan's rivers has been significantly altered by urbanisation. Sewerage works in Japan have progressed slower than other public utilities, with only 47% of the nation's population being serviced in 1993 (Ministry of Construction, 1993a). However, municipal and prefectural governments are investing heavily in programmes to increase this rate. Sewerage rates are much higher in cities. In the Tokyo Ward area the percentage of the population serviced by sewerage rose from around 35% in 1965 to 98% in mid-1994 (Sewerage Bureau, 1994). Improvements in living standards and the construction of multi-storey residences have dramatically increased the volumes of wastewater flowing through the sewer systems and waterways. In the Tokyo Ward area in 1974 the average sewerage flow per capita was $810 \, l \, day^{-1}$, having increased from $167 \, l \, day^{-1}$ in 1924 (Sewerage Bureau, 1989).

The present stormwater drainage facilities in Tokyo were designed to cope with a rainfall intensity of $50 \, mm \, h^{-1}$, but this was assuming a runoff coefficient of 0.5 (Fujita, 1994). Over much of Japan, the soils have a high capacity for infiltration, so another effect of urbanisation, whereby roads and buildings have increased the cover of impervious surfaces, has been to dramatically increase runoff coefficients. In the Tokyo Ward area, as the area of paved road as a percentage of the total area increased from 4.4% in 1960 to

14.4% in 1985, the runoff coefficient for Ginza increased from 0.7 to 0.9, for Ichigaya from 0.5 to 0.75 and for Shakujii from 0.4 to 0.65 (Sewerage Bureau, 1989). The traditional approach to managing urban runoff has been to drain stormwater as quickly as possible through sewer pipes and channels. Hydrological changes due to urbanisation have increased the risk of inundation, but this cannot be simply dealt with in the traditional manner because there is now little scope for further enlarging the capacity of urban channels (Fujita, 1994). Another consequence of increased surface runoff due to urbanisation has been lowering of water-tables, so that many streams in western Tokyo now cease to flow for much of the year (Fujita, 1993). These hydrological changes have adversely affected aquatic ecosystems (Ishizaki et al., 1993).

While much of the river regulation in Japan has been driven by flood mitigation and sediment control objectives, flows have also been regulated in order to generate electricity and to provide a more reliable supply of water. The gradient and discharge characteristics of rivers in Japan make them ideally suited to hydroelectricity production, with the first hydropower dam being constructed in 1890. Many hydropower plants were built during the post-war construction period, and the energy crises of the 1970s provided another impetus. Most dams constructed from the 1950s onwards were designed for multipurpose use: hydropower, flood mitigation, urban water supply, and agricultural water supply. Demand for water in urban areas grew rapidly in the 1960s, and it was at this time that land subsidence problems associated with groundwater extraction forced a change to dependence on surface water supplies (Hukunari and Hiroki, 1993).

Although Japan has a wet climate, the precipitation per capita is about one-fifth of the world average (Iyama, 1993). Annual precipitation per capita in the Tone River Basin, which includes the Tokyo metropolis, is only 1/25th of the world average (Hukunari and Hiroki, 1993). Despite the construction of ambitious water resources control schemes, the water supply situation is precarious in low rainfall years (Hukunari and Hiroki, 1993). The high degree of flow regulation imposed by impoundment, diversion and abstraction has grossly altered the natural regimes of rivers, but there is little information available on the consequences for the aquatic biota.

The expansion of sewerage services has improved the water quality of Japan's waterways, but there are still many rivers that are too polluted to support fish life. In Tokyo, the average biological oxygen demand (BOD) level in the Ayase River at the Takumi Bridge peaked at 55 mg l^{-1} in 1970, and in 1985 was still 25 mg l^{-1}. The BOD in the Sumida River at Odaibashi Bridge peaked at 35 mg l^{-1} in 1965, but now, like the Edo River and the Tama River, the levels are close to or below 5 mg l^{-1} (Sewerage Bureau, 1989). Some rivers are still highly polluted, with the annual average BOD in the Karabori River between 1982 and 1988 fluctuating between 30 and 60 mg l^{-1} (River Division, Bureau of Construction, 1990, p. 19). Only streams in the less populated, mountainous Saitama Prefecture to the west of Tokyo have BOD levels below 2 mg l^{-1} (River Division, Bureau of Construction, 1988). Ecological rehabilitation of in-stream habitats will only be successful where water quality is adequate to support healthy populations of flora and fauna.

THE APPROACH TO WATERWAY MANAGEMENT IN JAPAN

Serious floods in the 1890s prompted enactment of the River Law in 1896 (Ikebuchi, 1994). This law proclaimed rivers having some economic value as national property, to be managed by prefectural governments for the national good, especially with respect to flood mitigation (Uzuku and Tomita, 1993). The Act revoked all private ownership of these rivers, their riparian area and their water. Because of its emphasis on flood control, the River Law was reformed in 1964 to cover water utilisation (Kawasaki, 1994). The new law divided waterways into Class A, being major rivers or streams important for conservation or economic purposes, and Class B, being tributaries of these rivers, or other lesser streams, also important for the public interest. Major Class A rivers are managed by the Ministry of Construction, and other rivers are managed by prefectural governments. Central government bears two-thirds of the cost of work on Class A rivers and half of the cost for Class B rivers, with the prefectural governments paying the balance. However, some prefectural governments have recently chosen to carry out special works at their own expense (Iyama, 1993).

Flood disasters in the early 1900s led to the creation of the First Flood Control Plan in 1911. A Second (1921) and a Third (1933) Flood Control Plan were enacted before the Second World War. In response to the devastation caused by a series of intense post-war typhoons, the Erosion and Flood Control Emergency Measures Law was enacted in 1960. Five-year Flood Control Programs were launched, with the Eighth Flood Control Program running from 1992 to 1996 (Ikebuchi, 1994). The *'93 Construction Whitepaper* (Ministry of Construction, 1993b) proposed policies and objectives for flood mitigation in the long term. For large rivers the goal is to provide, by the beginning of the 21st century, security against flooding by the largest post-1945 flood or the 1 in 200 year flood, whichever is greater. For main cities on lesser streams, the goal is to provide, by the beginning of the 21st century, security against flooding and debris damage associated with rainfall events of $80\,\mathrm{mm\,h^{-1}}$, and for smaller cities and rural streams, events of $50\,\mathrm{mm\,h^{-1}}$. There is considerable work to do in this regard, as protection against flood damage associated with rainfall events of $50\,\mathrm{mm\,h^{-1}}$ has been achieved in only 45% of cases and protection against damage by debris in only 20% of cases. The Ministry of Construction is committed to a programme of modifying waterways to improve their capacity to convey flood flows. It is ironic that while past flood protection work has resulted in degradation of riverine habitats, the current and future upgrading programmes actually present an ideal opportunity to incorporate rehabilitation schemes.

Increasing interest in nature conservation by the Japanese public has created a need for waterways which not only have aesthetic and recreational values, but also have ecological value. In December 1981 the River Council advocated a unified approach to river basin management which considered not only public safety against floods, but also maintenance of the natural environment and provision of recreational space. By the end of 1991 the basic plans related to river space management were prepared for all Class A and Class B rivers (Imaoka and Mitsuishi, 1993). The pillar of the Eighth Flood Control Program was 'conservation and creation of rich and beautiful river system environments' (Ikebuchi, 1994).

The movement to rehabilitate Japan's waterways was initially inspired in the late 1980s by the promising results being achieved by stream rehabilitation projects being conducted in Switzerland. On 6 November 1990 the River Bureau, Ministry of Construction, served a notice to all authorities involved in river channel works, encouraging them to adopt the approach of *Naturnaher Wasserbau* (*Ta Shizen Gata Kawa Dukuri*). The literal English translation of this term is 'close-to-natural river construction', but the more common translation used in Japanese literature is 'close-to-nature river construction'

(Kochi Work Office, 1994), or 'multi-natural style river construction' (*Kin Shizen Kasen Koho*) (River Division, Hiroshima Prefecture, 1993). The general approach has also been described as 'creation of rivers rich in nature' (Seki and Takazawa, 1993), and 'naturally diverse construction methods' (Suzuki, 1993). The notice encouraged river managers to recognise the concept of rehabilitation for nature conservation when planning river improvements, and to follow guidelines in *The Operation Manual of Naturnaher Wasserbau*. The new approach has been adopted enthusiastically, with the number of project sites increasing from 606 in 1991 to 971 in 1992, and then further increasing to 1500 in 1993 (Suzuki, 1993).

In 1992 a plan was proposed to create 'model rivers'. Rivers were selected on the basis that the project had local community support; conservation, tourism and angling were currently active; there was potential to provide fish passage past barriers; and the river must be rehabilitated from headwaters to mouth (Anonymous, 1992). For example, the 5.3 km long Uji River, Kochi Prefecture (Figure 17.1), was selected as part of the 'model rivers' programme for rehabilitation under the theme 'The river which preserves the history and culture of our hometown (Ino Town)'. Government waterway engineers are being advised by a community group called the Committee for Improvement of the Uji River. The river is perched a few metres above the adjacent floodplain land and floods cause serious inundation damage to the town. At a cost of ¥110 million (approx. US$1 million), in 1992 the capacity of a 1.6 km reach of the channel, 2 km upstream of the Niyodo River junction (Figure 17.1), was improved by excavation. A natural appearance was preserved by covering the concrete toe protection and groynes with stones, and revegetating the banks. Gabions, rip-rap, artificially constructed bars and pools, and macrophytes were used to create favourable habitat (Kochi Work Office, 1994, p. 17; Sugihara, 1994).

Close-to-natural river construction work is a movement away from the hydraulically smooth, hard-lined canals having a very artificial appearance, towards the creation of waterways that appear and function more naturally but have similar or improved flood capacity. Even though the waterways are still highly engineered and stable structures, the designs include physical habitat structures, fishways, riparian vegetation, layers of stone or vegetation covering concrete revetments, and restored baseflows. The designs also encourage multiple-use of river environments by providing facilities and access for recreational pastimes such as fishing, swimming, and appreciation of the aesthetic qualities of the environment. Initial assessments of the biological response to stream rehabilitation are promising. For four rehabilitated sites, Suzuki (1993) found that while fish and insect diversity and biomass decreased in the year immediately following rehabilitation, these indices generally increased by the second year.

Kochi Work Office has been particularly active in applying close-to-natural construction techniques. The budget for the Waterway Section in the 1993 financial year was around ¥18 billion (approx. US$180 million). Flood mitigation works are being constructed with aesthetic, recreational, ecological and local cultural values in mind. Community use of river environments is promoted under the theme of 'friendly rivers', by constructing walkways on the banks, stones that allow entry to the water, and natural-looking swimming pools (Kochi Work Office, 1993). Rehabilitation of waterways in Toyota City, Aichi Prefecture (Figure 17.1), is consistent with the city's development theme of 'vitality, nature and humanity' (City of Toyota, 1989, p. 18). The city is becoming rapidly urbanised and plans are under way to provide recreational waterway areas that also have favourable ecological habitats. Near the junction of the Yahagi River (Figure 17.1), in the vicinity of Arai City, the Kago River contains habitats favourable for several fish species, including ayu. The Prefecture has allocated ¥870 million (approx. US$9 million) for rehabilitation of a 2 km reach of the river to improve flood defence, control erosion, provide access and recreational facilities, plant riparian vegetation native to the area, and protect in-stream habitats.

Waterway rehabilitation is sometimes facilitated because of an association with points of special cultural interest. The Momijidani River, on the island of Miyajima (Figure 17.1), required rehabilitation after the area around the famous Itsukushima Shrine was buried by debris flows during a typhoon in 1945. The works predate the current close-to-natural approach by 50 years, but because of the importance of the site (one of 'Japan's three greatest views'), the stabilised river channel was given a natural appearance and function by covering concrete surfaces with local stone (River Division, Hiroshima Prefecture, no date). There is a point of public interest on the Enokuchi River, Kochi Prefecture, at 133° 33′ 33″ E and 33° 33′ 33″ N called 'The Earth Number 33' (Figure 17.1). A bridge was built over the river to enable people to stand on the exact point of intersection of these lines of longitude and latitude. The popularity of the site encouraged application of the close-to-natural construction technique when flood improvement works were carried out in the vicinity. Aquatic habitats were created, and the foreshore was given an appearance that harmonised with the old-fashioned buildings in the vicinity (Kadota, 1994; Kochi Work Office, 1994, p. 15). In association with maintenance works on the historical site of the Matsudairas (Figure 17.1), it was planned to run the Soure River, Aichi Prefecture, through a closed conduit under a visitors' parking area. Because of the steep gradient, this plan was abandoned in favour of a stream rehabilitation project, completed in 1992. This involved infilling the incised stream and floodplain, sealing the channel with a fixed rock bed overlying a concrete sill, and creating a meandering stream with pools and chutes (River of Toyota-city, 1993).

Engineers have traditionally shouldered most of the responsibility for waterway management in Japan. With the trend towards providing aesthetically pleasing and ecologically sustainable river environments, landscape architects are now becoming involved in the planning process (Aso, 1991). Another development has been the establishment of some educational foundations by the national and prefectural governments. These include the Construction College established by the Ministry of Construction to re-educate their national and regional based waterway engineers, the Foundation of River Management which publishes booklets, provides technical support for river works and manages funds, the Foundation of Riverfront Adjustment Center, and the Foundation of Construction Engineering Center of Hiroshima-Prefecture. For some rehabilitation projects the Ministry of Construction has assembled a multidisciplinary team of specialist advisers (biologists, ecologists, chemists, hydrologists, geomorphologists, etc.). Community groups take an active interest in waterway management in Japan. For example, in the Kochi Prefecture there are 100 000 people involved in 148 riverine and coastal area protection groups. Each July, various clean-up and improvement activities are organised (Kochi Work Office, 1993). Community participation in waterway management in the Toyota City area, Aichi Prefecture, is exercised through the activities of the Yahagi Waters Cleanup Society (City of Toyota, 1989, p. 27).

CASE STUDIES OF WATERWAY REHABILITATION

While river rehabilitation has been actively promoted and attempted in Japan in recent years, very little information has been documented in the English language. The case studies presented below are intended to illustrate the current situation in Japan. As there are now over 1500 sites in Japan where river rehabilitation has been attempted, the review is not comprehensive. Rather, the case studies were selected to illustrate the range of problems being approached and the methodologies employed, with a special emphasis on projects involving ecological appraisal of the work. Information was gathered from several sources: two important English language reviews published by Seki and Takazawa (1993) and Tamai *et al.* (1994); other available English language literature; selected Japanese language literature that was translated; and a special (Japanese language) questionnaire that was sent in late 1993 to waterway managers in order to gather details of rehabilitation projects, with returned information later translated into English.

RESTORATION OF BASEFLOWS IN URBAN STREAMS

Over the past 30 years many streams in Tokyo (Figure 17.1) have effectively dried up due to lowering of groundwater levels associated with increased runoff from impervious urban surfaces. Two methods are being employed to restore baseflows: pumping purified waste water to streams, and groundwater cultivation (Fujita, 1993).

Flows of $1 \text{ m}^3 \text{ s}^{-1}$ are being provided in the Nomi and Meguro Rivers, Tokyo, by pumping treated sewerage water through tunnels up to 17 km long. The cost of the project is approximately ¥20 billion (approximately US$200 million) and the government has allocated ¥400 million (approximately US$4 million) annually for maintenance. The purpose of the project is to provide a flow in the waterways for recreational and ecological benefit; an economic effect is not expected (Ohgaki and Sato, 1991; Fujita, 1992, 1993). Flow in the Nobidome Water Supply channel was similarly restored in 1984 with $0.2 \text{ m}^3 \text{ s}^{-1}$ of water pumped from the Tamagawa-Joryu Wastewater Treatment Plant. Baseflow was restored in the Tamagawa Water Supply Channel in 1986, and in the Senkawa Water Supply Channel in 1989. Flows have been directed from these supply channels to the Zenpukuji and Kanda Rivers (River Division, Bureau of Construction, 1988, p. 45; Sewerage Bureau, 1989, p. 51; Bureau of Sewerage, 1993, p. 5).

Groundwater cultivation involves the construction of facilities that infiltrate stormwater runoff. These can be large-scale sewerage systems, trenches, permeable pavement for road, footpath and parking areas, and inlets accepting road and roof runoff (Fujita, 1992). During the 1980s, infiltration facilities were implemented with the primary objective of runoff reduction, but now the main objectives are recovery of urban stream flow, maintenance of groundwater as a water source, and prevention of land subsidence (Fujita, 1992, 1994; Ishizaki *et al.*, 1993).

The first rainwater infiltration system in Japan was installed in 1981 in an apartment complex covering 27.8 ha at Akishima City, Tokyo. Observations over 10 years, which compared 40 storm event hydrographs from the complex with those from a nearby control area, demonstrated that the infiltration facilities absorbed 68–100% of event runoff (Ishizaki *et al.*, 1993). In Koganei City, Tokyo, in 1991 it was estimated that 3.1% of the annual rainfall infiltrated through inlets installed on residential properties (Fujita,

1992, 1993). Groundwater now issues from springs to maintain baseflow in the No River, a tributary of the Tama River. The community prefers that river flow is restored by natural groundwater discharge, rather than pumping of treated sewerage. Experience has shown that when encouraging residents to construct infiltration facilities (at their own expense), cultivating groundwater to restore nature is a more persuasive argument than reducing storm runoff (Fujita, 1993).

In 1981 an experimental sewerage system combining infiltration and storage facilities was installed in an area of Tokyo at a cost of ¥59 billion (approximately US$590 million). Over the experimental area of 1329 ha, the system halved total surface runoff and reduced peak storm runoff by 40%. The discharge reduction lowered the annual suspended solids load by 55% (Fujita, 1992).

INNER-CITY REHABILITATED STREAM PARKS

The streams and rivers in Edogawa Ward, Tokyo, were once used for transport, irrigation, recreation and domestic proposes. However, urbanisation and industrialisation in the Meiji Era resulted in the rivers becoming polluted, and some ceased to function with development of the sewerage system. In the early 1970s the residents proposed rehabilitation of the flow, physical structure and water quality of their local streams.

The Furukawa River Shinsui ('familiarisation with water') Park was opened in 1974. It is a 1.2 km reach of rock-lined stream that has clean water suitable for bathing, and vegetated riparian areas. The more ambitious Komatsugawa-Sakaigawa Shinsui Park was finished in 1985 and is now regarded as one of the 100 scenic sites in Tokyo with up to one million visitors each summer (Edogawa Ward Office, 1986). This 3.2 km long reach has a channel constructed from boulders, stone and aggregate that has the appearance of a mountain stream. The stream comprises a *Seseragi* ('murmuring') zone, a *Mizu no Teien* ('water garden') zone, a *Mizu Shibuki* ('water splash') zone and a *Tayutai* ('rippling and flickering') zone. There is a plaza, and the stream is flanked by a vegetated riparian zone designed to attract wildlife (Edogawa Ward Office, 1986). Although the channel has a close-to-natural appearance, it is principally a physical environment rather than a biological one.

TRADITIONAL CHANNEL STABILISATION TECHNIQUES

The steep gradients and high flow velocities typical of rivers in Japan means that their bed and bank morphology is prone to be dynamic. Concrete, stone and prefabricated interlocking blocks are the popular choice for modern stream stabilisation in Japan, but there has been a revival of traditional methods in some areas because they provide historical links, use local materials and labour, and are seen to provide ecological benefit (Shimoda Work Office, no date). Suzuki (1993) accepted the benefits of these traditional techniques, but also stressed that close-to-natural river rehabilitation required an innovative approach.

Seigyuu ('holy cow') are tetrahedron-shaped, log-framed structures, 2–5 m high and 3–6 m long and wide, with one side weighted down to the bed with gabions. They were originally used in the Nara Period (710–794) to control water flow in irrigation ditches. During the Period of National Reunification (1568–1603), Shingen Takeda promoted the use of *Seigyuu* in river control works. They are stable in high-velocity flows and are normally placed in fields of up to 20 structures. Hydraulically they act like pile-fields to slow velocity and stabilise bed and bank sediments. *Seigyuu* have been used in recent projects on the Takatu River, Masuda City, Shimane Prefecture (Figure 17.1), and in the Ooi River, Shimada City, Shizuoka Prefecture (Figure 17.1) (Shimoda Work Office, no date).

Mokkoo Chinshoo, first used in the Meiji Era, are overlapping, rock-filled, log frames ('wooden mattress') that provide bank protection. They have recently been employed in the Niyodo River, Tosa City, Kochi Prefecture (Figure 17.1) (Kochi Work Office, 1994, pp. 21–22), and also in the Takatsu River project.

In carrying out recent improvement works on the lower Yahagi River, Aichi Prefecture (Figure 17.1), traditional 'willow-twig' revetment was the chosen method of stabilising the banks. With the willow-twig technique, stones of 10–50 kg are held in a gridded frame of braided willow twigs. The frame is fixed onto the bank using willow stakes which eventually root, adding extra strength to the bank (Tomita, 1989; Ishida *et al.*, 1994).

BANK REVETMENT AND IN-CHANNEL HABITAT IMPROVEMENT DEVICES

Prefabricated hollow concrete blocks (Tamai *et al.*, 1994) or 'Hume' pipes (Suzuki, 1993) have been incorporated into revetment walls to provide resting and refuge habitat for fish (NFIWFC, 1987). The blocks have inner spaces of approximately 0.5 m^3. Observations on the Yoshii River and the Asahi River (Figure 17.1), by the Okayama Prefectural Fisheries Laboratory, found that 15 species of fish used the blocks.

Suzuki (1993) described a method of creating fish resting and refuge spaces between bundles of willow twigs that were incorporated into stone revetment. Broken rock and large boulders are more appropriate for ayu, as this fish relies on algae attached to rock as a food source.

Although the current emphasis in bankside improvement is provision of ecologically favourable habitat, most of the previous work on the banksides of urban rivers was designed to improve recreational amenity. These works have included shaping banks to have moderate slopes, providing steps for access to the channel, and revetments made of concrete blocks and log frames that allowed vegetation growth. A survey of these improved banksides on six rivers by the NFIWFC (1993) showed that, compared with neighbouring natural zones, they had a substantially lower biomass of benthos, and generally produced a lower catch of fish per unit effort.

EXCAVATION OF POOLS

Ayu fishing was improved on a section of the Ohta River, Hiroshima Prefecture (Figure 17.1) by excavating a pool which allowed fish to shoal. The pool is 10 000 m^2 in area and 3 m deep. Large rocks were placed around the perimeter of the pool to stabilise its shape (NFIWFC, 1987).

In July 1981, a pool 3 m deep, 10 m wide and 30 m long was excavated from a flat section of the Maruyama River, Hyogo Prefecture (Figure 17.1). Within three months of creating the pool, fish numbers had at least doubled for most species of fish present, and for ayu and Japanese dace (*Leuciscus* (*Tribolodon*) *hakonensis*) the results were even more marked (Mizuno, 1985).

REHABILITATION OF BACKWATERS

Heavy settlement has resulted in the loss of most of Japan's floodplain wetland habitat, and regulation has removed backwater pool habitat (*wando*) from within many channels. Remaining backwater pools have a high conservation status because they are recognised as providing unique habitat for many species of plants, fish and shellfish. In 1990 an artificial backwater pool system was constructed on the Yodo River at Toyosato in Osaka City (Figure 17.1) to replace one that was buried by river improvement works (Seki and Takazawa, 1993). The backwater consists of two interconnected pools 2 m deep which cover a total area of 6000 m^2.

An oxbow lake on the Kiso River at Muchokuji, Gifu Prefecture (Figure 17.1) is renowned as an ideal habitat for dragonflies and wetland vegetation. To compensate for partial loss of the lake due to river improvement, an artificial lake modelled on the original one was constructed nearby. The works required an excavation covering 5800 m^2. Seki and Takazawa (1993) report that the pond is now inhabited by a diverse community of biota which includes dragonflies.

FULLY ARTIFICIAL CHANNEL

The Nohgu River, Nagano Prefecture, is located in the central part of Honshu Island (Figure 17.1). The river drains a catchment of 17 km^2, is 5–10 m wide, and has a stable flow from a series of lakes of 0.5–2 m^3 s^{-1}. As part of a plan to improve the surrounding land for irrigation agriculture, part of the channel was straightened and widened in 1967. The river is important for angling and contains populations of ayu. Since 1978 various devices have been installed in the artificial stream to enhance the fish habitat. The habitat improvement devices included boulders in log frames, grade reducing structures with fish blocks in the revetment, and pools in the middle of the channel. Nakamura and Tsukisaka (1994) recently reported on the outcomes of the project. Various indices of channel morphology variability and flow heterogeneity were measured for the different types of channel structure. The study area also included some concrete-walled and semi-natural (undisturbed post-1967) sections of channel.

Nakamura and Tsukisaka (1994) found that fish populations were highest in the semi-natural stream section, but that reasonable populations of fish inhabited the artificial channel. Even the concrete-walled channel had a relatively high population of fish, but only in the meandering section; fish were not found in the straight section. In the artificial channel with habitat improvement devices, the highest fish populations were positively correlated with the most variable channel morphology. Not surprisingly, the semi-natural channel section had the most variable channel morphology and the highest flow heterogeneity. Large-sized fish preferred morphological habitat with a high depth to width ratio.

The narrow, high-gradient, Asahi River in Okayama Prefecture (Figure 17.1) was recently stabilised by totally covering the bed and banks with concrete blocks. An attempt was made to provide artificial habitat for *Megalobatrachus japonicus*, a species of salamander that is important from a conservation viewpoint. Hollow blocks were used to line the banks in the lower part of the channel, and boulders and stones were placed in voids between blocks in the bed (Ishida *et al.*, 1994).

FISH PASSAGE

The most common type of fishway in Japan is the fish ladder, but other types which have been used are the vertical slot type, sloped channelways filled with cobble stones, and the lock type which are used on barrages in the lower reaches of large rivers (Mochizuki and Jikan, 1993). Fish ladders of various designs have been tried, including bending-slope (or fan-shaped) (Hara, 1993; Nakamura, 1994), boulder and pool (Seki and Takazawa, 1993), ladder with attraction flow flume, and spiral type (Mochizuki and Jikan, 1993).

The morphological heterogeneity of a natural channel is reflected in the wide range, and evenly spread distribution, of flow depth. In contrast, regulated channels tend to be uniformly shallow. When pools, rocks and fishways were placed on a grade control structure on a tributary of the Niyodo River, Kochi Prefecture, Mizuno (1989) reported only a modest increase in the spread of depth distribution. However, while the habitat around an ordinary grade control structure was found to support only a small population of Japanese dace, the modified structure supported a diverse population of fish similar to that found in a natural section of channel.

The success of a bending-slope fishway installed on the Hatta Weir, Niyodo River (Figure 17.1), encouraged the construction of several similar fishways on other rivers (Seki and Takazawa, 1993; Nakamura, 1994). The fishways were difficult to construct, and their performance was disappointing, probably because they were designed with only a rudimentary understanding of their hydraulic characteristics. Nakamura (1994) addressed these problems in a series of flume experiments using live fish. He found a complex flow pattern over the fishways that was dependent on the flow rate. Fish passage could not be guaranteed even when boulders were embedded into the slope surface. To provide passage during all flow conditions, Nakamura (1994) recommended that bending-slope fishways be supplemented with an adjacent alternative fishway of a different design.

DISCUSSION AND CONCLUSION

Japan's small areas of highly valuable floodplain land are extensively developed. In order to overcome naturally high flood susceptibility and channel instability, Japan's waterways have suffered modification for centuries. Given that providing flood protection and channel stability will always be the major consideration, restoration of rivers to a pristine hydrological, geomorphological and biological condition is not a realistic option. The challenge for rehabilitation of Japan's waterways is to establish, or improve the level of, biological functioning in systems which are hydrologically and geomorphologically highly regulated, and in places highly polluted.

Improvement in sewerage treatment will continue to improve water quality; the biggest problem is the degraded condition of the physical habitat, and in some places, shortage of water. The enormous pressures on water resources in Japan make environmental releases from impoundments difficult to secure. It is encouraging that some progress has recently been made towards legislation of minimum environmental flow releases from dams (Tamai et al., 1993). This issue presents a dilemma that must be faced in many countries: provision of more generous environmental releases in the future could partly rely on the construction of additional dams that will ease the pressure from urban, industrial and agricultural water demands.

Despite extensive river regulation efforts, the risk of flood damage has actually increased in many cities. This can be attributed to the continued concentration of assets in flood-prone areas, combined with the higher flood peaks that have resulted from urbanisation and works to reduce the hydraulic resistance of channels. Ironically, it is the demand for hydraulic upgrading of waterways which has created the opportunity to incorporate some in-stream habitat, riparian habitat and fish passage in the designs. In promoting a European-style close-to-natural construction approach, the Ministry of Construction is responding to community demands for rivers to be rehabilitated for recreational, aesthetic and ecological benefit. The response has been rapid and earnest, with the number of sites being rehabilitated now numbering over 1500. Early results of rehabilitation works show that while close-to-natural techniques can create favourable habitats, the artificial habitats are generally ecologically inferior to natural ones.

European methodologies have been adopted and adapted with great alacrity, but one of the survey respondents admitted that over-enthusiasm has sometimes led to creating an artificial 'garden-like' effect. This temptation is not all that surprising, given that management of the natural environment in Japan has traditionally involved a tendency to 'improve' on nature by imposing a kind of order (Kornhauser, 1982, p. 144). This concept is exemplified in the traditionally popular arts of garden architecture, bonsai and ikebana. Thus, compared to river rehabilitation projects conducted elsewhere in the world, Japanese examples have so far tended to have a more contrived and artificial appearance. Regardless of this cultural influence, the scope for restoring river environments to a natural condition is constrained by the necessity for flood defence and channel stability.

Economic and safety considerations deem it necessary to prevent channel mobility and flooding, but they are both important natural processes. The channel improvement (flood mitigation) programme currently under way will exacerbate the ecological alienation of floodplains from their rivers. This will not have a drastic ecological impact, as most natural floodplain habitats were lost long ago. Current rehabilitation projects generally do not include floodplain habitats, so it is unlikely that the rehabilitated ecology will resemble that which occurred long ago prior to regulation. The long-term ecological response to this partial, and relatively artificial, physical rehabilitation is awaited with great scientific interest, as there are many places in the world with these or similar constraints.

ACKNOWLEDGEMENTS

The contribution of Chris Gipel was funded by an Australian Vice-Chancellors Committee Bicentennial Fellowship and a British Council Academic Links and Interchange Scheme Grant. Information on individual river rehabilitation projects was kindly provided by officers of Aichi, Kochi, Shimoda, Ootagawa, Hamada, Fukuoka and Hiroshima Prefectures. Additional information and suggestions were kindly provided by Professor Shunroku Nakamura, Toyohashi University of Technology, Toyohashi, Dr Soichi Fujita, Organization for Sewerage Works New Technology, Tokyo and K. Iwai, Nishinihon Institute of Technology. Translations were done by Yoshi Abe, Melbourne.

REFERENCES

Anonymous (1992) Towards creation of a model river that allows fish to swim easily upstream. *Water Ecology*, Usuikyo, Tokyo, **4**, 6 (in Japanese).
Aso, Y. (1991) The river which makes a planner suffer. *Kawaraban: The Book of River*, Waterways Information Center, Tokyo, **8**, 2–7 (in Japanese).
Bureau of Sewerage (1993) How Can Sewerage Make the World More Beautiful? Brochure, Tokyo Metropolitan Government, Shinjuku-ku, Tokyo.
City of Toyota (1989) *Enjoy Toyota City*. Toyota City, Aichi Prefecture, Japan.
Edogawa Ward Office (1986) Nature in a Big City, the Real Home Land. Shinsui Park (Familiarisation with Water) in Edogawa Ward. Brochure, Edogawa Ward Office, Tokyo.
Fujita, S. (1992) Infiltration facilities in Tokyo: their purpose and practice. In *Proceedings of the Fifth European Junior Scientist Workshop on Urban Stormwater Infiltration*, 1–4 October 1992, Klintholm Havn, Møn, Denmark, pp. 179–192.
Fujita, S. (1993) Infiltration in congested urban areas of Tokyo. In Marsalek, J. and Torno, H.C. (eds) *Proceedings of the Sixth International Conference on Urban Storm Drainage*, Vol. 1, Niagara Falls, Ontario, Canada, IAHR/IAWQ Joint Committee on Urban Storm Drainage, pp. 993–998.
Fujita, S. (1994) Japanese experimental sewer system and the many source control developments in Tokyo and other cities. In *Proceedings of the Standing Conference on Stormwater Source Control*, 6 June, Coventry, UK.
Hara, Y. (1993) Sabo facilities harmonised into their environment. *Journal of Hydroscience and Hydraulic Engineering, Special Issues*, **51–4**, Research and Practice of Hydraulic Engineering in Japan, River Engineering, Committee on Hydraulics, Japan Society of Civil Engineers, pp. 209–217.
Hirano, M. (1993) Flood control works in the Ishikari River. *Journal of Hydroscience and Hydraulic Engineering, Special Issues* **SI–4**, Research and Practice of Hydraulic Engineering in Japan, River Engineering, Committee on Hydraulics, Japan Society of Civil Engineers, pp. 23–43.
Hukunari, K. and Hiroki, K. (1993) Water resources development and planning – the Tone River. *Journal of Hydroscience and Hydraulic Engineering, Special Issues* **SI–4**, Research and Practice of Hydraulic Engineering in Japan, River Engineering, Committee on Hydraulics, Japan Society of Civil Engineers, pp. 103–119.
Ikebuchi, S. (1994) River management in Japan. In *River Management and Planning in Japan* Technical Memorandum of PWRI No. 3265, Environment Section, River Department, Public Works Research Institute, Ministry of Construction, Asahi, Japan, pp. 1–19.
Imai, K. (1993) Sabo planning in Tenryu River system – case study on Yotagiri River. *Journal of Hydroscience and Hydraulic Engineering, Special Issues* **SI–4**, Research and Practice of Hydraulic Engineering in Japan, River Engineering, Committee on Hydraulics, Japan Society of Civil Engineers, pp. 175–181.
Imaoka, R. and Mitsuishi, S. (1993) Basic plan for river environment management. *Journal of Hydroscience and Hydraulic Engineering, Special Issues* **SI–4**, Research and Practice of Hydraulic Engineering in Japan, River Engineering, Committee on Hydraulics, Japan Society of Civil Engineers, pp. 133–143.
Ishida, R., Nakamura, S., Mizuno, N., Tamai, N. and Mayama, H. (1994) Stream restoration for fishes in Japan. In Armantrout, N.B. (ed.) *Condition of the World's Aquatic Habitats*. Proceedings of the World Fisheries Congress, Theme 1, 14–19th April 1991, Athens, Greece, Oxford IBH Publishing Co., New Delhi, pp. 225–237.
Ishizaki, K., Seiji, M., Kagawa, A., Mochizuki, T. and Imbe, M. (1993) Rainwater infiltration technology for urban areas. *Journal of Hydroscience and Hydraulic Engineering, Special Issues* **SI–4**, Research and Practice of Hydraulic Engineering in Japan, River Engineering, Committee on Hydraulics, Japan Society of Civil Engineers, pp. 72–85.
Iyama, S. (1993) Profile of Japanese rivers – background to river engineering in Japan. *Journal of Hydroscience and Hydraulic Engineering, Special Issues* **SI–4**, Research and Practice of Hydraulic Engineering in Japan, River Engineering, Committee on Hydraulics, Japan Society of Civil Engineers, pp. 1–4.

Kadota, T. (1994) Environmental maintenance around 'The Earth Number 33'. In *Lectures on Rivers, TOSA '93, International Forum for Waterside Environment*. Kochi Work Office, River Division, Civil Engineering Department, Kochi City, Kochi Prefecture, Japan, pp. 21–22.

Kawasaki, H. (1994) River administration and institutional aspects in Japan. In *River Management and Planning in Japan*. Technical Memorandum of PWRI No. *3265*, Environment Section, River Department, Public Works Research Institute, Ministry of Construction, Asahi, Japan, pp. 20–38.

Kikkawa, H., Tawara, T. and Ishizaka, K. (1975) Change of runoff characteristics due to river improvement. In *Proceedings of the Tokyo Symposium, IASH Publication No. 117*, pp. 571–578.

Kochi Work Office (1993) 1993 Project Summary. Brochure, River Division, Civil Engineering Department, Kochi City, Kochi Prefecture, Japan (in Japanese).

Kochi Work Office (1994) *Examples of Close-to-Nature River Construction, Excursion Notes, TOSA '93, International Forum for Waterside Environment*. Close-to-Nature Construction Technique Research Group, River Division, Civil Engineering Department, Kochi City, Kochi Prefecture, Japan.

Kornhauser, D. (1982) *Japan: Geographical Background to Urban–Industrial Development*, 2nd edition, The World's Landscapes, Longman, London.

Ministry of Construction (1993a) History and present state of sewerage works in Japan. In *Sewerage Works in Japan*. Japan Sewerage Works Association, Chiyoda-ku, Tokyo, pp. 58–79 (in Japanese).

Ministry of Construction (1993b) *'93 Construction Whitepaper*. Printing Bureau of the Ministry of Finance, Tokyo (in Japanese).

Mizuno, N. (1985) Channel alteration and bank revetment in the middle reaches of streams. *Tansuigyo 11*, Tokyo (in Japanese).

Mizuno, N. (1989) Report of the investigation on stream forms and environment for fish. Fisheries Agency of Japan, Tokyo (in Japanese).

Mochizuki, T. and Jikan, S. (1993) Environment-friendly water resources development facilities. *Journal of Hydroscience and Hydraulic Engineering, Special Issues* **SI-4**, Research and Practice of Hydraulic Engineering in Japan, River Engineering, Committee on Hydraulics, Japan Society of Civil Engineers, pp. 153–174.

Nakamura, S. (1994) Hydraulics of bending-slope fishway. In *Proceedings of the 1st International Symposium on Habitat Hydraulics*, 18–20 August 1994, Trondheim, Norway, pp. 143–153.

Nakamura, S. and Tsukisaka, M. (1994) Field investigations on artificial habitat in a Japanese channelized stream. In *Proceedings of the 1st International Symposium on Habitat Hydraulics*, 18–20 August 1994, Trondheim, Norway, pp. 305–315.

NFIWFC (National Federation of Inland Water Fisheries Cooperatives) (1987) Report of an investigation of the river environment for fish production in Japanese rivers, Tokyo (in Japanese).

NFIWFC (National Federation of Inland Water Fisheries Cooperatives) (1993) Report on the influence of ameliorated facilities for recreation on fish production, Tokyo (in Japanese).

Ohgaki, S. and Sato, K. (1991) Use of reclaimed wastewater for ornamental and recreational purposes. *Water Science Technology*, Kyoto, **23**, 2109–2117.

River Division, Bureau of Construction (1988) *River Improvement Projects in Tokyo*. Tokyo Metropolitan Government, Tokyo.

River Division, Bureau of Construction (1990) *Aiming for a Charming River: Planning the River Environment in Tokyo*. Brochure, Tokyo Metropolitan Government, Tokyo (in Japanese).

River Division, Hiroshima Prefecture (1993) *A Note Concerning Multi-Natural Style River Construction in Hiroshima Prefecture*. Civil Engineering Department, Hiroshima City, Hiroshima Prefecture, Japan (in Japanese).

River Division, Hiroshima Prefecture (no date) Erosion Control Works for the Park at Miyajima's Momijidani River. Brochure, Sand Control Section, Civil Engineering Department, Hiroshima City, Hiroshima Prefecture, Japan (in Japanese).

River Division of Toyota-city (1993) 5. Experimental cases. Document by River Division of Toyota-city, Aichi Prefecture (unpublished).

Seki, K. and Takazawa, K. (1993) Projects for creation of rivers rich in nature – toward a richer nature environment in towns and on watersides. *Journal of Hydroscience and Hydraulic Engin-*

eering, Special Issues **SI-4**, Research and Practice of Hydraulic Engineering in Japan, River Engineering, Committee on Hydraulics, Japan Society of Civil Engineers, pp. 86-101.

Sewerage Bureau (1989) *Sewerage in Tokyo*. Tokyo Metropolitan Government, Chiyoda-ku, Tokyo.

Sewerage Bureau (1994) *Monthly Sewerage*, **33**(4), 24, Tokyo Metropolitan Government, Tokyo (in Japanese).

Shimatani, Y., Kayaba, Y., Oguri, S. and Suzuki, O. (1993) Impacts of stream modification on habitat component and biota in Tagawa River. In *Proceedings of the XXV Congress, IAHR Vol. VII*, pp. 53-60.

Shimoda Work Office (no date) *How to Apply Close-to-Natural River Construction to Practical Works: Traditional Japanese Techniques for Rivers*. Unpublished, Shimoda, Shizuoka Prefecture, Japan (in Japanese).

Sugihara, N. (1994) Naturnaher Wasserbau in Uji River. In *Lectures on Rivers, TOSA '93, International Forum for Waterside Environment*. Kochi Work Office, River Division, Civil Engineering Department, Kochi City, Kochi Prefecture, Japan, pp. 33-35.

Suzuki, O. (1993) Effect by naturally diverse river construction methods. River Environmental Research Department, Public Works Research Institute, Ministry of Construction, Tsukuba City.

Tamai, N., Mizuno, N. and Nakamura, S. (1993) *Environmental River Engineering*. University of Tokyo Press, Tokyo (in Japanese).

Tamai, N., Nakamura, S., Mizuno, N. and Shimatani, H. (1994) Recent experiences on stream restoration in Japan. In *Proceedings of the 1st International Symposium on Habitat Hydraulics*, 18-20 August 1994, Trondheim, Norway, pp. 316-330.

Tomita, N. (1989) On the traditional 'willow-twig revetment works' in the Yahagi River. In *Proceedings of the Annual Technical Conference of the Chubu Construction Bureau*. Ministry of Construction, pp. 157-162 (in Japanese).

Uzuka, K. and Tomita, K. (1993) Flood control planning - case study of the Tone River. *Journal of Hydroscience and Hydraulic Engineering, Special Issues* **SI-4**, Research and Practice of Hydraulic Engineering in Japan, River Engineering, Committee on Hydraulics, Japan Society of Civil Engineers, pp. 5-22.

Index

Aboriginal people 270
abiotic 34, 60, 71, 88
abiotic/biotic 91
abiotic/stochastic 91
Achelse Kluis, The 116–20, 122, 125–6
acid rain 24, 126–7, 129
acidification 24, 40, 43, 47, 66
acidity 15,
Acque Alte Canal, Italy 251–3, 260, 263–4
Adelaide 2
aerial deposition 107
aesthetics 176, 264, 303, 306–8, 314
agricultural fields 110
agricultural irrigation 24
agricultural land 109
agricultural landscape 98
agricultural waters 215, 252
agriculture 26, 39, 57, 62, 89, 91, 95, 97–9, 108, 110, 162, 218, 252, 271, 280, 286, 301, 305
Aims of rehabilitation 140
air pollution 181
air temperature 106
alder 47, 90
algal blooms 215, 256, 274, 276, 290
alien pasture 271
alien species 287, 291, 293–4 (*see also* exotic species)
alkalinity 47
allochthonous organic 47, 89
ammonia 162
amphibians 120
analyses 47, 105
analysis programme 105
angiosperms 45
animal migration 61
animal(s) 1–2, 66, 71, 75, 79, 98, 180
annual rainfall 303
anthropogenic pollution 39
Aotearoa 285 (*see also* New Zealand)
appraisal 309
aquatic life 289–90
Aque Alte Canal 2
archipelago 301
architects 74
Area of Concern (AOC) 174–5, 177, 183–4

Association of Drainage Authorities (UK) 134
Astel 4
Australia 2, 269–84
Austria 6
automated monitoring stations 184
autonomous nature development 128–9
average daily flow 218

backwater pools 312
bacteria 180
Baden-Wurttemberg 31, 32, 34–6
bank erosion 218, 288–9
bank protection 73, 224, 288, 311
bank stabilization 224, 228, 311
bank vegetation 13, 23
bankfull discharge 42
banks 92
bank structure 93
barbel rivers 94
barbellus fish 73
barley 101
Barmah Forest 228
beach closings 176
bed erosion 282, 289
bed sediment 307
bed stabilization 311
Belgium 20, 23, 116, 120
benefits 76
benthic algal communities 90
benthic macroinvertebrates 90, 180
benthos 176, 311
bioaccumulation 179–81
bioavailability 179–80
biochemical oxygen demand 162
biodiversity 47, 88, 126–7, 129, 160, 252, 262
biogeochemical processes 67
biogeographical provinces 172
biographical 74
biological functioning 313
biological monitoring 179
biological processing 88, 92
biomanipulation 61, 263
biomass 89, 90
biomonitor-defined 179
biomonitoring 179–80

biomonitoring suites   181
biomonitors   179, 181
biotic communities   91
biotic factors   34
biotic indices   68
biotic structure   91
biotopes   78
birds   216, 264, 290, 292
Black Forest   39, 41, 43, 47
BOD   289–90, 305
bog streams   34
boulders   39, 43
Boundary Water Treaty   174
branches   93
bream rivers   94
Breda   115
Brenderup   101
brown tides   178
Buffalo   172
buffer strips   251, 263–4, 266, 290
buffer zones   76
Bunter Sandstone   47
Burgenland   6
bushes   81, 93

calcium   39, 40
Campaspe River, Australia   270, 277
Canale Emiliano-Romagnolo, Italy   252–3, 263
carcinogenesis   179
cascades   39
catchment   2, 3, 20, 24, 89, 95, 97
catchment area   14, 32, 43, 99
catchment erosion   91
catchment level   16
catchment management   76, 216, 274, 287, 306
Catchment Management Plans   157, 167
catchment rehabilitation   13, 124–6
catchment-scale project   99
cattle   77, 87, 88
channel habitats   62
channel modification   302, 312
channel riparian zones   3
channel stability   314
channel stabilisation   87, 224, 230, 310, 313
channel structure   229, 232
channelised rivers   1, 20, 67, 88, 93
channelisation   77, 84, 240–1, 262, 271, 286, 293, 301, 304
chaos   177
chemical monitoring   179, 183
chemical monitors   184
Chicago   172
Chicago River   173
Chinese philosophies   303
chironomids   181

chi-square test   197
chlorine   174
Chowilla floodplain   224, 230, 234
chromosome breaks   182
clay   107
clay soil   101, 106
Cleveland   172
climate   20
coal mining   163
cobbles   39, 53
Coliban River, Australia   270, 277
Columbia River   171
combined sewer overflows (CSOs)   177–8
comet test   182
community involvement   76, 234, 292, 307–8
community perception   142
complex mixtures   179, 183
complex systems   177–8
    chaos in   177
    emergence in   177
    order in   177
complexity   177
conductance   102
Confederation of British Industry   134
conservation   1, 2, 17, 26, 54, 77, 149–50, 153, 155, 159, 307
conservation enhancement   149
conservation management   7
construction   76
constructions costs   61
constructors   58
contamination   97
coordinated management   75, 77, 80, 83, 229, 232
corridors   20, 98
cost assessment   76, 78–9
cost-benefit assessment   138
cost-benefit analysis   76, 228
costs   1, 8, 72
costs of maintenance   58, 218
Country Landowners Association (UK)   134
countryside   97–8, 100
Countryside Commission (UK)   133, 135, 153–4
Countryside Council for Wales   135, 160
countryside rehabilitation   98–100
countryside rehabilitation project   109
Countryside Stewardship Scheme   163
cover   304
CPOM   261
crop   100
crustacean   47
cultivated land   97–8, 100–1, 107–10
cultural interest   308, 314
culverts   164
current   15–16

INDEX

Cuyahoga River RAP 180

Dales area 162
dam building 1
dams 66, 93, 219, 239–40, 271, 273, 280, 290–1, 301–2, 304–5, 313–14
Denmark 97, 100, 105–6, 107–9
Darling River, Australia 276
Darlington Borough Council 140, 146
Darlington, UK 140, 144–5
database 61
deadwood 53
debris flow 302–3, 308
decision-making 113
deflectors 80–1, 83, 230–1, 244
deforestation 285
degradation 14, 218, 270–4, 292, 294, 313
demonstration project 109
dentrification 107, 109
Department of Agriculture Northern Ireland 134
Department of the Environment (UK) 159
Department of the Environment Northern Ireland 135
Derby 155–6
Derby City Council 155
Derbyshire Wildlife Trust 156
desalination 65
de-snagging 271, 273, 282
detritus 39
Detroit 172
Detroit River 173, 180
Detroit River RAP 180
development 1, 18, 23, 26–7
development plans 20
developmental instability 182
diffuse leaching 109
dikes 68, 251–2
dimensions 15, 16
Dinkel system 24, 27
dioxin 179
dipteran 47
discharge of pollutants 162
Disher Creek Disposal Basin 57, 59, 61, 230–1
dissolved oxygen 57, 62–3, 65, 67, 84, 162, 256
diversion canal 23
diversion channel 120
domestic effluent 254
Dommel, The 20, 113
dragonflies 312
drain tiles 97
drain water 101–2
drainage 20, 24, 66, 97, 99–100, 108, 252
drainage consents 166
drainage water 109–10

dredging 59, 60–1, 98, 110, 227, 240, 271
drop structures 278
drought 15, 24, 126–9
dry weather flow 60
drying 64, 67
Duffins Creek 189, 199–201
Dutch 14, 16–17, 20, 23, 27, 59, 61–3, 66–71, 73–5, 77–84
Dutch water management 28

EC 12
EC Nitrate Directive 163
EC Urban Wastewater Directive 162
ecotechnology 8
Ecological corridors 17, 114
ecological degradation 276
ecological development 88
ecological issues 31
ecological main structure 17
ecological recovery 24, 31
ecological sustainability 114, 294, 308
ecological value 306, 309–10, 314
economic compensation 108
economic factors 67–70, 71, 79, 220, 227, 266, 301
economic value 68, 227, 303
economy 58, 218
ecosystems 87
ecotones 88, 91, 95, 251
ecotopes 74
EC-regulation 'environmental friendly production methods' 263–4
EC-regulation 'set-aside' 263–4
Edo River, Japan 301
education 63, 79, 293, 308
educational component 8
effluent 39
Eifel 47
Eindhoven 115
electric power industry 171
electrical screens 291
electricity 305
electromagnetic flow meter 102
embankments 1, 301, 304
emergence 177
emergent properties 178
energy 91
engineers 23, 87
English Nature 135, 150, 160
enhancement 8
environment 17, 26, 64, 66, 73, 98–9
Environment Agency (UK) 134, 144, 149, 157, 159–60, 163, 166–7
environmental assessment 54
environmental conservation 68
Environmental Impact Assessment 164

environmental projects 28
environmental quality standards 61, 218, 220, 224, 252
Enz river 134
EPA 171, 179
erosion 20, 116, 118, 129, 224, 303, 306
erosion control 308
estuary 71, 76, 79–84, 91
Europe 1, 3, 6
European Centre for River restoration (ECRR) 137
European settlement 218, 232, 269, 271–2, 285
European Union Habitats and Species Directive 160
European Union 'Life' Programme 136–7, 139–40, 146
eutrophic conditions 128
eutrophic water 23
eutrophication 24, 91, 95, 97, 126–8, 218, 220, 224
evaluation 224
evapotranspiration 105
exotic animals 218, 220, 224, 287
  Cyprinus carpio 224
  Oncorhynchus mykiss 273
  Oryctolagus cuniculus 60, 224, 273
  Phyla nodiflora 57, 226, 229–30, 232
  Salmo trutta 273, 287
  Salmon spp 287
exotic plants 218, 220, 224, 273
  Eichornea crassipes 273
  Lupinus spp 291
  Rubus fruiticosus 273
  Salix alba 57, 224, 230, 273, 288
  Salix rubens 67, 71, 224, 273, 288
  Salix spp 291
exotic species 162, 165

fallow areas 109, 110
farmers/farming 23–4, 27, 78, 98–9
farming organizations 109
farmland 97–8, 108–10
fauna and flora 2, 5, 40, 46–7, 53, 73, 98–9, 109
fencing 291
fertilizers 97, 99, 127, 272
field survey 179–80
field trials 87–8, 218, 226, 229–30
fields 100, 110
filters 99
financial aspects 16
financial resources 57, 59, 60, 61
financial support 20
fish 47, 71, 88–92, 94, 100, 164, 167, 180, 218, 239, 246, 258, 260, 262, 273–4, 280–2, 286–7, 290, 293, 304–5, 308, 311–12

associations 190, 194–200, 202–3
biomass 47, 88, 91
caged 180
carnivorous 180
communities 91, 93, 95
consumption 176, 181
diversity 307
fauna 91
food 88
habitat 243, 273, 281, 290, 303–4, 308, 311–12
ladder 313
macro invertebrates 36
migration 278, 303
planktonivorous 180
species
  American Signal crayfish 166
  Astacus pallipes 151
  barbel 93
  Bidyanus bidyanus 274
  bream 93
  Cottus gobio 247
  Cottus poecilopus 247
  eels 291
  Gadopsis bispinosis 281
  Gadopsis marmoratus 273, 280
  Galaxiid spp 291
  grayling 92, 93, 94
  Leuciscus (*Tribolodon*) *hakonensis* 311, 313
  Maccullochella peelii 57–8, 87, 216, 219, 221, 223, 227, 230, 273
  Macquaria novemaculeata 274
  Phoxinus phoxinus 247
  Plecoglossus altivelis 304, 308, 311–12
  Prototroctes maraena 274
  Pseudodorasbora parva 260
  Salmo salar 189
  Salmo trutta 243–4, 246–7, 281, 290–2
  Salmo trutta fario 121
  Salmon spp 243, 291
  salmonids 121
  Salvelinus fontinachus 189
  Thymallus thymallus 121, 247
  trout 47
  zander 166
stocking 62, 229, 303–4
tumours 176, 180
fisheries 135, 140, 153, 159
fishing 62, 65, 73, 75–6, 303, 312
fishing places 79
Fishpass Regulations 291
flood alleviation 133
flood alleviation schemes 218, 220, 224, 230
flood banks 58
flood control 171

flood defence   1, 153, 159, 164, 308, 313
flood defence works   149
flood mitigation   301–2, 305
flood protection   98
flood risk/defence   136, 139, 142, 144
flood storage   159
flood waters   102, 303, 306
flooded meadows   109
floodplain forest   125
floodplain habitats   133
floodplain vegetation   36
floodplains   2, 3, 8, 31, 34, 36, 39, 43, 47, 62, 91, 93, 95, 97, 108, 144, 218, 301, 303
floods   227, 272, 280, 293, 301–2, 305–6, 314
flow   13, 43, 45, 73, 91, 93, 98–101, 109–10, 216, 219, 221, 223, 227, 242, 290, 309
flow heterogeneity   312
flow management   62, 229, 279–80, 284
flow rate   313
flow regulation   283, 291, 293
flow velocity   91
flow meter   102
fluvial geomorphology   134
fluvial landscape   8
food chain   91, 218
food gathering   270
food webs   91
footpaths   59, 60, 226
forest (mixed and deciduous)   45, 90, 301
forestry   272
forestry commission   4
funding   140
Funen   97–101, 105, 107–10
Funen watercourses   108, 110

Gedlintg Borough Council   153
geology   34, 269, 272, 284, 301, 303
geomorphological audits   142
geomorphological processes   2, 36, 75
geomorphologists   2
geomorphology   32, 34, 36, 39, 76, 153
German/Dutch border   76
Germany   24, 31, 67, 68–9, 71, 73–4
global radiation   106
goals   126
Gordon hydroelectric scheme   283
government   306
Graben   32
gradient   243, 310
grass   100, 107
grasses   101
grasslands   100, 220–4
gravel beaches   39, 93
grazing   77, 79, 100, 218–20, 222, 224–5, 230, 233 256, 271, 289, 291
Great Lakes   171, 176

Great Lakes Water Quality Agreement   174
Great Lakes–St Lawrence watershed   171
Green Bay and Fox River RAP   180
Green belt   159
Greenpeace   174
Greenwood Community Forest Project   154
Gronau   24
groundwater   34, 62, 87–8, 97, 102, 119–20, 305, 309–10
groundwater abstraction   24
groundwater flow   13
groundwater levels   13, 24
groundwater resources   110
guidelines   100, 292
Gulf of Mexico   171
Gulf of St Lawrence   172
gully erosion   277–8
Gwydir River, Australia   280–1

habitat   176, 220, 239, 247, 263, 292, 308, 312–13
habitat diversity   87, 91, 226, 312
habitat improvement   167
habitat structure   91
habitat(s)   42, 47, 53, 67, 91, 94, 98–9
heavy metals   15, 179
Heisei Era (1989–present)   303
Her Majesty's Inspectorate of Pollution (HMIP)   159
herbicide   66, 292
herbivores   129
herbs   101, 107
heterogeneity   69, 88
historic value   252
human activities   1, 33–4, 40, 45, 302
hume pipes   311
Huron–Eerie corridor   178
hydraulic
  conductivity   101
  diversity   281
  efficiency   252
  modelling   141–2
  performance   138
hydrodynamics   230, 232
hydroelectric power   1
hydrogen carbonate   39
hydrograph   222, 224–5, 233–4, 303, 309
  conditions   219, 221, 230
  factors   221, 230
  management   254
  regime   87, 220–4, 252
hydrology   220–1
hypertrophic conditions   128

IFIM   280, 290
impact assessments   138

impaired uses   174
impeded drainage   110
implementation   18, 23
impoundment   271, 292–3, 313
Index of Biotic Integrity (IBI)   192, 194
indicator species   84
industrialization   286, 301–2, 310
infiltration   106–7, 310
inflow   105
infrastructure   264
inlet   107
inlet water   102
insect diversity   307
insects   256, 312
instream and bankside habitats   73, 221, 224
instream benefits   292
integrated pollution control   159
integrated water management   14, 18, 20
Integrated Watering Strategy Project   230, 232
integrating   57, 58
inter-basin transfer schemes   1
interdisciplinary research   53, 254, 292, 294, 308
Internal Drainage Boards (UK)   163
International Association for Great Lakes Research   175
International Joint Commission (IJC)   174, 176–7
internet   184
introduced species   87, 224, 233–4 (*see also* exotic plants & exotic animals)
inundation   20, 23, 219, 220–1, 271, 280, 302, 305, 307
invertebrate biomass   47, 224
invertebrates   5, 47, 57, 88–9, 93, 100, 221, 230, 246, 256, 280, 282, 292
investigations   97, 101, 110
ions   105
irrigation   13, 87, 98, 224, 252, 271, 274, 301, 310, 312
islands   221, 224
Italy   2
Itsukushima Shrine   308

Japan   301–14
Joint Nature Conservation Council   160
Jutland   107

Kalgan River, Australia   270, 277
Kammbach   134
Keersop Project   116, 120–5
Kerbel streams   42, 45
kilometres   3

lagooning   264
Laguna Madre   178

lake classification   232
Lake Erie   172, 175, 180
lake effluent   34
Lake Huron   172, 175
Lake Michigan   172, 175
Lake Ontario   172, 175, 189
Lake St Clair   178, 180
Lake Superior   172, 175
Lake Victoria   77, 221, 224
lake volume   172
land drainage   1, 224, 251–2
land reclamation   13
land subsidence   309
landowners   17, 99, 108, 110
landownership   20
landscape   3, 8, 33, 43, 98, 241, 264
landscape architects   308
landscape assessments   142
landscape ecology   2
landscape geomorphology   36
land-use   8, 20, 99
land–water buffer zone   91
laws   17
leaching   98–9, 110
Leen   2
legal instrument   71–3
legal issues   240
legislation   1, 2, 100, 275, 287–8, 292–3, 301, 303, 305–6, 313
legislators   31
Leitbild   126–7, 157
lentic   91
levee (banks)   79–80, 87, 91, 220
levee construction   271
licences   23
lined channel   304, 310, 312
livestock damage   271
livestock effluent   254
Local Environment Agency Plans (LEAPs)   157, 160, 167–8
location   102
log-jams   39
longitudinal profile/zonation   40, 43, 45, 47
low-flow   75, 215, 301
lowland   292
lowland streams   13–14, 18, 20
lowland streams rehabilitation   14, 17–19, 27
lowland water systems   18
Lubrzanka River   66, 91–3, 97

machine mowing   121, 125
macroinvertebrates   75, 78–9, 82–3, 125, 224, 230, 244, 246, 256
macrophytes   82, 121, 221, 230, 224, 261
   Azolla ficuloides   166
   Bolboschoenus fluviatilis   73, 232

INDEX

Bolboschoenus maritimus 261
Carex eleata 261
Carex riparia Curtis 261
Ceratophyllum 182
Ceratophyllum demersum 261
Crassula helmsii 165–6
Elodea 182
Fallopia japonica 165–6
Heracleum mantegazzianum 165–6
Impatiens glandulifera 165–6
Iris pseudocorus 261
Phragmites australis 261, 263
Polygonum amphibium 261
Typha angustifolia 261
Typha latifolia 261
macrozoobenthic invertebrates 256, 261
Ischnura elegans 256
maintenance 8, 19, 54, 60–1, 81–2, 98–9, 108, 110, 216, 218, 232–3
maintenance of waterways 109
Makkink 106
mammals 81–2, 224
Manchester 3
manure 97
margins 26
marshes 216, 229
marshland 62, 71–2
mayfly 47
meadows 90, 97–9, 108–9
meanders 13,
measurements 109
measures 16
Medloc River 3–4
Meiji Era (1868–1912) 301, 310–11
meltwater sand 107
methodology 36
methods 69
Michigan 179
microhabitats 24
Microtox 182
migration zones 124
Milwaukee 172
mineralisation 91
minimum acceptable flows 225, 233, 279–80, 290, 293, 313
mining
  alluvial 289
  catchment 173
  gold 271, 286
Ministry of Agriculture, Fisheries and Food (UK) 134, 159
Mississippi River 171
Missouri River 171
mitigation measures 60
Model Partnerships 136–7
modelling 292, 294

models 219, 233, 247
  CREAMS 292
  Dissolved Oxygen 292
  IFIM 292
  SEGMENT 292
Mokkoo Chinshoo (wooden mattress) 311
monitoring 75, 100, 178, 230, 245–8, 292
  benthic fauna 246
  fish 246
  spawning grounds 247–8
  water quality 245–6
monitoring stations 99
monitoring water quality 99
Montreal 172
morphodynamics 60, 215
morphological 42, 78–9
morphological heterogeneity 313
morphology 32, 40, 43, 45, 53, 312
mosses 20
mouthparts 182
municipalities 43
murdental streams 57, 71
Murray-Darling Basin Commission 216, 229, 269, 278
Murray-Darling Basin, Australia 215–6, 218, 232–3, 273–4
Murrumbidgee River, NSW, Australia 273
Mussel – *Margaritifera margaritifera* (L) 246
mutagenesis 179
Mutatox 182

Nara Period (710–794) 310
National Carp Task Force 62, 73, 221, 224, 233–4
National Landcare Program 276, 277
National River Health Program 276
National Rivers Authority (UK) 135, 149, 152–4, 156, 159, 163, 166–7
National Trust (UK) 134, 139
natural properties 33
Natura 2000 160
natural beauty 1
natural environment 71
natural habitat 2, 23, 40, 98
natural lowland streams 14–16, 27
  rehabilitation 16, 20, 24, 28
natural nutrient filter 109
natural processes 8, 20, 67, 76, 225, 233
natural regeneration 219, 233, 264
natural river system 77
natural stream channels 26, 31–2, 36, 39
natural structures 31
natural watercourses 108
Nature Conservancy Council, UK 150
nature conservation 1, 14, 17, 24, 27, 68, 98, 109, 306–7

nature development 13, 17, 24, 81
nature development project 20
nature reserves 118, 127, 151–2
navigation 215, 271, 301, 310
nesting grounds 291
Netherlands 4, 13–14, 18–19, 24, 27, 82
　Ministry of Agriculture, Nature Management & Fisheries 113, 116, 124
　National Ecological Network 114–5, 123
　Nature Policy Plan 113, 115, 124
New Zealand 269, 284–93
nitrate analysis 105–7
nitrate filters 108
nitrate inflow 109–10
nitrate input 109
nitrate leaching 110
nitrate runoff 110
nitrate transformation 109–10
Nitrate Vulnerable Areas (NVZs) 163
nitrate year 108
nitrate(s) 14, 24, 97–9, 108–10
nitrogen 89, 97, 108–9
nitrogen fertilizers 97
nitrogen leaching 108
nitrogen run-off 95, 97, 108
non-point source pollution 181
Noord-Brabant Province 113, 115, 126
North East Region 160, 163, 165–8
North Sea 113
north-east Twente 24, 26
Northern Ireland 7
Northern Jutland 100
Northumbria Area 161, 164
Northumbrian Water plc 146
Northwest Planning Council 171
Northwest Power Act 171
Norway Spruce 5
Nottingham 2
Nottingham City Council 153–4
Nottingham County Council 153
Nottinghamshire Wildlife Trust 151, 154
nursery habitat 7
nutrient(s) 13, 15, 24, 26, 39, 40, 43, 74, 89, 91, 97–100, 109
　concentrations 109
　discharge 26
　filter 109
　leaching 109
　levels 40, 88, 215, 257–8, 286, 290, 293
　load 15
　spiralling 91
　transformation 108, 110
nutrients 251, 272, 292

Oakland County 173
Oberrheinebene 32–4, 41

objectives 16, 88
Observed Maximum Stream Habitat (OMST) 190, 193, 199
obstructions 100
ochre precipitation 138
ochrous mine water 163
Odenwald 32–4, 37, 39, 41–2, 46, 47, 53
Ohio River 171
oligotrophic 43
Oostvaarderplassen 114, 128
organic matter 15, 47, 93, 106
organic micro-pollution 15
organic pollution 40
organic sediments 107
Organization, Planning and Reporting 59, 71, 75
organisms 77, 78
ornithological 100
Otter Habitat Enhancement Project 149
outflow 109
outlet 109
outlet water 102
over-abstraction 1
overbank flow 20, 79, 218–19, 221, 224
over-engineering 1
Overijsselse Vecht 24, 66
ownership 24
oxbow lake 93, 312

PAC 174
PAHs 179, 181
Panaro River 260
parameters 92
participants 20
PBBs 179
PCBs 179, 181
peak flows 303, 310
peat soil 106
Pennines 160
people 91, 99, 109, 228
permits 23
pesticides 105, 109, 166, 181
PHABSIM 247
phosphate removal 116
phosphorous 97
phosphorous runoff 108
phosphorus stripping 163
physical conditions 102
physical diversity 284
physical habitat 313
phytoplankton 180, 256, 258–60, 263
　Chlorophyceae 256, 258–9
　Cyanophyceae 256, 258–9
　Diatomeae 256, 258–9
　Peridineae 256, 258–9
phytosociological Methodology 43

INDEX

piezometers 102
plankton 176
planning 8, 303
planning document 20
plant community 107
plant species 107
plant uptake 109
plants 1, 2, 43, 89, 98, 180, 197, 100
point source pollution 181
Poland 89, 90
policies 306
policy concept 18
policy networks 14, 16, 20, 24, 26
policy process 23
politicians 31
pollutants 91, 228, 254
pollutants – organic 255–6, 286
pollution 1, 82, 97, 287, 305, 310, 313
    non-point source 289
    point source 289
    sources 254
polybromiated biphenyls (PBBs) 179
polychlorinated biphenyls (PCBs) 179, 181
polycyclic aromatic hydrocarbons (PAHs) 179, 181
Polynesian migrants 285
pool-riffle 45
pools 42, 84, 93, 221, 281, 304
pool-step 45
population(s) 91, 100, 173
post-project monitoring 134
precipitation 61, 218, 220, 303, 305–6, 309
predation 127, 129
pre-regulation 77, 218–9
preservation 57, 218–19, 224
price-setting 62, 231
Processing 71, 73
project monitoring 59, 60, 73
Project Riverlife 155–6
project(s) 26, 100
public 31, 66, 67, 87, 225
Public Advisory Council (PAC) 174
public and scientific communities 1
public budgets 31
public education 8

Rabis Stream 107
rainfall 173
Ramsar 87, 228
RAP 174
raptors 180
reclamation 67–8, 224, 248 (*see also* rehabilitation)
recombinogenesis 179
recreation 87, 91, 149, 153, 155, 264, 274, 282, 289, 303, 307–8, 310–11

recreational fisheries 290–1, 304, 307
recreational use 218, 228, 264
recreational value 306, 309, 314
refuge 304, 311
regional economic development 13
regional surface waters 19
regional water systems 14
regulation 218, 220
rehabilitated process 3
rehabilitated schemes 3, 8, 9
rehabilitation 2–3, 13, 18, 20, 28, 58–9, 91, 94, 98–100, 157, 218–9, 224, 251–64, 269, 292, 308
    cost 241, 277, 279, 281, 290, 292, 306–10
    examples 215–16, 221, 223, 228, 230
        Asahi River, Japan 312
        Bjelke-Peterson Dam, Australia 279
        Edogawa Ward, Tokyo, Japan 310
        Emu Creek, Australia 281
        Furukawa River Shinsui ParkTokyo, Japan 310
        Heathcote River, New Zealand 289
        Lower Latrobe River, Australia 279
        lower Manawatu River, New Zealand 289
        Lower Yarra River, Australia 279
        Mangatangi Stream, New Zealand 290
        Maruyama River, Japan 311
        Meguro River, Japan 309
        Nohgu River, Japan 312
        Nomi River, Japan 309
        Ohau River, New Zealand 290
        Project River Recovery, New Zealand 291–3
        Rangitaiki River, New Zealand 291
        River Iijoki, Finland 239–50
        River Murray, Australia 229–32, 278
        Thomson River, Australia 280
    floodplain wetlands 66, 215–6, 225, 280–1
    nature conservation 303, 307
    objectives 241–2, 254, 293, 309
rehabilitation planning process 8
rehabilitation policies 13, 14, 16, 27
rehabilitation project 18–19, 20, 23–4, 27–8, 97–9, 109
rehabilitation schemes 306
rehabilitation techniques
    artificial channel 312
    backwaters 144, 165, 312
    bank re-profiling 144
    bank revetment 311
    baseflow restoration 309–10
    berms 165
    boulder dam 243
    boulders 244, 262, 278, 281, 289, 312
    channel cross-section 263, 265

rehabilitation techniques (*cont.*)
  channel deepening   140
  channel diversion   133
  channel stabilization   310
  channel straightening   133, 140
  channel widening   133
  culverting   151
  deflectors   144, 244
  demonstration projects   134
  feasibility study   136
  fish habitat   243
  fish passage   278–9, 291, 307, 313–14
  fish predators   263
  fish re-stocking   165
  gabions   307
  impoundment   263
  in-channel habitat   311, 314
  log weir   281
  logs   289
  macrophyte planting   165, 307
  marsh   263, 265
  masterplans   150
  meander reinstatement   121, 279
  physical habitat   293, 307, 310
  point bars   144
  pool-riffle sequences   118, 144, 262
  pools   311, 312
  rapids/slows   242, 307–8
  re-meandering   142, 144, 165
  revetments   144
  riparian vegetation   263, 277, 288–90, 294, 307, 314
  rip-rap   307
  rock weir   282
  sinuosity   262, 265–6, 308
  spawning habitat   244–5, 281
  topsoil stripping   144
  water quality   282, 310
  weirs   154–5, 164
  wetland restoration   142
  woody debris   263
rehabilitated lowland streams   13
relief channels   57, 75
Remedial Action Plan (RAP)   174
renewable resources – harvesting   162
Reno River, Italy   264
reptiles   62, 224
  Megalobatrachus japonicus   312
reservoirs   87, 91, 304
residence   102
resources   98, 99
restoration   2, 8, 60, 87, 91, 93, 94, 95, 100, 133–46, 218, 254, 262–6, 269
Restoration Network   137
restoration strategy   91
revetments   87, 304, 311

Rhine Graben   32
Rhine Valley   36
Ribe   100
Ridings Area   160–1, 163, 167
riffles   42, 47, 57, 87–8, 93, 290, 304
riparian   43, 45, 47, 87, 91–2, 98, 107
Riparian Ecotones   59, 62, 65, 66, 67, 89, 90, 91, 94, 224
riparian forest   107, 251
riparian habitats   224, 264, 306, 311
riparian management   292, 304
riparian vegetation   121, 206–7, 226, 264, 288, 291, 308
riparian vegetation   5, 48, 62–4, 66, 68–71, 87, 89
riparian zones   97, 108–10, 122
riparian/floodplain ecotones   89, 91
risk   140
river   8, 13, 23, 34, 66, 91–3, 97–8
River Avon   149, 157
river bank management/managers   57–8, 62
river banks   18, 58, 62, 67, 68
river basin   62, 183
river bed   88
River Breda   134
River Bush   7
River Cam   3
river catchment   5
river channel   93
River Cole, Birmingham   150
river conservation   7, 95
river corridor   1, 114, 127, 133, 166
River Credit   189, 199
River Dearne   162, 164
River Derwent, Derby   150, 155–7
River Dinkel   24, 26
River Don, Canada   189
River Exe   133
River Gelsa   134
River Hull   160, 162–3
River Humber, Canada   189
River Iijoki, Finland   239–40, 243, 247–9
River Ijssel   134
River Leen, Nottingham   2, 4, 150–2, 154, 156–7
  materplan   153–4
River Leitha   6, 134
river maintenance   108
river management   1, 2, 301, 303, 308
river management policies   1
River Meuse   57–8, 113
River Murray, Australia   59, 215–6, 221, 223, 228, 230, 271, 273, 278
River Murray Commission   58, 67–8, 215, 216
River Po, Italy   251–2
river purification   93

INDEX

river regulation 1
river rehabilitation 1, 8–9, 54
river restoration 4, 6
River Restoration Centre 137, 146
River Restoration Project (RRP) 134–40, 146
River Rhine 57, 73, 82, 113
River Scheldt 113, 222
River Severn 149
River Skerne 140–4, 145
River Thames 146
River Trent 149, 152–3
River Waal 66
riverine (see also riparian) 272
riverine ecosystem 66, 68, 91
riverine fish 62, 64, 67
riverine habitats 59, 91, 94
Rouge River 189
Royal Society for the Protection of Birds (UK) 134
run-off 91, 100, 106, 272, 303–4, 309–10
run-off coefficient 305
rural areas 40
rural context 3

salination 64
saline groundwater 74, 224, 273
salinity 215, 272–3, 280, 293
salmon fishery 171
salmonids 7, 92
salt water intrusion 289, 291
Samoggia River, Italy 264
sampling 102, 246
Sandbach 134
sandstone 36, 39
Scottish Natural Heritage 135, 160
sea 98–9, 109
sediment 181, 247, 272, 305
sediment monitoring 138
sediment transport 13
sedimentation 2, 20, 23, 39, 230, 116, 118, 129
Seigyuu (holy cow) 310–11
self-organizing complex systems 177
self-regulation 128, 129
semi-natural lowland streams 13
SERCON 160
set-aside 263–4, 266
Severn–Trent Region 149–50
sewage 77, 232
sewage effluent 162, 164
sewage outfalls 192
sewage plants 24
sewage treatment 31
sewerage 252, 276, 289, 304–5, 309–10, 313
shading 87, 261, 277, 290, 292, 304
shellfish 312
shipping 87

Showa Era (1926–1989) 304
shredders 47
silicate 39, 47
siltation 61, 91
sinuosity 100
smart surface chemistry 184
snails 47
social factors 224
softwood plantation 272, 286
sohlental streams 42, 53
soil erosion 91, 271–2, 286, 302
soil structure 20
soil water 109
soils 66, 68, 272, 304
South Australia 75, 222
South Jutland Council 136, 139
Southern Jutland 100
Southern Yorkshire Area 162
south-west Germany 32, 32
spawning 100, 291, 304
spawning grounds 241, 244, 273, 303
species 107
Species Association Tolerance Index (SATI) 192, 194, 196
species diversity 23
Species Tolerance Score (STS) 195
SSSI 159
St Lawrence watershed 171
St Louis River 172
stagnant 102
stations 102
Stoke 150
Stora 97, 100, 107, 109
Stora Watercourse 101
stormwater 305, 309
stormwater drainage 304
Straatsbosheheer 20
strategic approach 150
Strategic Urban Rivers Initiative 150
stream and floodplain vegetation 32
stream and river rehabilitation 31
stream beds 39, 45
stream classification 32, 34
stream ecology 7
stream ecosystems 31
stream gradient 92
stream management 31, 32, 54
stream planform 40, 41
stream rehabilitation 27, 31
stream storage 39
stream systems 13, 14
stream/channel 74, 89, 91–2, 95
stream/river biota 89
streams 23, 24, 31–3, 36, 40, 74, 88, 90, 92–3
  trout 94
stress proteins 182

stressed and degraded river systems   1
substrate   43
subsurface water   102, 106
succession   120
surface water   97, 100, 102, 106, 108–9
surveillance   77
suspended solids   62, 246, 310
sustainable agricultural growth   98
sustainable development   81
Swindon, UK   140
Syvs Stream   106

Ta River, Japan   304
tailings   178
TAMAR 2000 Project   139
tannin   89
targets   123
technical demands   66
technical solutions   77
technicians   77
terms of payment   76
Thames Region   168
thymol   105
Tibooburra, New South Wales, Australia   271
timber floating   239–40, 286
Tokugawa (Edo) Period (1603–1868)   301
Tokyo   309
Tone River, Japan   301, 305
Tongelreep Brook   20, 22–3, 26–7, 116, 118, 120, 126
Tongolreep   24
topography   226
Toronto   172
Toronto Area Watershed (TAR)   189, 191–2, 194, 199–200, 208–12
toxic substances   65, 218
toxicity   179–80
Tradescantia test   182
TRAD-MN test   182
transformation   100
tree roots   53
trees   60– 61, 92, 93, 215
   Alnus glutinosa   120, 251
   Betula pubescens   120
   Carpinus betulus   251
   deciduous and coniferous   89
   Eucalyptus camaldulensis   64, 73, 230, 232, 271
   Eucalyptus coolabah   60–1, 232
   Eucalyptus largiflorens   73, 224
   Fraxinus oxycarpa   251
   Populus alba   251, 264
   Populus nigra   251
   Quercus robur   251, 264
   Salix alba   251
   Salix aurita   120
   Salix cinerea   251
   Salix eleagnos   251
   Salix triandra   251
   Ulmus minor   251
   Willows   76, 218, 224
Triassic Bunter Sandstone   32, 34, 37, 39, 91
tributaries   24
trophic conditions   91
trout   92, 93
turbidity   65–6, 91, 215–16, 219, 224, 272, 274, 293
typhoons   301, 303, 306, 308

Uji River, Japan   307
UK Biodiversity Action Plan   160
uncultivated riparian zones   109
unpolluted streams   40
unregulated discharge   178
upland streams   39, 42–3, 47
urban development   3, 20, 24, 71, 218, 252, 271, 282, 301–2, 305, 312
urban environment   292, 309
urban streams   309
urbanization   302, 304, 310, 314 (*see also* urban development)
US Army Corps of Engineers (USACE)   171
US Environmental protection Agency (EPA)   171
USA   107

vacuum pump   102
vacuum, pumped   105
Valkernswaard   27
valley density   37
vegetation   20, 34, 43, 67, 107, 218, 246
vegetation management   88, 292
Vejle   100
velocity   15, 16
Viborg   100
Victorian State Government   276
vision plans   142, 146

Wando (backwater pool)   312
Waste Regulation Authorities (UK)   159
waste water   20
water   2, 23, 101–2, 109, 110
water abstraction   1
Water Act 1989   149, 159
water balance   102, 109
water beetles   47
water chemistry   32, 36
water demand   314
water environment   99
water flow   39, 101, 106
water fowl   180

water levels   41, 224
water management   14, 226, 275–6
water mining   173, 178
water plants   87
water policy   218
Water Pollution Control Plants   192
water purifying   31
water quality   23–4, 28, 31, 99–100, 108–9, 121, 128, 135, 141–2, 153, 162–7, 194, 199, 245, 255–7, 263, 276–7, 280, 282, 287, 289–90, 292–3, 305, 313
water quality data   99
water quantity   13, 24, 28
water recipients   109
water resources   1
Water Resources Act 1991   159
water retention time   172
water rights   274, 287, 293
water samples   102
water storage   39
water supply   215, 274, 276, 289, 292, 305, 314
water systems   13, 99
water tailings   173, 178
water temperature   92, 192, 194, 196, 198–200, 204–5, 255, 257, 280, 290, 292
water usage   173
water velocity   88, 310
water volume   26
watercourse system   99
watercourse(s)   97–100, 107–10
waterfowl   218, 224, 273, 280
  Anarhynchus frontalis   291
  Himantopus novaezelandiae   291–2
  Hymenolaimus malacorhynchos   290–1
  Paspalum disticum   232
watertable   305
waterways   19, 98–100, 108–10
  dynamic   301, 310
  unstable   301
weed cutting   98
Weighted Species Association Tolerance Index   192–3, 196–212
weirs   1, 13, 93, 97, 116, 120–1, 124–5, 215–6, 219, 242, 273, 289–91
West Germany   34
wet meadows   100, 108–10
wetlands   6, 8, 97, 218, 221, 264, 280, 312
wild flora   109
Wild Rivers Project   139
Wildlife and Countryside Act 1981   165–6
wildlife corridor   151
wood cutting   110
woodland areas   20
woody debris   224
World Wide Fund for Nature   139

XAD resins   184

Yodo River, Japan   312
Yorkshire   162, 167

Zealand   106
zebra mussels   178
Zena Ditch   264
zero discharge   174
zooplankton   180, 256, 260–1, 263
  Rotatoria   256